Topics in Plant Population Biology

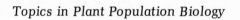

TOPICS IN PLANT POPULATION BIOLOGY

edited by

OTTO T. SOLBRIG
SUBODH JAIN
GEORGE B. JOHNSON
PETER H. RAVEN

COLUMBIA UNIVERSITY PRESS
NEW YORK   1979

LIBRARY OF CONGRESS CATALOGING IN PUBLICATION DATA

MAIN ENTRY UNDER TITLE:

TOPICS IN PLANT POPULATION BIOLOGY.

BIBLIOGRAPHY: P.
INCLUDES INDEX.
1. PLANT POPULATIONS. I. SOLBRIG, OTTO THOMAS.
QK910.T66   581.5′24   78-27630
ISBN 0-231-04336-8

COLUMBIA UNIVERSITY PRESS
NEW YORK AND GUILDFORD, SURREY
COPYRIGHT © 1979 COLUMBIA UNIVERSITY PRESS
ALL RIGHTS RESERVED
PRINTED IN THE UNITED STATES OF AMERICA

To George Ledyard Stebbins
Scientist, Teacher, Friend

CONTENTS

LIST OF CONTRIBUTORS

Mark W. Angevine. Section of Ecology and Systematics, Cornell University, Ithaca, N.Y. 14850.

James A. Bunce. Light and Plant Growth Laboratory, U.S. Dept. of Agriculture, Beltsville, Md. 20705.

Martyn M. Caldwell. Department of Range Science, Utah State University, Logan, Utah 84322.

Brian F. Chabot. Section of Ecology and Systematics, Cornell University, Ithaca, N.Y. 14850.

Robert E. Cook. Department of Biology, Harvard University, Cambridge, Mass. 02138.

Thomas Givnish. Department of Biology, Harvard University, Cambridge, Mass. 02138.

Leslie D. Gottlieb. Department of Genetics, University of California, Davis, Calif. 95616.

Sherry L. Gulmon. Department of Biology, Stanford University, Stanford, Calif. 94305.

James L. Hamrick. Department of Botany, University of Kansas, Lawrence, Kan. 66044.

James C. Hickman. Department of Biology, Swarthmore College, Swarthmore, Pa. 19081 (present address: Dept. of Botany, University of California, Berkeley, Calif. 94720).

Henry S. Horn. Department of Biology, Princeton University, Princeton, N.J. 08540.

Subodh Jain. Department of Agronomy, University of California, Davis, Calif. 95616.

George B. Johnson. Department of Biology, Washington University, St. Louis, Mo. 63130.

Donald L. Levin. Department of Botany, University of Texas, Austin, Tex. 78712.

Phillip C. Miller. Systems Ecology Group, San Diego State University, San Diego, Calif. 92182.

Harold A. Mooney. Department of Biology, Stanford University, Stanford, Calif. 94305.

Peter H. Raven. Missouri Botanical Garden, St. Louis, Mo. 63110.

Otto T. Solbrig. Department of Biology, Harvard University, Cambridge, Mass. 02138.

G. Ledyard Stebbins. Department of Genetics, University of California, Davis, Calif. 95616.

William A. Stoner. Systems Ecology Research Group, San Diego State University, San Diego, Calif. 92182.

James A. Teeri. Barnes Laboratory, The University of Chicago, Chicago, Ill. 60637.

Patricia A. Werner. W. K. Kellogg Biological Station, Michigan State University, Hickory Corners, Mich. 49060.

INTRODUCTION

POPULATION BIOLOGY is concerned with the documentation and explanation of changes in gene frequency, number of individuals, and phenotypic characteristics of members of populations over time. Population biology includes aspects of population genetics, demography, population ecology, plant physiology, and adaptive morphology; it is primarily a synthetic discipline.

The two main objectives of population biology are: to understand in precise detail how natural selection operates (this involves not only determining that heritable changes take place but establishing precisely when and where in the life cycle these changes occur) and to understand rigorously the mechanisms of adaptation. It is not sufficient, therefore, to establish a correlation between the presence of a given phenotypic characteristic and a set of environmental factors, but it is necessary to provide the causal factors. Consequently, the population biologist tries to establish the patterns of genetic and phenotypic variation in time and space, hoping to understand how these patterns arise by the action of natural selection and in exactly what ways they are adaptive, in order to develop predictive models and theories.

Biological phenomena are, however, so complex, so multidimensional, that they cannot be understood by the patient accumulation of data, no matter how accurately and carefully gathered. Theoretical notions are necessary in the quest for knowledge, notions that often are deemed "tautological, Panglossian, teleological speculations" (discussed further by Horn, article 2 in this volume). Since counterexamples to any theory can usually be found, biological theories are not necessarily disproven by the performance of a single crucial experiment as in the physical sciences (where this model of proof may also be invalid in a large number of cases; Kuhn, 1970) but stand or fall in direct measure to their general predictive power. Good theories are those that lead to new insights, or new twists, and thereby help in the gathering of new and appropriate data.

Plant population biology is no different in its objectives than animal population biology (Harper, 1977). However, because plants as organisms have special characteristics, the investigation of the population biology of plants has certain unique advantages and disadvantages that require special approaches and techniques.

The most obvious characteristic of plants, which applies to all land plants and a large number of aquatic plants, is their sessile nature. This is shared by a small number of marine animals, most notably corals. The sessile nature of plants makes them very good material for demographic studies. Although the individual plant— the ramet—is truly sessile, the genotype—the genet—has a certain degree of mobility, which appears in the form of runners, stolons, or other forms of asexual reproduction, that allows an individual to produce exact genetic replicates. Although occasionally this form of movement is appreciable—as in the case of the aspen (*Populus tremuloides*), where an individual genotype is capable of occupying an entire hillside of several acres—vegetative reproduction, although complicating demographic studies, is not an overwhelming problem.

Interactions between sessile plants, both intra- and inter-specific, as well as interactions with the physical environment, are largely individual and very localized processes. Each plant has a unique physical environment, a unique set of competitors, and a unique set of herbivores and parasites, from which there is no escape. Therefore, by necessity, any models in plant population must include a vector indicating this neighborhood effect (Mack and Harper, 1977). As Schaffer and Leigh (1976) have pointed out, however, spatial relations are very difficult to model and to measure.

Another advantage of plants over animals in population biology studies is the relative ease of handling live plants and the facility with which they are reproduced vegetatively. This allows the design of replicate experiments with identical genetic individuals, without the necessity of inbreeding or special genetic manipulations. On the other hand, plants as a group have long generations, and it is seldom possible to obtain more than one or two generations a year; plants also tend to be bulkier, requiring more space, and for a longer time, than experiments with *Drosophila*, or even mice.

Plant species show a great degree of variation in life-cycle characteristics, longevity, breeding systems, and physiological and morphological traits. This facilitates observations and experiments on adaptation. Plants, however, are phenotypically very plastic, which makes studies of adaptation, and especially of selection, quite difficult. The plasticity of the phenotype almost mandates a determination of the genetic and environmental components of the variance in studies of plant populations, a time-consuming procedure that is very seldom followed.

Finally, plants are simply built, and their development and function should be easier to comprehend than those of most animals, particularly vertebrates, with their complex nervous and endocrine systems. On the other hand, plants have a very complex chemistry owing to the presence of a very large number of so-called secondary compounds (flavonoids, tannins, complex carbohydrates, etc.). The presence of these compounds not only makes the study of intermediate metabolism in plants very difficult, it makes the application of one of the most powerful techniques of modern population genetics—determination of allozyme phenotypes—very difficult.

Plants, therefore, present advantages and disadvantages compared to animals for population biology studies. However, it is the unique role of green plants as harvesters of solar energy that makes their study imperative as well as fascinating.

The last 20 years have seen the development of many novel and powerful concepts and techniques that have propelled population biology forward at an accelerating pace. In the realm of genetics, the introduction of allozyme techniques by Harris (1966) and Hubby and Lewontin (1966) has provided a powerful tool to ascertain gene and genotype frequencies in populations and their change over time. Developmental genetics, that is, efforts to map genotypes into phenotypes, has made slower progress. The present inability to predict phenotypic composition from genotypic structure of a population, and vice versa, is probably the largest lacuna in our understanding of populations. Development of a more rigorous theoretical analysis of the components of natural selection, owing mostly to Prout (1969), and the introduction of notions of optimality into evolutionary theory (Rosen, 1967) have materially aided in increasing the level of rigor with which problems of selection and adaptation are attacked today. A more detailed discussion is found in Part 1 of this book.

Largely owing to the activities of John Harper, Anthony Bradshaw, and their students, plant population biologists have become very aware of the importance of understanding the meaning of the various aspects of the life cycle of plants. The concept is of course not new: Schimper (1898), in his classical treatise, attempted to explain the characteristics of the various ontogenetic stages of a plant in ecological and adaptive terms. However, demographic approaches in plant evolution and plant ecology have not been exploited until recently. As is pointed out by several contributors to this volume (Cook, article 9; Hickman, article 10; Werner, article 12), it is becoming increasingly clear that plant evolution is influenced enormously by the fate of seeds and seedlings. Traditionally, these stages have been neglected by taxonomists as well as ecologists. We need studies that record genetic and demographic changes simultaneously during these early stages of the life cycle.

Another interesting area is life-history theory (Gadgil and Bossert, 1970; Schaffer, 1974a,b; Schaffer and Gadgil, 1975; Bell, 1976; Schaeffer and Rosenzweig, 1977). When lifespan, competition, and reproduction are considered in an integrated context, the diversity of reproductive strategies becomes understandable, and general predictions are possible (see Jain, article 7; Cook, and Hickman).

Great strides in our understanding have also been made in the area loosely known as "physiological ecology," or "functional morphology." Although botanists have always tried to interpret form in terms of function, it is only in the last 20 years that this ideal has been translated into reality in populational studies. Advances in our understanding of basic phenomena such as the biochemistry of photosynthesis (Zelitch, 1969) and the mechanics of water and nutrient transport (Zimmermann, 1974), together with the development of reliable instrumentation for field use, have totally revolutionized this area. In Part 3 of this book, some of these advances, as well as the new and fascinating problems that have arisen from these approaches, are documented.

Studies of the genetic structure of populations need now to be integrated with demographic studies and with the more functional insights derived from physiological ecology. Since great lacunae exist in our knowledge of each of these three approaches, it may be premature to expect much integration at this time. However, as the papers in the three parts of this volume show, the genetic composition of the population is determined by the forces that structure the life history parameters through the morphological and physiological characteristics of the phenotype, which are in turn determined by the genotype and, ultimately, by the genetic structure of the population.

In choosing the topics for this book, the editors deliberately avoided the issues of the influence of community structure on population and species coevolution, and, in particular, plant–herbivore interactions. This was done in part because others are presently investigating these problems very competently (May, 1974; Gilbert and Raven, 1976) and in part because unless some boundaries, no matter how artificial, are set, we are in danger of becoming lost in a sea of data and general statements.

Because we do stand on the shoulders of our predecessors, it was felt important to present a review of the historical development of plant population biology (Stebbins, "Fifty Years of Plant Evolution"). Plant population biology as it is developing in the United States has three historical roots. First there is the "biosystematic" root, whose historical development is described by Stebbins. Starting with Turesson in the 1920s, the biosystematists documented the importance of breeding populations as the unit of evolution, the local nature of adaptation, and the genetic structure of species. A second historical root is the tradition of population ecology, which is concerned with the changes in numbers of indiviudals in populations. Although some classical work with plants was done in the 1920s (Clements et al., 1929), this approach was developed by animal ecologists and survived mostly among botanists in the applied fields (Harper, 1977). There has been a tremendous surge of interest in this area in the last 20 years, spurred largely by the drive and imagination of John Harper in Great Britain. Finally, the physio-logical–ecological approach also has a long tradition that reaches back to the German ecologist Andreas Fran Schimper and comes to us through Maximov and Walther. But it has only been in the last 20 years that plant physiology has been integrated into population studies, mainly as a result of a new and better understanding of the process of photosynthesis and the development of reliable field instrumentation. Although clearly defined conceptual approaches, techniques, and intellectual schools can be discerned within these three traditions, there has always been a fair amount of intellectual exchange between these areas, so that today's synthetic approach does not represent a radical departure.

The papers that follow represent an attempt by a generation of American plant population biologists in their 30s and 40s to present their view of the field. Some papers are defenses of particular con-cepts (Horn, article 2); others represent critical evaluations of ideas and approaches (Jain and Hickman). Most papers synthesize the present status of theory and research on specific topics within plant population biology. Collectively, the book is our scientific mani-festo, in which we indicate what we see as the present problems and future directions of this field.

The editors were solely responsible for choosing the general topics and assembling the participants. A conscious decision was made to favor members of the new generation. In so doing, we do not mean to denigrate in any way the great contribution of some of the more senior population biologists. Their work is the foundation of ours. Financial considerations excluded the participation of researchers from outside the United States. This we regret very much. Finally, not all invited persons could participate.

All contributors but one (M. Caldwell) attended a preliminary conference held at Ithaca College, in Ithaca, New York, on June 16–18, 1977. The conference was attended also by a number of guests (see the list at the end of the book). We wish to thank all participants for their comments and contributions during the lively discussions that took place. We all benefited greatly. The papers were sent for review to the following persons, whose critical comments and edito-rial advice are acknowledged by editors and authors: P. Cavers, M.

Clegg, B. Chabot, H. Lewis, R. Loomis, J. Mitten, R. Ornduff, H. Wilbur, and M. Zimmermann. Last but not least, we wish to thank the authorities of Ithaca College, especially S. LaMotte, and Brian Chabot of Cornell University, for local arrangements, and Kathryn W. Rollins and Anita Fahey of Harvard University for able editorial and secretarial assistance and organization help. Joe Ingram and Maria Caliandro of Columbia University were unfailing in their confidence in this project and their editorial advice.

GEORGE LEDYARD STEBBINS

Otto T. Solbrig

GEORGE LEDYARD STEBBINS was born on January 6, 1906, in the town of Lawrence, New York. He attended Harvard University, where he obtained an A.B. degree in 1928, an A.M. the same year, and a Ph.D. in biology in 1931. At Harvard he worked first with A. L. Fernald on the taxonomy of grasses and the Compositae and then with E. C. Jeffrey, with whom he did his thesis work on the cytology, taxonomy, and evolution of the American species of the genus *Antennaria*.

From 1931 to 1935, Stebbins was Instructor in Biology at Colgate University in Hamilton, New York. In 1935, he joined the Department of Genetics of the University of California at Berkeley, as Junior Geneticist. Here he became part of the team that E. B. Babcock assembled to study *Crepis* and related genera. In Berkeley, Stebbins rapidly ascended the academic ladder, becoming Professor of Genetics in 1947.

These were years of great intellectual ferment in the San Francisco Bay Area, and Stebbins joined other geneticists, such as Babcock, Richard Goldschmidt, and Jens Clausen of the Carnegie Institution, and systematists interested in evolutionary questions, among whom were David Keck, Lincoln Constance, and Herbert Mason, in founding an informal discussion group known as the "Biosystematists." In monthly gatherings, this group (still extant) reviewed the then-emerging ideas regarding ecotypic variation, hybridization, and polyploidy and how the new concepts would affect the purely morphological concept of the species that then predominated. In 1938, Stebbins, in collaboration with Babcock, published "The

North American species of *Crepis*: Their relationships and distribution as affected by polyploidy and apomixis." In this work, they advanced the concept that polyploids that combine the genetic patrimony and ecological amplitude of two species become more widespread than either parent. This study was followed by two papers in which Stebbins developed these ideas further: "The significance of polyploidy in plant evolution" (1940) and "Apomixis in the angiosperms" (1941). Later (1947), he summarized his work on polyploid complexes in his review "Types of polyploids: Their classification and significance."

Another area where Stebbins did pioneering work is the relationship between plant habit and chromosome number. He showed that polyploidy is most frequent in herbaceous perennials with some means of vegetative reproduction, whereas annuals, especially colonizing species, tend to have low chromosome numbers. These ideas were developed in a number of papers starting in 1938 and culminating in his "Longevity, habitat, and release of genetic variability in the higher plants" (1978). Still another theme that Stebbins developed in his research relates to the details of the evolution of the genetic system of plants. In 1957 he published "Self-fertilization and population variability in the higher plants," in which he set forth the concept—now generally accepted—that self-fertilization is a derived condition in the angiosperms. This paper was followed in 1960 by "The comparative evolution of genetic systems."

In 1950 Stebbins published *Variation and Evolution in Plants*, the expanded text of the Jessup Lectures he had delivered at Columbia University in October and November of 1946. This book masterfully applies the tenets of the synthetic theory of evolution to plants. It is a seminal book that immediately attracted the attention of botanists and evolutionists throughout the world. In 1974, *Flowering Plants: Evolution above the Species Level* saw the light of day. In this book, which resulted from his presentation of the Prather Lectures at Harvard University in 1958, Stebbins summarized and documented his views regarding the origin and evolution of the angiosperms. This book, too, has attracted wide attention. In addition, Stebbins has published a half-dozen textbooks, of which *Chromosomal Evolution in Higher Plants* is probably the best text in its area.

Professor Stebbins moved in 1951 from Berkeley to the new campus of the University of California at Davis to start a new genetics department, which he directed until 1963. From 1960 to 1964, he served as secretary-general of the International Union of Biological Sciences.

In his long career, Stebbins' achievements have received wide recognition at home and abroad. He was elected to membership in the National Academy of Sciences (1952), the American Academy of Arts and Sciences (1953), and the American Philosophical Society (1961). He was elected president of the Society for the Study of Evolution in 1948 and of the Botanical Society of America in 1961. He has served on the editorial boards of the *American Naturalist, Evolution*, and the *American Journal of Botany*. He was awarded the John F. Lewis Prize of the American Philosophical Society in 1960 and the Addison Emery Verrill Medal of Yale University in 1967. In 1962 he was awarded the degree of Docteur, Honoris Causa, by the Faculté des Sciences of the University of Paris.

Stebbins has a variety of interests outside of his purely academic pursuits, and has always been concerned with the social problems of his time. He has, for example, been deeply involved in defending the teaching of evolutionary concepts in the California schools. He also has been very interested in California plant life and conservation, which has resulted in his being elected president of the California Native Plant Society. In a lighter vein, Stebbins' love of music and singing, especially the music of Gilbert and Sullivan's operettas, is well known.

Stebbins is an effective and colorful lecturer and teacher. He has always shown keen interest in all facets of botanical research, and his imagination and enthusiasm are boundless and very infectious. Those who have had any contact with him have been amazed by his great energy, enthusiasm, and knowledge. The editors, in their name and that of the authors, take great pleasure in dedicating this book to George Ledyard Stebbins in honor of his seventieth birthday and to celebrate his innumerable contributions to plant population biology. May the future provide him with many more opportunities for further research and personal fulfillment.

PUBLICATIONS BY GLS

BOOKS

GLS and C. W. Young. *Human Organism and the World of Life.* 1938. New York: Harpers.

GLS. *Variation and Evolution in Plants.* 1950. New York: Columbia University Press.

GLS and H. G. Baker, eds. *The Genetics of Colonizing Species.* 1965. New York: Academic Press.

GLS. *Processes of Organic Evolution.* 1966; 2nd ed., 1971; 3rd ed., 1977. Englewood Cliffs, N.J.: Prentice-Hall.

GLS. *The Basis of Progressive Evolution.* 1969. Chapel Hill, N.C.: University of North Carolina Press.

GLS. *Chromosomal Evolution in Higher Plants.* 1971. Reading, Mass.: Addison-Wesley.

GLS. *Flowering Plants: Evolution Above the Species Level.* 1975. Cambridge, Mass.: Harvard University Press, Belknap Press.

Dobzhansky, Th., F. J. Ayala, GLS, and J. W. Valentine. *Principles of Evolution.* 1977. San Francisco: W.H. Freeman and Co.

ARTICLES

GLS. 1929. Further additions to the Mt. Desert flora. *Rhodora* 31: 81–88.

GLS. 1929. *Lomatogonium rotatum* (L.) Fries in Maine. *Rhodora* 31: 143.

GLS. 1929. Some interesting plants from Mt. Katahdin. *Rhodora* 31: 142–143.

GLS. 1930. An interesting form of *Eupatorium perfoliatum. Rhodora* 32: 132–33.

GLS. 1930. Contribution from the Gray Herbarium of Harvard University—No. LXXXVII. III. A revision of some North American species of *Calamagrostis*. *Rhodora* 32: 35–57.

GLS. 1932. Some interesting plants from the north shore of the St. Lawrence. *Rhodora* 34: 66–67.

GLS. 1932. Cytology of *Antennaria*. I. Normal species. *Bot. Gaz.* 94: 134–51.

GLS. 1932. Cytology of *Antennaria*. II. Parthenogenetic species. *Bot. Gaz.* 94: 322–45.

Hicks, G. C. and GLS. 1934. Meiosis in some species and a hybrid of *Paeonia*. *Amer. J. Bot.* 21: 228–41.

GLS. 1935. Some observations on the flora of the Bruce Peninsula, Ontario. *Rhodora* 37: 63–74.

GLS. 1935. A new species of *Antennaria* from the Appalachian region. *Rhodora* 37: 229–37.

GLS. 1936. A note on species differentiation in *Antennaria*. *Rhodora* 38: 367–69.

GLS. 1936. Two new species of *Lactuca* from tropical Africa. *Bull. Jard. Bot. L'Etat* 14: 223–26.

GLS. 1937. Critical notes on *Lactuca* and related genera. *J. Bot.*, January 1937:12–18.

GLS. 1937. Critical notes on the genus *Ixeris*. *J. Bot.* February 1937: 43–51.

Babcock, E. B., GLS, and J. A. Jenkins. 1937. Chromosomes and phylogeny in some genera of the Crepidinae. *Cytologia Fujii (Jubilee Volume)*, pp. 188–210.

GLS. 1937. The scandent species of *Prenanthes* and *Lactuca* in Africa. *Bull. Jard. Bot. L'Etat* XIV: 333–51.

Saunders, A. P., and GLS. 1938. Cytogenetic studies in *Paeonia*. I. The compatibility of the species and the appearance of the hybrids. *Genetics* 23: 65–82.

GLS. 1938. Cytogenetic studies in *Paeonia*. II. The cytology of the diploid species and hybrids. *Genetics* 23: 83–110.

GLS. 1938. A bleaching and clearing method for plant tissues. *Science* 87: 21–22.

GLS. 1938. An anomalous new species of *Lapsana* from China. *Madroño* 4: 154–57.

GLS. 1938. Cytological characteristics associated with the different growth habits in the dicotyledons. *Amer. J. Bot.* 25: 189–98.

GLS. 1938. The Western American species of *Paeonia*. *Madroño* 4: 252–60.

Babcock, E. B. and GLS. 1938. The American species of Crepis: Their relationships and distribution as affected by polyploidy. Carnegie Inst. Washington Publ. 504.

GLS. 1939. Notes on some systematic relationships in the genus Paeonia. Univ. Calif. Publ. Bot. 19: 245–66.

GLS and S. Ellerton. 1939. Structural hybridity in Paeonia californica and P. brownii. J. Genet. 38: 1–36.

GLS and E. B. Babcock. 1939. The effect of polyploidy and apomixis on the evolution of species in Crepis. J. Hered. 30: 519–30.

GLS and J. A. Jenkins. 1939. Aposporic development in the North American species of Crepis. Genetica 21: 191–224.

GLS. 1939. Notes on Lactuca in Western North America. Madroño 5: 123–26.

GLS. 1940. The significance of polyploidy in plant evolution. Amer. Nat. 74: 54–66.

GLS. 1940. Studies in the Cichorieae: Dubyaea and Soroseris, endemics of the Sino-Himalayan Region. Mem. Torrey Bot. Club 19: 5–76.

GLS. 1939. Notes on some Indian species of Lactuca. Indian For. Rec. 1: 237–44.

GLS and R. M. Love. 1941. A cytological study of California forage grasses. Amer. J. Bot. 28: 371–82.

GLS. 1941. Additional evidence for a holarctic dispersal of flowering plants in the Mesozoic era. Proceedings of Sixth Pacific Science Congress, pp. 649–60.

GLS. 1941. Apomixis in the angiosperms. Bot. Rev. 7: 507–42.

GLS and R. M. Love. 1941. An undescribed species of Stipa from California. Madroño 6: 137–41.

GLS. 1942. Polyploid complexes in relation to ecology and the history of floras. Amer. Nat. 76: 36–45.

Babcock, E. B., GLS, and J. A. Jenkins. 1942. Genetic evolutionary processes in Crepis. Amer. Nat. 76: 337–63.

GLS. 1942. The role of isolation in the differentiation of plant species. Biol. Symp. 6: 217–33.

GLS. 1942. The genetic approach to problems of rare and endemic species. Madroño 6: 241–58.

GLS. 1942. The concept of genetic homogeneity as an explanation for the existence and behavior of rare and endemic species. Chron. Bot. 7: 252–53.

Babcock, E. B., and GLS. 1943. Systematic studies in the Cichorieae. *Univ. Calif. Publ. Bot.* 18: 227–40.

GLS. 1944. Review of "Vegetation and Flora of Mount Diablo, California." *Ecology* 25: 481–82.

GLS and H. A. Tobgy. 1944. The cytogenetics of hybrids in *Bromus*. I. Hybrids within the section Ceratochloa. *Amer. J. Bot.* 31: 1–11.

GLS, H. A. Tobgy, and Jack R. Harlan. 1944. The cytogenetics of hybrids in *Bromus*. II. *Bromus carinatus* and *Bromus arizonicus*. *Proc. Calif. Acad. Sci.* 25: 307–22.

GLS and Masuo Kodani. 1944. Chromosomal variation in *Guayule* and *Mariola*. *J. Hered.* 35: 163–72.

GLS. 1945. Role of isolation in the differentiation of plant species. *Nature (London)* 155: 150–51.

GLS. 1945. Review of "Plant Evolution through Amphiploidy and Autoploidy." *Ecology* 26: 420–21.

GLS. 1945. The cytological analysis of species hybrids. II. *Bot. Rev.* 11:463–86.

GLS. 1945. Evidence for abnormally slow rates of evolution, with particular reference to the higher plants and the genus *Drosophila*. *Lloydia* 8: 84–102.

GLS, J. I. Valencia, and R. Marie Valencia. 1946. Artificial and natural hybrids in the Gramineae, tribe Hordeae. I. *Elymus, Sitanion,* and *Agropyron*. *Amer. J. Bot.* 33: 338–51.

GLS, J. I. Valencia, and R. Marie Valencia. 1946. Artificial and natural hybrids in the Gramineae, tribe Hordeae. II. *Agropyron, Elymus,* and *Hordeum*. *Amer. J. Bot.* 33: 579–86.

GLS. 1947. Improved forage grasses to be put in field trials. *Calif. Agr.* 1(4).

GLS. 1947. Types of polyploids: Their classification and significance. *Adv. Genet.* 1: 403–29.

GLS. 1947. Evidence on rates of evolution from the distribution of existing and fossil plant species. *Ecol. Monogr.* 17: 149–58.

GLS, E. B. Matzke, and C. Epling. 1947. Hybridization in a population of Quercus marilandica and Quercus ilicifolia. *Evolution* 1: 79–88.

GLS. 1947. The origin of the complex of *Bromus carinatus* and its phytogeographic implications. *Contrib. Gray Herb.* 165: 42–55.

GLS. 1948. Review of "A Study of the genus *Paeonia*." *Madroño* 9: 193–99.

8   GEORGE LEDYARD STEBBINS

GLS. 1948. The chromosomes and relationships of Metasequoia and Sequoia. Science 108: 95–98.

GLS. 1949. Asexual reproduction in relation to plant evolution. Evolution 3: 98–101.

GLS and Marta Sherman Walters. 1949. Artificial and natural hybrids in the Gramineae, tribe Hordeae. III. Hybrids involving Elymus condensatus and E. triticoides. Amer. J. Bot. 36: 291–301.

GLS. 1949. The evolutionary significance of natural and artificial polyploids in the family Gramineae. Proceedings of the Eighth International Congress of Genetics; Hereditas (Suppl.): 461–85.

GLS. 1949. Rates of evolution in plants. In Glenn L. Jepsen, George Gaylord Simpson, and Ernst Mayr (eds.), Genetics, Paleontology, and Evolution, p. 14. Princeton, N.J.: Princeton University Press.

GLS. 1949. Speciation, evolutionary trends, and distribution patterns in Crepis. Evolution 3: 188–93.

GLS and Elton F. Paddock. 1949. The Solanum nigrum complex in Pacific North America. Madroño 10: 70–81.

GLS. 1949. Reality and efficacy of selection in plants. Proc. Amer. Phil. Soc. 93: 501–13.

GLS and Marta Sherman Walters. 1949. The evolutionary significance of two synthetic allopolyploid species of Bromus. Port. Acta Biol. Ser. A, pp. 106–36.

GLS. 1950. Evolution in the Soviet Union. Evolution 4: 99–100.

GLS and Ranjit Singh. 1950. Artificial and natural hybrids in the Gramineae, tribe Hordeae. IV. Two triploid hybrids of Agropyron and Elymus. Amer. J. Bot. 37: 388–93.

GLS. 1950. New grasses, drought-resistant strains of perennials developed for dry range lands. Calif. Agr., September 1950, 3 pp.

GLS. 1951. Push-button evolution. Quart. Rev. Biol. 26: 191–93.

GLS. 1951. Review of "Problems of Cytology and Evolution in the Pteridophyta." Science 113: 533–35.

GLS. 1951. Cataclysmic evolution. Sci. Amer. 184: 54–59.

GLS. 1951. Natural selection and the differentiation of angiosperm families. Evolution 5: 299–324.

GLS. 1952. Pastos resistentes a la sequia. 1 p. La Hacienda, N.Y.

GLS. 1952. Comments on literature in plant evolution. Evolution 6: 131–33.

GLS. 1952. Species hybrids in grasses. Proceedings of the 6th International Grassland Congress, August 17–23, vol. I, pp. 247–53.

GLS. 1952. The evolution of cultivated plants and weeds. *Evolution* 6: 445–48.

GLS. 1952. Aridity as a stimulus to plant evolution. *Amer. Nat.* 86: 33–44.

Duara, B. N., and GLS. 1952. A polyhaploid obtained from a hybrid derivative of *Sorghum halepense* × *S. vulgare* var. *sudanense. Genetics* 37: 369–374.

GLS. 1952. Organic evolution and social evolution. *A University at Work* 11: 3–7, December 1952.

GLS. 1953. Heterosis and evolution. *Evolution* 7: 90–92.

GLS, James A. Jenkins, and Marta S. Walters. 1953. Chromosomes and phylogeny in the Compositae, tribe Cichorieae. *Univ. Calif. Publ. Bot.* 26: 401–30.

GLS. 1953. A new classification of the tribe Cichorieae, family Compositae. *Madroño* 12: 33–64.

GLS and Fung Ting Pun. 1953. Artificial and natural hybrids in the Gramineae, tribe Hordeae. V. Diploid hybrids of *Agropyron. Amer. J. Bot.* 40: 444–49.

GLS and Fung Ting Pun. 1953. Artificial and natural hybrids in the Gramineae, tribe Hordeae. VI. Chromosome pairing in *Secale cereale* × *Agropyron intermedium* and the problem of genome homologies in the Triticinae. *Genetics* 38: 600–608.

GLS. 1953. Plant phylogeny and evolution. *Evolution* 7: 281–84.

GLS. 1953. *Asplenium viride* in California. *Madroño* 7: 1.

GLS. 1953. Les processus de l'évolution aux hautes montagnes.

GLS. 1954. Review of "The Major Features of Evolution." *Science* 119: 699–701.

GLS and Antero Vaarama. 1954. Artificial and natural hybrids in the Gramineae, tribe Hordeae. VII. Hybrids and allopolyploids between *Elymus glaucus* and *Sitanion* spp. *Genetics* 39: 378–95.

Anderson, E., and GLS. 1954. Hybridization as an evolutionary stimulus. *Evolution* 8: 378–88.

Stokes, Susan G., and GLS. 1954. Chromosome numbers in the genus *Eriogonum. Leafl. West. Bot.* 7: 228–33.

GLS. 1955. Review of "Synthetische Artbildung." *Quart. Rev. Biol.* 30: 384–85.

GLS. 1955. Memorial, Ernest Brown Babcock. *Madroño* 13: 81–83.

GLS and L. Ferlan. 1956. Population variability, hybridization, and introgression in some species of *Ophrys. Evolution* 10: 32–46.

GLS. 1956. New look in Soviet genetics. *Science* 123: 720–22.

Sarkar, P., and GLS. 1956. Morphological evidence concerning the origin of the B genome in wheat. Amer. J. Bot. 43: 297–304.

GLS and L. A. Snyder. 1956. Artificial and natural hybids in the Gramineae, tribe Hordeae. IX. Hybrids between western and eastern North American species. Amer. J. Bot. 43: 305–12.

GLS. 1956. Taxonomy and the evolution of genera, with special reference to the family Gramineae. Evolution 10: 235–45.

GLS. 1956. Artificial polyploidy as a tool in plant breeding. Genetics in Plant Breedings: Brookhaven Symp. Biol. 9.

GLS. 1956. Cytogenetics and evolution of the grass family. Amer. J. Bot. 43: 890–905.

GLS. 1957. Regularities of transformation in the flower. Evolution 11: 106–8.

GLS. 1957. O gibridnom proiskhozhdenii pokrytosemennykh [On the hybrid origin of angiosperms]. In Russian. Bot. J. Russ. Acad. 42: 1503–6.

GLS. 1957. The hybrid origin of microspecies in the Elymus glaucus complex. Proceedings of the International Genetics Symposia, 1956, pp. 336–40.

GLS. 1957. Self-fertilization and population variability in the higher plants. Amer. Nat. 91: 337–54.

GLS. 1957. The inviability, weakness and sterility of interspecific hybrids. Adv. Genet. 9: 147–215.

GLS. 1957. The use of plant breeding to increase the world's food supply. Indian J. Genet. Plant Breed. 17: 121–28.

GLS. 1957. Genetics, evolution and plant breeding. Indian J. Genet. Plant Breed. 17: 129–41.

GLS. 1957. The use of experimental data in floras and monographs. VIII Congres International de Botanique, pp. 186–192.

Popov, M. G. (posthumous) and GLS. 1958. Comments on the origin and phylogeny of the angiosperms, and on the hybrid origin of the angiosperms. Evolution 12: 266–70.

GLS. 1958. Longevity, habitat, and release of genetic variability in the higher plants. Cold Spring Harbor Symp. Quant. Biol. 23: 365–78.

GLS. 1959. The role of hybridization in evolution. Proc. Amer. Phil. Soc. 103: 231–51.

GLS. 1959. Genes, chromosomes, and evolution. Vistas Bot., pp. 258–90.

GLS. 1959. The synthetic approach to problems of organic evolution. *Cold Spring Harbor Symp. Quant. Biol.* 24: 305–11.

GLS and D. Zohary. 1959. Cytogenetic and evolutionary studies in the genus *Dactylis*. I. Morphology, distribution and interrelationships of the diploid subspecies. *Univ. Calif. Pub. Bot.* 31: 1–40.

GLS. 1959. Seedling heterophylly in the California flora. *Bull. Res. Council Isr.* 7D: 248–55.

GLS. 1959. Differences between the process of speciation in higher animals and plants. American Society of Zoologists, Refresher Course for 1959, Pennsylvania State University.

GLS. 1960. The comparative evolution of genetic systems. In *Evolution after Darwin*, pp. 197–226. Chicago: University of Chicago Press.

GLS and S. K. Jain. 1960. Development studies of cell differentiation in the epidermis of monocotyledons. I. *Allium, Rhoeo*, and *Commelina*. *Devel. Biol.* 2: 409–26.

McKell, Cyrus M., Eugene R. Perrier, and GLS. 1960. Responses of two subspecies of orchard grass (*Dactylis glomerata* subsp. *lusitanica* and *judaica*) to increasing soil moisture stress. *Ecology* 41: 785–90.

GLS and Gurdev S. Khush. 1961. Variation in the organization of the stomatal complex in the leaf epidermis of monocotyledons and its bearing on their phylogeny. *Amer. J. Bot.* 48: 51–59.

GLS and K. Daly. 1961. Changes in the variation pattern of a hybrid population of *Helianthus* over an eight-year period. *Evolution* 15: 60–71.

Khush, Gurdev S., and GLS. 1961. Cytogenetic and evolutionary studies in *Secale*. I. Some new data on the ancestry of *S. cereale*. *Amer. J. Bot.* 48: 723–30.

GLS. 1961. A diploid subspecies of the *Dactylis glomerata* complex from Portugal. *De Flora Lusitana Commentarii* 14: 9–15.

GLS and Beecher Crampton. 1961. A suggested revision of the grass genera of temperate North America. In *Recent Advances in Botany*, pp. 133–45. Toronto: University of Toronto Press.

GLS and S. S. Shah. 1962. Differences in free amino acid content of seedlings of awned and hooded barley and their alteration by chloramphenicol. *Proc. Nat. Acad. Sci. USA* 48: 1513–19.

GLS. 1962. Toward better international cooperation in the life sciences. *Plant Sci. Bull.*, pp. 8–10.

GLS. 1962. International horizons in the life sciences. *Amer. Inst. Biol. Sci. Bull.*, pp. 13–19.

Ariyanayagam, David V., and GLS. 1962. Developmental studies of cell differentiation in the epidermis of monocotyledons. III. Interaction of environmental and genetic factors on somatal differentiation in three genotypes of barley. *Devel. Biol.* 4: 117–33.

GLS. 1963. The dynamics of evolutionary change. *Lectures in Biological Sciences*, pp. 39–62. Knoxville: University of Tennessee Press.

GLS. 1963. Perspectives. I. "Animal Species and Evolution" by Ernst Mayr, a review. *Amer. Sci.* 51: 362–70.

GLS. 1963. Identification of the ancestry of an amphiploid *Viola* with the aid of paper chromatography. *Amer. J. Bot.* 5: 830–39.

GLS and E. A. Yagil. 1963. Environmental factors affecting the development and expression of the hooded phenotype in barley (Abstract). *Amer. J. Bot.* 50: 619.

GLS. 1964. Four basic questions in plant biology. *Amer. J. Bot.* 51: 220–30.

GLS. 1964. The evolution of animal species. *Evolution* 18: 134–37.

GLS. 1964. Modern evolutionary theory (review of Verne Grant, "The Origin of Adaptations"). *J. Hered.* 55: 44.

GLS and R. T. Wijewantha. 1964. Developmental and biochemical effect of the agropyroides mutation in barley. *Genetics* 50: 65–80.

GLS. 1965. The experimental approach to problems of evolution. *Folia Biol.* 11: 1–10.

GLS and Jack Major. 1965. Endemism and speciation in the California flora. *Ecol. Monogr.* 35: 1–35.

GLS. 1965. From gene to character in higher plants. *Amer. Sci.* 53: 104–26.

GLS. 1965. Evolution of crop plants (review of Sir Joseph Hutchinson, ed. "Essays on Crop Plant Evolution"). *J. Hered.*, 56.

GLS. 1965. Pitfalls and guideposts in comparing organic and social evolution. *Pac. Sociol. Rev.* 8: 3–10.

GLS. 1965. The probable growth habit of the earliest flowering plants. *Ann. Mo. Bot. Gard.* 52: 457–68.

GLS. 1965. Some relationships between mitotic rhythm, nucleic acid synthesis, and morphogenesis in higher plants. In *Genetic Control of Differentiation; Brookhaven Symp. Biol.* 18: 204–21.

Day, Alva, and GLS. 1965. Cytogenetic evidence for long-continued

evolutionary stability in the genus Plantago (abstract). Science
150: 371.
GLS and Peter Jura. 1965. Differential synthesis of nucleic acids
associated with cellular differentiation in the leaf sheath epidermis of barley (abstract). Science 150: 385–86.
GLS. 1965. Colonizing species of the native California flora. In The
Genetics of Colonizing Species, pp. 173–191. New York: Academic Press.
GLS. 1966. Chromosomal variation and evolution. Science 152:
1463–69.
GLS. 1966. Variation and adaptation in Galapagos plants. In The
Galapagos: Proceedings of the Symposia of the Galapagos International Scientific Project, Robert I. Bowman, ed., pp. 46–54. Berkeley, Calif.: University of California Press.
GLS and Ezra Yagil. 1966. The morphogenetic effects of the hooded
gene in barley. I. The course of development in hooded and awned
genotypes. Genetics 54: 727–41.
Dempster, Lauramay T., and GLS. 1965. The fleshy-fruited Galium
species of California (Rubiaceae). I. Cytological findings and some
taxonomic conclusions. Madroño 18: 104–13.
GLS. 1966. Polarity gradients and the development of cell form. In E.
G. Cutter (ed.), Trends in Plant Morphogenesis, pp. 115–26. Long-
manns, Green.
GLS. 1967. Two symposiums on chromosomes ("Chromosomes
Today," C. D. Darlington and K. R. Lewis, 1964, and "Chromo-
some Manipulations and Plant Genetics," Ralph Riley and K. R.
Lewis, 1964). Science 155: 184–85.
GLS, Suryakant S. Shah, Denise Jamin, and Peter Jura. 1967.
Changed orientation of the mitotic spindle of stomatal guard cell
divisions in Hordeum vulgare (abstract). Amer. J. Bot. 54: 71–80.
GLS. 1967. The place of botany in a unified science of biology.
BioScience 17: 83–87.
GLS. 1967. Adaptive radiation and trends of evolution in higher
plants. In Th. Dobzhansky, M. K. Hecht, and W. C. Steere (eds.),
Evolutionary Biology, vol. I, pp. 101–42. New York: Appleton-
Century-Crofts.
GLS and Alva Day. 1967. Cytogenetic evidence for long-continued
stability in the genus Plantago. Evolution 21: 409–428.
GLS. 1969. From gene to character in higher plants. In Science in

Progress, Sixteenth Series. New Haven, Conn.: Yale University Press.

GLS. 1967. Gene action, mitotic frequency, and morphogenesis in higher plants. *Devel. Biol. (Suppl.)* 1: 113–35.

Dempster, Lauramay T., and GLS. 1968. A cytotaxonomic revision of the fleshy-fruited *Galium* species of the Californias and southern Oregon (Rubiaceae). *Univ. Calif. Publ. Bot.* 46: 1–52.

GLS. 1968. Integration of development and evolutionary progress. In R. C. Lewontin (ed.), *Population Biology and Evolution,* pp. 17–36. Syracuse, N.Y.: Syracuse University Press.

GLS. 1968. The impact of modern genetics upon our understanding of life and of the future of mankind. *J. Mysore Univ., Sect. B.,* Golden Jubilee vol., 1–6.

GLS. 1968. Present and potential contributions of developmental genetics to our understanding of plant evolution. *Proceedings of the XII International Congress of Genetics,* vol. II, p. 222 (abstract).

GLS. 1969. The significance of hybridization for plant taxonomy and evolution. *Taxon* 18: 26–35.

GLS. 1969. The effect of asexual reproduction on higher plant genera with special reference to Citrus. *Proc. Int. Citrus Symp.* 1: 455–58.

Gupta, Vimal, and GLS. 1969. Peroxidase activity in hooded and awned barley at successive stages of development. *Biochem. Genet.* 3: 15–24.

Whittingham, Alva Day, and GLS. 1969. Chromosomal rearrangements in *Plantago insularis* Eastw. *Chromosoma* (Berlin) 26: 449–68.

Yagil, E., and GLS. 1969. The morphogenetic effects of the hooded gene in barley. II. Cytological and environmental factors affecting gene expression. *Genetics* 62: 307–19.

GLS. 1969. Comments on the search for a "Perfect System." *Taxon* 18: 357–59.

GLS. 1969. Developmental genetics and plant evolution. *Japan. J. Genet.* (Suppl 1) 44: 344–50.

GLS, and V. K. Gupta. 1969. The relation between peroxidase activity and the morphological expression of the hooded gene in barley. *Proc. Nat. Acad. Sci.* 64: 50–56.

GLS. 1970. Prospects for spaceship man. *Saturday Review,* pp. 48–66, March 7, 1970.

GLS. 1970. Variation and evolution in plants: Progress during the past twenty years. In M. K. Hecht and W. C. Steere (eds.), *Essays in*

Evolution and Genetics in Honor of Theodosius Dobzhansky. New York: Appleton-Century-Crofts; (1970) Evol. Biol. (Suppl.), pp. 173–208.

GLS. 1970. Botanizing in California's nooks and crannies. Calif. Native Plant Soc. Newsletter, October 1970.

GLS. 1970. The natural history and evolutionary future of mankind. Amer. Nat. 104: 111–26.

GLS. 1970. Biosystematics: An avenue towards understanding. Taxon 19: 205–14.

GLS. 1970. Transference of function as a factor in the evolution of seeds and their accessory structures. Isr. J. Bot. 19: 59–70.

GLS. 1970. Adaptive radiation of reproductive characteristics in angiosperms. I. Pollination mechanisms. In Ann. Rev. Ecol. Syst. 1.

GLS. 1971. Relationships between adaptive radiation, speciation and major evolutionary trends. Taxon 20: 3–16.

GLS. 1971. A Review of "Genetic Resources in Plants." Science 172: 1018–19.

Murr, Sandra M., and GLS. 1971. An albino mutant in Plantago insularis requiring thiamine pyrophosphate. I. Genetics. Genetics 68: 231–58.

GLS and H. J. Price. 1971. The developmental genetics of the Calcaroides gene in barley. I. Divergent expression at the morphological and histological level. Genetics 68: 527–38.

Price, H. J., and GLS. 1971. The developmental genetics of the Calcaroides gene in barley. II. Peroxidase activity in mutant and normal plants at progressive stages of development. Genetics 68: 539–46.

GLS. 1971. Adaptive radiation of reproductive characteristics in angiosperms. II. Seeds and seedlings. Ann. Rev. Ecol. Syst. 2: 237–60.

GLS and Eduardo Zeiger. 1972. Developmental genetics in barley: A mutant for stomatal development. Amer. J. Bot. 59: 143–48.

GLS. 1972. The Evolution of the Grass Family. In The Biology and Utilization of Grasses, Chap. 1. New York and London: Academic Press.

Smith, A. T., and GLS. 1971. A morphological and histological study of the tomato mutant "curl." Amer. J. Bot. 58: 517–24.

GLS. 1972. Ecological distribution of centers of major adaptive radiation in angiosperms. In D. H. Valentine (ed.), Taxonomy,

Phytogeography, and Evolution, pp. 7–34. New York: Academic Press.

GLS. 1972. Research on the evolution of higher plants: Problems and prospects. *Canad. J. Genet. Cytol.* 14: 453–62.

GLS and R. Lewontin. 1971. Comparative evolution at the levels of molecules, organisms and populations. In *Darwinian, Neo-Darwinian, and Non-Darwinian Evolution. Proc. Sixth Berkeley Symp. Math. Stat. Prob.* 5: 23–42.

GLS. 1973. The evolution of design. *Amer. Biol. Teach.* 35: 57–61.

GLS. 1972. Edgar Anderson: Recollections of a long friendship. *Ann. Mo. Bot. Gard.* 59: 373–79.

GLS. 1973. Morphogenesis, vascularization and phylogeny in angiosperms. *Breviora, Mus. Comp. Zool.,* no. 418.

GLS. 1973. Adaptive radiation and the origin of form in the earliest multinuclear organisms. *Syst. Zool.* 22: 478–85.

GLS. 1973. Evolution of morphogenic patterns. *Brookhaven Symp. Biol.,* no. 25. (reprinted fom *Basic Mech. Plant Morphogenesis*).

GLS. 1974. Building bridges between evolutionary disciplines. *Taxon* 23: 11–20.

GLS. 1974. Adaptive shifts and evolution novelty: A compositionist approach. In F. J. Ayala and Th. Dobzhansky (eds.), *Studies in the Philosophy of Biology.* London: Macmillan.

GLS. 1974. The evolutionary significance of biological templates. In A. D. Breck and W. Yourgrau (eds.), *Biology, History, and Natural Philosophy,* pp. 79–102. New York: Plenum Publications.

GLS. 1974. The role of polyploidy in the evolution of North American grasslands. *Taxon* 24: 91–106.

GLS. 1974. A California botanist in Chile. *Fremontia* 2: 8–13.

Heckard, L. R., and GLS. 1974. A new *Lewisia* (Portulacaceae) from the Sierra Nevada of California. *Brittonia* 26: 305–8.

GLS. 1975. Deductions about transspecific evolution through extrapolation from processes at the population and species level. *Ann. Mo. Bot. Gard.* 62: 825–34.

Frias, L. D., R. Godoy, P. Iturra, S. Koref-Santibanez, J. Navarro, N. Pacheco, and GLS. 1975. Polymorphism and geographic variation of flower color in Chilean populations of *Eschscholzia californica. Plant Syst. Evol.* 123: 185–98.

GLS. 1975. Shrubs as centers of adaptive radiation and evolution. *Proc. Symp. Workshop on Wildland Plants,* pp. 120–40.

GLS. 1976. L'Écologie comparative de quelques espèces de legumineuses de la flore méditerranéenne. *Colloq. Int. CNRS* 235: 361–68.

GLS. 1976. Seeds, seedlings, and the origin of angiosperms. In Charles B. Beck (ed.), *Origin and Early Evolution of Angiosperms,* pp. 300–311. New York: Columbia University Press.

GLS. 1976. Seed and seedling ecology in annual legumes. I. A comparison of seed size and seedling development in some annual species. *Ecol. Plant.* 11: 321–31.

GLS. 1976. Seed and seedling ecology in annual legumes. II. Stem growth, seed production and mechanisms for transport. *Ecol. Plant.* 11: 333–44.

GLS and R. D. Hoogland. 1976. Species diversity, ecology and evolution in a primitive angiosperm genus, *Hibbertia* (Dilleniaceae). *Plant Syst. Evol.* 125: 139–54.

GLS. 1976. Ecological islands and vernal pools. In S. K. Jain (ed.), *Vernal Pools: Their Ecology and Conservation.* *Proc. Symp. Inst. Ecology,* University of California at Davis.

GLS. 1977. In defense of evolution: Tautology or theory? *Amer. Nat.* 111: 386–90.

1 FIFTY YEARS OF PLANT EVOLUTION

G. LEDYARD STEBBINS

THE LAST HALF-CENTURY has seen a revolution in the thinking of biologists about evolution. During the 1920s, when I became a biologist, the study of evolution as a separate discipline of the life sciences did not exist. Anatomists and systematists were using evolutionary phylogeny to interpret the facts that they observed. Paleontologists were clarifying the course of evolution, particularly in vertebrate animals, by describing successions of fossils. Most geneticists believed that the principal mechanism of evolution was in the origin of mutations having large and conspicuous effects. Darwinian natural selection was under a cloud, the prevailing belief being that it had only the negative effect of eliminating unfavorable mutations. The burgeoning field of chromosomal cytology was concerned chiefly with analyzing and understanding the mechanisms responsible for the chromosomal cycle, particularly meiosis and crossing-over, and with recording the effects of such chromosomal changes as translocations, inversions, trisomics, and polyploidy. The principal contribution of cytogenetics to evolution, before 1930, was the demonstration that new species can arise suddenly through hybridization—accompanied, preceded, or followed by doubling of the chromosome number. Population genetics did not become a viable discipline until the 1930s, and its marriage with ecology, which produced the modern, dynamic approach to the processes of evolution, was not consummated until the 1960s.

This final integration of the evolutionary discipline was preceded by research having three different kinds of motivation. First, systematists hoped that a better understanding of evolution would help them to produce more clear-cut, stable classifications of species, genera, and families. They were primarily concerned with developing a valid species concept and separating related species from each other. Since the constancy of chromosome number within a species was a generally accepted dogma, chromosome counting and analysis of chromosome behavior in hybrids appeared to be the best way of attaining this goal. Second, many botanists were fascinated with the rich diversity of form, structure, ecological and geographic distribution, and adaptation among plants and enjoyed analyzing this pattern of diversity for its own sake. Finally, a relatively small group of broadly trained geneticists—J. B. S. Haldane, R. A. Fisher, and S. C. Harland in England; E. B. Babcock, A. F. Blakeslee, and R. E. Clausen in the United States; as well as M. S. Navashin and N. I. Vavilov in the USSR—believed that the study of higher plants could contribute significantly to a better understanding of the principles of evolution in general. Although these three objectives were often combined, the threads of investigation governed by them are separate enough that they can serve as the framework about which I have organized this brief history of plant evolution.

CHROMOSOMES AND SPECIATION

Botanists who were concerned with using evolutionary principles as aids to classification concentrated upon two phenomena: chromosomal differences and the formation and nature of interspecific hybrids. The first demonstration that a chromosomal difference could have fundamental significance for the differentiation of species was Otto Rosenberg's (1909) analysis of *Drosera rotundifolia* and *D. "longifolia"* (now known as *D. anglica*). He showed that these two species have the chromosome numbers $2n = 20$ and $2n = 40$, respectively, and that, in sterile hybrids between them, the 30 chromosomes present form at meiosis 10 pairs and 10 single, or univalent, chromosomes. The ultimate resolution of this problem, the demonstration that *D. anglica* is an allopolyploid derived from the hybrid *D. linearis* × *rotundifolia*, was accomplished much later by Carroll Wood (1955). The next step forward was taken by Ojvind Winge (1917) in a long review article whose title, "The chromosomes: Their numbers and general importance," appears to us naively simple and general but accurately expresses the nature and position of chromosome cytology at that time. Winge first realized that if the sterility of interspecific hybrids is based upon failure of chromosomes from the parental species to pair with each other and the consequent irregularity of meiosis, this condition can be corrected by doubling the chromosome number, and thus a stable, fertile allopolyploid or amphidiploid can be produced. His conclusions were entirely theoretical since no examples were available to him. Actually, four cases of stable, fertile derivatives of interspecific hybrids, which were later recognized as allopolyploids, were already known. For three of these, *Triticale* (Müntzing, 1939), *Anemone janczewskii* (Gajewski, 1946), and *Spartina townsendii* (Goodman et al., 1969), chromosomal information was not obtained until much later. For the fourth, *Primula kewensis*, chromosome numbers were known; some information on chromosomal behavior was also available, but it had been misinterpreted (Digby, 1912). Probably because of poor communication between England and Denmark during the First World War, Winge did not know about *Primula kewensis* when he wrote his review. If he had, he would almost certainly have interpreted it correctly. The correct interpretation of *P. kewensis* was not made until 1929 (Newton and Pellew, 1929).

A number of investigations during the 1920s demonstrated clearly the validity and wide significance of Winge's hypothesis. The first artificial allopolyploid to be recognized as such was *Nicotiana digluta* (Clausen and Goodspeed, 1925). Then followed *Raphanobrassica* (Karpechenko, 1927) and species of the *Solanum nigrum* complex (Jorgenson, 1928). The demonstration by Müntzing (1930a,b) that the Linnean species *Galeopsis tetrahit* is of allopolyploid origin opened the way to verifiable hypotheses regarding the phylogeny of the numerous genera of plants that were already known to possess polyploid series of chromosome numbers. Verifications of such hypotheses were made much easier by the discovery, on the part of three separate research workers in the late 1930s, that polyploids could be synthesized with the aid of colchicine (Dustin, 1934; Dustin et al., 1937; Eigsti, 1938; Nebel and Ruttle, 1938; Blakeslee, 1939). Polyploid research became an important part of evolutionary studies in plants. Its significance in understanding the diversity of plants is discussed later.

Paradoxically, the recognition and analysis of polyploidy has not lightened the task of classifying and recognizing species in groups that are dominated by this phenomenon. Allopolyploids derived from two clearly distinct species or genera are easy to recognize and define but form only a small proportion of natural polyploids. The autopolyploids that arise in garden cultures were recognized in *Oenothera* as early as 1907 (Lutz, 1907), and their nature was well analyzed before allopolyploids were known to exist (Winkler, 1916). Based upon the genome concept as developed by Kihara (Kihara and Ono, 1926; Kihara and Nishiyama, 1930), the earlier cytologists regarded autopolyploids and allopolyploids as sharply distinct classes, so that the only major taxonomic problem they recognized was whether or not to recognize autopolyploids as distinct species, separate from their diploid ancestors. Later, conditions intermediate between autopolyploidy and allopolyploidy were recognized (Stebbins, 1950), and the concept of the polyploid complex was developed. In many genera, polyploids were found to form a whole series of intermediate forms between otherwise distinct diploid entities. These results showed that the difficulty of recognizing species boundaries in many genera characterized by a high proportion of polyploids is intrinsic and is not due to lack of perception on the part of the taxonomist, which did not ease the task of producing a clear-cut division of a genus into easily recognized species. Particularly in those genera that combine polyploidy with asexual reproduction, such as *Hieracium, Taraxacum, Rubus, Potentilla,* and *Poa,* the "nightmare for the taxonomist" will never be removed, and species boundaries will always be a matter of personal judgment.

In addition to chromosome number, the form and size of chromosomes were soon found to possess great taxonomic significance.

The pioneer comparative research in this field was by Babcock and his associates in the genus *Crepis* (Hollingshead and Babcock, 1930; Babcock and Cameron, 1934) and by a large group of workers in the USSR headed by G. A. Levitzky (1931a,b) and M. S. Navashin (1932). Both of these groups recognized the importance of structural rearrangements, as described by Belling and Blakeslee (1926) and McClintock (1931), and interpreted karyotypic differences on the basis of such changes. A similar approach, with additional emphasis on chiasmata in bivalents at meiosis and on "secondary pairing," was adopted by Darlington and his school (Darlington, 1937). Karyotype analyses were reconized to be of great importance for delimiting groups of species and genera.

The analysis of chromosomal translocations held a particular fascination for *Oenothera* cytology, since cytology could be combined with genetics to explain the nature and origin of the entities that were then recognized as separate species (Renner 1917, 1921; Cleland, 1936). Since these "species" were soon recognized as a peculiar form of chromosomal races (Darlington, 1929, 1931; Blakeslee and Cleland, 1930), *Oenothera* cytogenetics has suffered the same fate with reference to taxonomy as has the cytogenetics of polyploids: It has shown why delimitation of species in many subgenera of *Oenothera*, as well as in other genera, is intrinsically difficult, but it has not led the way to a clear-cut simple classification.

HYBRIDIZATION AND THE SPECIES PROBLEM

The formation of artificial hybrids was used for delimiting species even before Mendel, by Kölreuter and others. The Dutch geneticist J. P. Lotsy (1916) first postulated that hybridization could give rise to new species and to new morphological characteristics, but his hypothesis was based upon an ambiguous concept of the nature of species and was supported by little evidence other than anecdotal; as a consequence, it received little attention. Beginning soon after the rediscovery of Mendel's laws and continuing up to the present time, artificial hybridization, followed by estimates of fertility and analysis of chromosome behavior in the hybrids, has been a standard method of determining the distinctness of species and deciding whether two groups of populations should be ranked as species or subspecies.

The botanist who more than any other exploited natural hybridization as a source of new variation within species populations was Edgar Anderson. According to his own account, he arrived at his conclusions concerning the importance of introgression by careful observations of natural populations, using methods which came more naturally to a geneticist than to a taxonomist. He began by comparing variability within the populations of *Aster anomalus* (Anderson, 1929), a species which he chose because it belongs to a large and complex genus but which is, nevertheless, well defined and never confused with other species. He found that, in this species, each population has a range of variability comparable to that of the species as a whole; racial or ecotypic differences could not be recognized. He then directed his attention to the blue flag irises, at that time placed by most taxonomists in a single species, *Iris versicolor* (Anderson, 1936b). Using the earliest versions of the diagrammatic methods for which he later became famous, which reduce complex variation patterns to relatively simple elements, he recognized two kinds of variation. One kind existed within each population, and was of the same kind but greater when certain populations were compared. The other kind of variation separated the populations into two groups, which he later determined as having different chromosome numbers, $2n = 72$ and $2n = 108$. He realized not only that he was dealing with two different species, *Iris versicolor* ($2n = 108$) and *I. virginica* ($2n = 72$), but also that genetic variation within each species was of a different nature from that between the species. Using the principle of amphidiploidy, to which he added the method of extrapolation, he postulated that *I. versicolor* is an amphidiploid between *I. virginica* and an unknown species having $2n = 36$ chromosomes, which he first described hypothetically by extrapolation from the difference between *I. versicolor* and *I. virginica*. He later discovered the species in herbarium specimens from eastern Alaska, which had been classified as *I. setosa* var. *interior*.

After this initial success, he and his students developed the concept of introgression, based upon studies of other species of *Iris*,

as well as *Tradescantia, Oxytropis, Juniperus,* and other genera. He
showed that hybridization and introgression are of great importance
for taxonomic determinations by an ingenious and highly original
device—the use of his friend and taxonomic colleague, Robert
Woodson (Anderson, 1936a). Since Woodson was a recognized
authority on the Apocynaceae, the logical genus for the collaboration was *Apocynum,* in which several species and varieties
described by taxonomists were suspected to be of hybrid origin.
Anderson raised segregating progeny from a partly sterile F_1 hybrid
between *Apocynum androsaemifolium* and *A. cannabinum* and
sent herbarium specimens of them to Woodson for identification,
without indication of their source. Among siblings of the same
progeny, some were identified as *A. androsaemifolium,* some as *A.
cannabinum,* and still others as *A. medium,* which had been
regarded by some taxonomists as of hybrid origin. Of equal importance was the fact that the genetic factors responsible for the partial
sterility of the F_1 hybrid were segregating in the F_2 progeny, which
included plants having a range of fertility values from high fertility
to almost complete sterility. Those identified by Woodson as belonging to one or the other of the parental species included plants which
were just as sterile as, or more so than, some plants classified as the
supposed hybrid *A. medium* and were similar to the known F_1
parent.

The personality of Edgar Anderson, combining a capacity for
precise, careful observation and logical deduction with flamboyant
showmanship and a penchant for intemperate attacks on those
whom he regarded as misguided or stupid, was bound to make
implacable enemies as well as devoted friends. Perhaps it is still too
soon to estimate the ultimate value of his contributions to evolutionary science. Undoubtedly, he taught a whole generation of botanists
to observe variation in plant populations in an entirely new way.
The term introgression, which he coined, is firmly fixed in the
vocabulary of population genetics and evolution.

Many botanists have regarded the acknowledged prevalence of
natural hybridization between distinct populations of plants as a
reason for rejecting the biological species concept. Undoubtedly, the
claim that some zoologists have made for this concept, that it

removes subjectivity and arbitrariness from the delimitation of species, is not valid for plants. Not only do hybrids between various
plant species exhibit a whole spectrum of sterility, from almost
complete fertility through various degrees of semisterility to the
virtual absence of seed setting, but, in addition, the significance of
semisterility as a barrier to gene flow depends greatly upon other
characteristics of the species involved. If the plants are annuals,
even a small amount of sterility, combined with the adaptive inferi-
ority of many of the progeny that are produced, can restrict gene
flow to such an extent that the populations of the parental species
remain distinct. On the other hand, F_1 hybrids between species that
are long-lived perennials, particularly those that have effective
methods of vegetative, asexual reproduction, can serve as vehicles
for gene transfer between their parental populations even though
they produce only 1 percent or less of the seed produced by their
fertile parents. The example of *Elymus condensatus* × *triticoides*, in
which a highly sterile hybrid (produced artificially) is very similar to
equally sterile clones that are widely established under natural
conditions, has been discussed elsewhere (Stebbins, 1959). In other
examples, particularly species of oaks *(Quercus)* that are sympatric
over wide areas, the same pair of species can exist side by side in
some regions with little or no intergradation, whereas in other
regions their identity has become so blurred by hybridization and
introgression that the originally distinct populations are now unre-
cognizable (Burger, 1975).

Does this mean that all considerations of reproductive isolation
should be eliminated from delimitations of plant species and that we
should revert to purely morphological definitions, supported by
computer-based analyses? As an evolutionist, I regard this as a step
backward. Recent analyses of interspecific differences with respect
to allozymes (Ayala, 1975; Dobzhansky et al., 1977) strongly support
the notion that reproductive isolation serves as the starting point for
divergence of evolutionary lines from each other. Consequently,
species definitions based at least partly upon reproductive isolation
under natural conditions are of great value for understanding critical
steps of the evolutionary process. Definitions based only upon morphological data are useless to the evolutionist.

THE EXPLORATION OF VARIATION
WITHIN SPECIES

Many important research ventures have aimed only to understand the intricate pattern of variability found within plant species
and populations. In the 1920s, the questions most frequently asked
were: How much of this variation is genetic, and how much is based
upon environmental modification of the phenotype? Can phenotypic modifications be transformed into hereditary differences?
What proportion of the differences between individuals and natural
populations is due to differences with respect to genes having large
effects upon the phenotype, and what proportion is based upon
multilocus inheritance of quantitative characters? The pioneer
experiments of Turesson (1922a,b, 1925, 1931) developed methods
to answer these questions. He first showed that specialized habitats
within the same climatic zone, such as coastal bluffs, sand dunes,
and open fields, are inhabited by genetic races or ecotypes that can
be clearly recognized and differentiated when transplanted into a
uniform garden. He then showed that widespread species comprise
an extensive series of climatic ecotypes, each of which possesses
distinctive morphological and physiological characteristics. Turesson's techniques were exploited in Great Britain by Gregor (1938,
1939) and in the United States by Clausen, Keck, and Hiesey (1940,
1948). Similar transplantations of samples derived from various
provenances had already been used by foresters for interpreting
intraspecific variation (Engler, 1913), and these types of studies
received added stimulus from Turesson's research (Langlet, 1936;
Burger, 1941).

     This pioneer research on ecotypes brought forth two additional
questions from evolutionists: Does the concept of the ecotype as a
distinctive entity express accurately the intraspecific variation pattern in most plant species? Are ecotypes of plant species comparable
to subspecies as recognized by zoologists?

     Turesson maintained that ecotypes are distinctive populations
and that the interfaces between different ecotypes are occupied not
by populations having intermediate modes of variation but by heterogeneous populations, highly variable with respect to the distinctive

differences between recognizable ecotypes. Other workers, particularly Langlet (1936), Gregor (1939), Gregor, Davey, and Lang (1936), and Olmstead (1944), believed that distinctive, geographically separated ecotypes are largely an artifact of sampling. They regarded much of the variation within widespread species as consisting of genetical gradients, or clines, which vary with respect to quantitative characters. Langlet first developed the concept that the steepness of these clines, and therefore the distinctiveness of spatially separated "ecotypes," is largely a function of the nature—whether abrupt or gradual—of the transition between different climatic zones or conditions. This concept has now largely superseded the view which regards species as made up of a mosaic of distinctive ecotypes (Ehrlich and Raven, 1969).

The ecotype concept overlaps the subspecies concept to a considerable degree, but the two concepts differ from each other enough so that both are valuable aids to understanding variation within species. The subspecies concept is based primarily upon phenotypes, and the operational methods for recognizing subspecies are essentially the same as those by which taxonomists recognize species. Turesson and Jens Clausen, however, insisted that ecotypes can be recognized only by means of transplant experiments, in which uniform garden conditions reduce differential modification of the phenotype to a minimum. The more recent investigations (Clausen, Keck, and Hiesey, 1948), which have provided further evidence favoring the predominance of clinal genotypic variation, show that, in plants, the description of morphologically recognizable subspecies or "varieties" calls attention to only a small proportion of the adaptive differences among the populations of a species. On the basis of present knowledge, we may more appropriately refer to ecotypic variation, which is often quantitative or clinal in nature, rather than try to subdivide species into distinct ecotypes. Nevertheless, the recognition of geographically and edaphically different gene complexes on the basis of comparisons under uniform garden or laboratory conditions is an indispensable method of exploring differences between populations of a species. An extensive and illuminating review of this topic has been presented by Bennett (1964). The experimental research on ecotypic differentiation did much to dispel once and for all the beliefs in Lamarckian inheritance of acquired characteristics that many botanists still held during the first quarter of our century. The apparent demonstration of earlier workers that transplanting plants into new habitats brings about genetic adaptations to those habitats was not confirmed by the more careful experiments of Turesson and his followers. They showed that adaptation to new environments can be accomplished only by selection of appropriate gene complexes.

Nevertheless, the refutation of Lamarckian concepts does not eliminate the evolutionary importance of phenotypic plasticity, as Bradshaw (1965) has ably demonstrated. The phenomenon of genetic assimilation, which Waddington (1960) demonstrated experimentally in *Drosophila* and which can be explained by assuming quantitative genetic control of enzyme action (Stebbins, 1966), provides a bridge between the "capture" of an ecological niche on the basis of phenotypic plasticity and the subsequent acquisition of a permanent hold on the niche via selection of adaptive gene complexes. An example is the evolution of succulent races in maritime habitats. If a maritime niche is available for colonization by a neighboring inland species, genotypes that are nonsucculent in an inland environment but have enough plasticity that they can acquire partly succulent phenotypes under the influence of salt spray from the ocean will be able to enter the habitat. Their progeny will be subject to strong selection, both for genotypes having equal plasticity and for gene complexes that code for increasing succulence regardless of the environment. In this way, several generations of selection can convert a population of highly modifiable phenotypes into a population that has complete adaptation based upon new complexes of genes.

The question of the relative importance of genes having large effects and those having small effects was conclusively answered by the later research of Clausen, Keck, and Hiesey (1948) and Clausen and Hiesey (1958). With respect to plant height, progeny tests of open-pollinated individuals of the *Achillea millefolium* complex revealed a large amount of continuous variability, indicating the presence of heterozygosity and polymorphism for this quantitative character at many different gene loci. When alpine and lowland ecotypes of *Potentilla glandulosa* were crossed and F_2 populations of hundreds of individuals were analyzed, multiple-locus inheritance was found for nearly every character difference between the parents. For quantitative characters such as height, degree of dormancy, and leaf size, the distribution patterns of the F_2 segregants approached a normal curve. In many instances, correlations between pairs of characters were observed, but the extent to which these correlations were due to genetic linkage rather than developmental pleiotropy was not determined.

EVOLUTIONARY MONOGRAPHS OF MAJOR TAXA

The understanding of the evolutionary basis of diversity among higher plants was greatly aided by a series of monographs synthesizing findings from a variety of sources—taxonomy, genetics, chromosome cytology, anatomy, histology, geographic distribution, and paleontology. The most notable of these were by Avdulov (1931) on the Gramineae and by Babcock (1947) on the genus *Crepis*. Also important but of narrower scope were those of Goodspeed (1954) on *Nicotiana*, Blakeslee and co-workers (Avery, Satina, and Rietsema, 1959) on *Datura*, and several monographic revisions of smaller genera.

Avdulov's synthesis was remarkable in many respects. In the first place, it provided a sound basis for rearranging the tribes of grasses into a system that reflects geographic and ecological affinities to a greater extent than the classical system of Hackel, upon which all systematic treatments of the family were based at that time. Second, it showed that morphological characteristics of the spikelets, particularly the structure and arrangement of the glumes and lemmas, are far less valuable as indicators of true affinities than are characteristics of leaf anatomy, starch grains, seedling development, and chromosomes. Third, it showed that the most primitive herbaceous grasses are not temperate genera like *Bromus* and *Festuca*, as is implied by the Hackelian system, but tropical groups related to the bamboos. Avdulov's evidence indicates that the grass family probably originated in subtropical or tropical semiarid climates and radiated into temperate climates only after it was fully evolved. Finally, Avdulov was the first to observe a correlation between absolute size of chromosomes and climatic distribution of a major plant group. He pointed out the striking difference between the medium-sized or small chromosomes of nearly all tropical grasses and the larger chromosomes of most temperate genera. He noted a correlation between this size difference and the characteristics of the metabolic nucleus: Larger nucleoli and, frequently, heterochromatic chromocenters are found in species having small chromosomes, and a more general distribution of Feulgen-staining material is found in species having large chromosomes. Recent estimates of DNA content (Rees and Jones, 1972; Sparrow et al., 1972) permit us to interpret Avdulov's data as indicating an increase in DNA content in various evolutionary lines of grasses in association with occupation of cooler climates.

The importance of Avdulov's research has been emphasized in more recent years by additional investigations along the same lines, particularly those of Prat (1932) on the histology of the epidermis, Reeder (1957) on embryo morphology, W. V. Brown (1958) on leaf anatomy, and various workers on lodicule morphology (Stebbins and Crampton,1961). Because of the great economic importance of this family, a new synthesis that would constitute an amplification of the framework laid down by Avdulov would be of great value, both to evolutionists and to research workers in a variety of applied fields, particularly agronomy, pasture and range research, and weed control.

Babcock's monograph of *Crepis* differed from that of Avdulov on grasses chiefly in its far greater attention to taxonomic and cytogenetic differentiation of species. Valuable to botanists generally was his demonstration that morphological characters previously regarded as "fundamental," and, therefore, as sure indications of the existence of separate genera, could vary among subspecies of a species and could even appear as mutations within a species population. Such was the fate of the presence or absence of paleae on the receptacle, which, prior to Babcock's research, was used to separate the genus *Rodigia* from *Crepis* but is now recognized merely as the diagnostic character of a subspecies, since the only species of *Rodi-*

gia has been reduced on genetic grounds to a subspecies of Crepis foetida. Babcock and his associates were among the first cytogeneticists to demonstrate that morphological differences between karyotypes of related species, including heteroploid alterations of basic number, are due to simple or complex combinations of structural differences, particularly translocations and inversions (Babcock and Cave, 1938; Babcock and Emsweller, 1938; Müntzing, 1934; Sherman, 1946; Tobgy, 1943).

At a different level, Babcock's analysis of evolutionary lines in Crepis showed how this important genus of the Mediterranean region has been derived from ancestors that lived in central Asia during the Tertiary period. He also reinforced the demonstration, by Avdulov and others, that karyotype information is a powerful tool for determining relationships between genera. His monograph, plus his support of monographic and karyotypic research on related genera of Cichorieae, paved the way for a more logical and evolution-based revision of the tribe Cichorieae (Stebbins, 1953; Jeffrey, 1966).

During the 1930s and 1940s, plant evolutionists living in California had high hopes that a monograph of another group of Compositae would be produced that would have been even more significant than that of Babcock on Crepis. Along with their research on ecotypic variation in Potentilla, Achillea, and other genera, Clausen, Keck and Hiesey were obtaining extensive data on species hybrids in the tarweeds (family, Compositae; subtribe, Madiinae). Their research reinforced that of Babcock's group in showing that the conspicuous morphological differences used by taxonomists to separate genera were in some instances diagnostic only of subspecies and in other cases exhibited polymorphism within populations. They also showed that most karyotypic differences are brought about by segmental rearrangements of chromosomes. They worked out the phylogeny of polyploid complexes by artificially synthesizing some of the polyploid entities. They went beyond Babcock, Avdulov, and other monographers of that period in their attention to intraspecific variation. In particular, they noted the presence of reproductive isolating mechanisms between two populations that belong to the same species and are connected by other intermediate populations not reproductively isolated from each other. They had planned to assemble the enormous amount of data they had collected into a monographic volume similar to that published in 1940, which reported the research on ecotypes. Just as they were beginning this project, our country entered the Second World War. They considered that a war-related project would be more appropriate and so did research on variation in forage grasses, with the hope of increasing the productivity of western rangelands. A small part of the research on Madiinae appeared in the second volume of *Experimental Studies on the Nature of Species* (Clausen, Keck, and Hiesey, 1945), but its full scope is evident only from the general review in Clausen's admirable volume that appeared as a result of his Messenger Lectures at Cornell University (Clausen, 1951).

In Europe, the synthetic approach to the evolution of genera and families was developed with similar success by Ehrendorfer (1959, 1965), who concentrated on the genera *Galium* and *Achillea* and the family Dipsacaceae. The research on *Galium* provided concrete evidence for the concept of Sewall Wright (1932) that evolution is greatly speeded up by cycles of isolation and differentiation that alternate with reunion of partly isolated populations, with consequent hybridization and segregation of adaptive gene complexes as well as adaptively neutral differences. In his research on the Dipsacaceae, Ehrendorfer showed that drastic reorganization of the karyotype accompanies the morphological differentiation that taxonomists use to recognize distinct genera. Surveys by Smith-White (1959) of cytogenetics and variation in genera of shrubs belonging to the largely Australian families Epacridaceae, Myrtaceae, Proteaceae, and Casuarinaceae called the attention of evolutionists to the rich diversity of problems that can be explored in the Australian flora.

POLYPLOIDY AND PLANT EVOLUTION

Beginning with the research of Winkler (1916) and continuing into the 1950s, the role of polyploidy in plant evolution, aside from the origin of species, was explored in various ways. A number of workers compared the physiology of autopolyploids with that of closely related diploids (see Stebbins, 1950, pp. 301–5), finding that the effects of chromosome doubling depend to a large extent on the genotype of the original diploid. Despite their higher chromosome numbers and larger cells, autopolyploids do not usually become larger plants than their diploid progenitors, since the rate of cellular proliferation is usually lower (Stebbins, 1949). Autopolyploids nearly always have lowered fertility owing partly to chromosomal irregularities at meiosis and partly to physiological factors.

The effect of polyploidy on genetic segregation was first explored by Haldane (1930) and later studied by several other workers, as reviewed by Little (1945). They all concluded that the more complex nature of segregation in polyploids, plus the presence of duplicate genes, greatly reduces the ability of individual mutations to contribute to evolutionary change and, as a rule, can be expected to retard greatly the rate of evolution in these species.

The comparative geographic distribution of diploids and polyploids was first studied by Tischler (1935) and Hagerup (1932). Both of these authors concluded that polyploids are more resistant than diploids to cold and drought. These results were based upon statistical comparisons of relatively small proportions of the floras involved and, in the case of Tischler, did not take into account the fact that northern Europe differs from southern Europe not only in having a colder climate but also in having been subjected to much greater alteration by the Pleistocene ice sheet. More careful comparisons, particularly those of Favarger (1964), showed that the proportion of polyploids found in the high Alps, a climate even more severe than that of northern Europe, is no higher than the proportion of polyploids in the flora of the surrounding lowlands, which have a much milder climate. Later, high proportions of polyploids were found in the fern flora of tropical Ceylon (Manton and Sledge, 1954) and in the Cameroons Mountains of tropical Africa (Morton, 1955). The most recent review of the comparative distribution of diploids and polyploids (Stebbins, 1971) indicates that, although a number of factors probably contribute to differential distributions of polyploids as compared to their immediate diploid relatives, the most important of these reflects the greater ability of polyploids to colonize newly available ecological niches.

The analysis of polyploid phylogenies has been greatly affected by fuller understanding of the basis of chromosome pairing. The belief that the amount of pairing accurately reflects the degree of genic similarity or homology, which is implicit in the genome concept as developed by Kihara (Kihara and Ono, 1926; Kihara and Nishiyama, 1930; Kihara, 1940), was considerably weakened by the recognition that physiological factors can often prevent the pairing of fully homologous chromosomes (Gaul, 1954). Further difficulties arose when Riley (1960) showed that, in the classic allopolyploid bread wheat (Triticum aestivum), meiotic pairing can become "diploidized" by the effects of single gene. To completely reject chromosome pairing as a tool for analyzing the relationships and origin of polyploids, which some cytogeneticists have proposed, is to throw out the baby with the bath water. Nevertheless, evidence from pairing should always be used with great caution and combined with other evidence before conclusions are reached.

The relationship between chromosome pairing and the gene content of chromosomes has been greatly clarified by Sears's development of the technique of nullisomic–tetrasomic substitutions (Sears, 1952). By means of this technique, partly homologous or homeologous chromosomes belonging to different genomes can be compared with respect to their developmental and physiological effects. Extensions of this technique have made possible the transfer of genes for disease resistance and other valuable characters between species so distantly related to each other that their F_1 hybrids are completely sterile. This kind of chromosomal engineering, initiated by Dr. Sears and his associates, has proved to be the most important contribution that evolutionary knowledge of plants has made to practical problems of plant breeding.

More recently, analyses of differences with respect to phenolic compounds (Stebbins et al., 1963), and particularly with respect to seed proteins (Johnson and Thein, 1970), have been used with great success to clarify phylogenetic relationships. Polyploid phylogeny, like many other aspects of plant evolution, will in the future be greatly aided by the use of biochemical techniques.

THE CONTRIBUTION OF PLANTS TO THE
PROBLEM OF
THE ORIGIN OF SPECIES

Evidence from plants has helped to clarify general problems of evolution in a number of ways. It has had great impact upon our understanding of the speciation process. Among animals, analysis of genetic changes that accompany speciation has until very recently been confined to the few groups that can easily be raised in the laboratory and hybridized, such as *Drosophila* and small rodents. The number of plant groups that can be handled in this manner is far greater and includes a much wider spectrum of life cycles and growth habits. Furthermore, much of the research on plant hybridization, from small annuals to large trees, is of practical value to agronomists, horticulturists, and foresters, so that motivation and funding for large-scale projects is relatively easy.

Evidence from plants first made possible the concept that each species contains a complex system of numerous interacting genes and that speciation involves chiefly a revolutionary alteration of a part of this complex. This concept of the species was developed by S. C. Harland from his analysis of gene segregation in hybrids of cotton (*Gossypium*) and the behavior of individual genes when transferred from one species to another. He stated (Harland, 1939) that "the study of species differences now resolves itself into a study of genetical backgrounds or modifier complexes and how they are built up, in short, the study of the genetical architecture of the species . . . It is a huge interlocking series of modifier complexes that characteristically marks off one species from another in a given genus." East (1935a) reached a similar conclusion from his studies in *Nicotiana*. More recent studies of speciation have done little more than elaborate this hypothesis.

The fact that adaptive differences among populations that form races of the same species are not necessarily the basis of interspecific differences was already evident from Müntzing's (1930a) analysis of segregation of morphological characters and sterility-promoting factors in the F_2 progeny of *Galeopsis tetrahit* × *bifida*. Further evidence in this direction came from the observations of Babcock (1947) arising from comparisons of different species complexes of *Crepis* and from similar observations by Clausen, Keck, and Hiesey (1945; Clausen, 1951) on various species complexes in the Madiinae. These workers showed that in some groups, populations that are very different from each other morphologically and ecologically will cross to form partly or completely fertile hybrids, whereas in other groups, very similar populations are separated from each other by such strong barriers of reproductive isolation that they form clusters of sibling species. Stebbins (1950) first suggested that such differences may be associated with differences in growth habit. Woody plants and long-lived perennials are slower to develop barriers of reproductive isolation in association with ecotypic differentiation than are small, quick-growing annuals. This and related concepts were reaffirmed and developed further by Grant (1963, 1971).

Other contributions of plant evolutionists to general problems of speciation include the demonstration by Baker (1959) that the shift from cross- to self-fertilization can greatly further the splitting of populations into separate species. Approaching the problem from a still different direction, Lewis (1962) reviewed evidence showing that, particularly in *Clarkia* and other genera of the family Onagraceae, "cataclysmic" bursts of chromosomal alterations can bring about rapid speciation. As has been elegantly shown in the excellent review of plant speciation by Grant (1971) and reaffirmed for animals by Bush (1975), speciation does not result from a single kind of process but may involve any one of several different interactions between altered gene complexes, chromosomal rearrangements, and adaptations to different environments.

CONTRIBUTIONS OF
PLANT EVOLUTIONISTS TO PROBLEMS
OF POPULATION GENETICS

Because of the relative ease with which large populations can be followed over many generations under controlled conditions, *Drosophila* and microorganisms are far more suitable objects for experimental studies of population genetics than are any plant groups.

Nevertheless, plants have provided the first evidence in favor of several concepts that are basic to understanding evolution at the level of populations. The most important of these is the Mendelian interpretation of quantitative variation based upon multilocus (or "polygenic") inheritance. Methods for determining the Mendelian basis of this kind of inheritance were explained by East (1916), using corolla length in *Nicotiana*, long before its importance was recognized in *Drosophila* and other animals. Fisher (1930) clearly showed the primary importance of this kind of inheritance by presenting the dictum that the smaller the effect of a mutated gene on the phenotype, the greater the likelihood that it will be incorporated into the gene system. In Fisher's classic book the similarity, or even identity, of the multilocus concept with the modifier complex concept as developed by Harland was made clear by Fisher's use of Harland's discovery that the genetic action of the crinkled dwarf gene becomes modified in the backgrounds presented by different species of cotton. This example was the principal evidence used by Fisher to support his hypothesis of dominance. The development of these concepts caused E. M. East to remark (1935b): "The deviations forming the fundamental material of evolution are the small variations of Darwin. We return to the Darwinian idea, modified by the demonstration of alternative inheritance." Thus, evidence from plant genetics played a key role in reestablishing the importance of natural selection upon slight heritable variations with respect to quantitative characters, as postulated by Darwin, rather than mutations with conspicuous effects, which most geneticists between 1900 and 1930 regarded as the most important genetic material for causing evolution. The primary importance of quantitative variation and multilocus inheritance was established firmly by analyses of segregation in F_2 progeny of hybrids between ecotypes (Clausen and Hiesey, 1958).

The first attempt to analyze the gene composition and evolution of a natural plant population was carried out by Epling and his associates (Epling and Dobzhansky, 1942; Epling, Lewis, and Ball, 1960) on the desert annual *Linanthus parryae*. They showed that the blue-flowered genotype, which differs by a single Mendelian factor from the more common genotype having white flowers, is distributed in the southwestern part of the Mohave Desert of California according to a pattern that is difficult or impossible to explain on the basis of adaptive differences and is probably due to random fixation of the blue allele, perhaps at the time when the large desert population of this species was first established. From the evolutionary standpoint, a difficulty in this analysis lies in the fact that blue-flowered and white-flowered races of L. *parryae* are found in the inner South Coast Ranges, to the northwest of the Mohave Desert, and there a clear ecological separation exists between the two races (oral communication from H. Lewis and C. Hardham, and personal observations). The full story of the evolution of Linanthus *parryae* has not yet been told. It is still a challenging example which may provide the evolutionist with information about relationships between natural selection and the effects of chance sampling.

More recently, plant population genetics has contributed to a deeper understanding of the importance of the genetic structure of populations. This has been possible because plants exhibit a wide spectrum of easily recognizable differences with regard to obligate outcrossing, partial inbreeding, and complete or nearly complete uniparental reproduction via enforced self-fertilization or apomixis. The importance of this condition was first recognized by Darlington (1939), whose *Evolution of Genetic Systems* was the foundation upon which all subsequent concepts have been built. Relatively little attention was paid to this topic during the 1940s, but it was further elaborated by Stebbins (1950, chapter 5), and particularly by Grant (1964). We presented evidence indicating that predominant self-fertilization is a derived condition (Stebbins, 1957) and recognized three ways of achieving temporary reduction in the amount of genetic recombination—increased linkage through reduction in chromosome number and chiasma frequency; facultative, or pre-dominant, self-fertilization; and apomixis. These three processes function as alternative evolutionary strategies for colonizing tempo-rary habitats and habitats having uncertain environmental fluctua-tions (Stebbins, 1958). The same kind of reasoning was extended to produce a hypothesis for explaining the haploid condition of many algae and fungi and the adaptive value of diploidy in higher plants (Stebbins, 1960). More recently, the topic of the evolution of genetic recombination and sex has been developed in depth, particularly through the work of Maynard Smith (1971) and G. C. Williams (1975).

THE UNION OF ECOLOGY WITH POPULATION GENETICS

To many modern evolutionists, whose thinking is based largely upon the concept of alternative evolutionary strategies and the fundamental importance of population–environment interactions, the lateness of the union between ecology and population genetics may appear rather surprising. The explanation for this is twofold. Up to 1950, most geneticists were confined to their laboratories and gardens, and those who did look upon natural habitats as something more interesting than simply places to collect had few contacts with practicing ecologists. Second, ecology, as it was practiced during the first half of the twentieth century, consisted largely of describing associations or communities, and its theoretical content consisted chiefly of highly controversial concepts such as the integral unit concept of the ecosystem and the rigid viewpoint on succession and climax proposed by F. E. Clements.

An intellectual climate favorable to the essential union between population genetics and ecology was created by two series of events that took place in Britain during the late 1940s and 1950s. In the first place, in 1941 the British Ecological Society initiated a series of contributions published in the *Journal of Ecology* under the heading "The Biological Flora of the British Isles" (British Ecological Society, 1941). Each article dealt with a separate species or species group and presented information according to a standardized list of topics: abundance, altitudinal limits, description of habitat, substratum, light and humidity factors, communities in which the species occurs, response to disturbance, gregariousness, sensitivity to frost or drought, morphology of underground parts, vegetative reproduction, longevity and seed setting, ecotypic variation, chromosome number, pollination biology, seed production and dispersal, hybridization with other species, seed germination and seedling biology, relation to herbivorous insects, parasites, and diseases. This was exactly the kind of information needed for constructing an evolutionary synthesis.

In addition, a number of workers in Great Britain and elsewhere began to study competition among related species both under natural conditions and in artificial simulations of microhabitats. Before this time, studies on plant competition were carried out chiefly by agronomists, such as Harlan and Martini (1938) and Suneson and Wiebe (1942). The pioneer research of Sukatschew (1928), though frequently cited in review articles and books, was for a long time an isolated example of this kind of work. The work of Salisbury (1942) on *The Reproductive Capacity of Plants* was another isolated forerunner of the modern era. The research that prepared the ground for and led into the modern discipline of ecological population genetics of plants was that of John L. Harper and his associates (Harper, 1965) on seed germination and seedling establishment; this research showed that small differences in microhabitat can exert strong selective pressures that most probably affect greatly the gene composition of the adult population.

An important milestone in the development of ecological population genetics was the symposium sponsored by the International Union of Biological Sciences at Asilomar, California, in 1963 on "The Genetics of Colonizing Species." This symposium was the brainchild of C. H. Waddington, who, when he became president of IUBS, decided that an important function of the Union should be to foster interdisciplinary conferences between research workers. He felt that the most important unsolved problems having to do with evolution were those concerning competition and establishment and that solutions of these problems would help greatly the development of methods for dealing with ecosystems that are of necessity under human influence. He therefore asked his secretary general (Stebbins) to organize the symposium in collaboration with H. G. Baker of the University of California. Most of us who were at Asilomar regard it as one of the most stimulating meetings that we have ever attended. The housing arrangements were excellent, and the atmoshere was informal and cordial. Evolutionists like Dobzhansky, Mayr, and H. L. Carson exchanged ideas not only with population geneticists like R.

Lewontin, R. W. Allard, and A. S. Fraser but also with botanists such as H. G. Baker, F. Ehrendorfer, J. L. Harper, C. B. Heiser, and D. Zohary, and, in particular, with experts in applied ecological disciplines, like M. H. Bannister, Frank Fenner, Paul DeBach, W. E. Howard, F. H. W. Morley, J. W. Purseglove, and K. Wodzicki. The volume that emerged from this symposium (Baker and Stebbins, 1965) achieved a popularity far beyond the predictions of its editors and still serves as a starting point for research in this field.

The present symposium can in many respects be regarded as a lineal descendant of Asilomar in 1963, even though the participants are almost completely different. Perhaps it is time for another gathering of ecological geneticists of both plants and animals, who will join with ecologists and particularly applied biologists to determine the most important contributions that our discipline can make toward achieving more balanced ecosystems under human influence.

Since 1965, the discipline of ecological population genetics has matured rapidly. The discovery of isozyme differences and their importance as indicators of genetic variability has provided the greatest stimulus on the side of genetics. At the same time, the concepts and quantitative methods developed by R. H. Mac Arthur and his associates have transformed ecology into a discipline that complements and furthers genetic research. Concepts such as alternative evolutionary strategies and K-determined vs. r-determined selection have provided a conceptual framework on which population geneticists and ecologists are building an entirely new structure. The future flourishing of the field is assured. I therefore conclude with an optimistic prediction: The next half century of research on processes of evolution will be just as exciting as the last 50 years have been. The coming years may be even more satisfying since they may see the advent of significant contributions made by evolutionists toward solving the world's problems.

1 ADAPTATION AND GENETIC VARIATION IN POPULATIONS

INTRODUCTION

EVOLUTION is fundamentally a genetic phenomenon. Genes provide the continuity of living matter over time: seeds, sporophytes, gametes, and gametophytes all have limited lifespans, even when viewed as populations of dividing cells. But survival of genes depends on the characteristics of the various phenotypes that the plant assumes throughout ontogeny and phylogeny and the ability of those phenotypes to capture resources and convert them into offspring in the face of competition and environmental constraints. Consequently, genes and the phenotypic structures they indirectly produce are intimately connected through a series of feedbacks during development and through the action of natural selection.

The genetically inclined population biologist is interested in ascertaining the frequency of genes and genotypes throughout the life cycle, the changes that take place in those frequencies, and the underlying causal factors (Figure 1). Such a goal, however, is unattainable at present for a variety of technical and conceptual reasons. Two factors are paramount. The first is the very limited knowledge that we presently possess regarding the process of development. Although it is possible in many instances to predict fairly accurately the genic composition of the population of gametes and zygotes from a knowledge of the phenotype of the parent plant population (provided the inheritance mechanism of the character is understood), it is not possible to predict the phenotype of a population of mature plants from knowledge of the genic composition of the zygote population, because we have a very limited understanding of how genes and environmental factors interact in development (Lewontin, 1974). The second reason is the technical problem of actually identifying genes and alleles. The technique of characterizing alleles by their primary gene products (allozymes) through gel

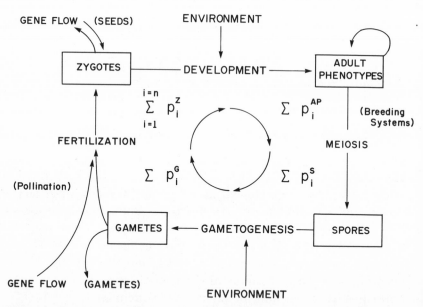

Figure 1. Generalized genetic model of the life cycle. The information of concern here is the gene frequency (Σp_i) at different stages of the life cycle and the effect of external factors (environment, breeding systems, pollination) on gene frequency change.

electrophoresis has provided a means of obtaining empirical data on gene frequencies in populations. However, as Johnson points out in this volume (article 3), the problems of biochemical regulation during development put in question the validity of evolutionary studies based on the monitoring of allele frequency changes of single genes in populations. Nevertheless, preliminary conclusions can be drawn in certain instances, as Hamrick (article 4) shows in correlating gene frequency with lifespan in plants.

Two factors influencing changes in gene frequency distributions are the breeding system and the details of pollination and fertilization. Solbrig (article 5) presents a cost–benefit analysis of recombination. One of the unresolved paradoxes of modern population biology is the apparent high level of heterozygosity in populations, which is impossible to account for under present single-gene models with selection. Plants, with their great variety of breeding systems, are ideal for testing costs and benefits of inbreeding and outbreeding.

Donald Levin (article 6) presents some novel ideas regarding pollinator behavior and interspecific hybridization, as well as the way that pollinators affect and are affected by the genetic structure of the population. The matter of gametic selection is not addressed in this section, but the interested reader is referred to Mulcahy's (1976) recent book.

But are models of plant evolution based on optimality arguments justified? Horn, in the introductory article (2), presents arguments in favor of an optimality view of evolution, although cautioning against circular and tautological thinking. He also presents the tantalizing view that botanists and zoologists have developed different outlooks as a result of being involved in the study of primarily sedentary and mobile organisms, respectively.

2 ADAPTATION FROM THE PERSPECTIVE OF OPTIMALITY

Henry S. Horn

ADAPTIVE and optimal arguments in botany are often scorned as tautological, Panglossian, teleological speculations about artifacts of history. Indeed, they are all that and more, but this is not as devastating a criticism as it might seem. In this article, I shall first make some idiosyncratic comments about the differences between plants and animals that seem to underlie some of the differences between botanists and zoologists in their use of adaptive and optimal arguments. I shall then review the formal argument of evolution by natural selection, distinguishing the circular from the noncircular parts and examining the role of optimality in generating testable statements about adaptive patterns. Various criteria for optimal models will follow, along with other techniques for generating adaptive predictions. Returning to the formal argument of adaptation, I shall show that evolutionary history plays a surprisingly small role in the origin and maintenance of contemporary adaptive patterns. Finally, I shall summarize some intriguing problems and the prospects for their solution, again emphasizing the peculiarities of plants that will not yield to metaphorical appeals to the ideas of population biology in animals.

Most of my philosophy is plagiarized from my mentors, Gordon H. Orians and Robert H. Mac Arthur, and from my literary heroes, Henry A. Gleason, Ronald A. Fisher, Robert H. Whittaker, and John

L. Harper. The only novelty in my perspective comes from my peculiar history of formal training as a zoologist, partial retreading as a botanist, and vicarious dabbling in theoretical population biology. I shall adopt an ecological, rather than evolutionary or genetic, perspective, partly from personal preference but mainly because optimal models are easier to frame unambiguously in an ecological context.

THE TROUBLE WITH PLANTS

A botanist presenting an optimal adaptive argument hears a litany of objections. Panglossian and teleological speculations are self-fulfilling and hence not testable. Adaptation to an environment including competitors requires stability of the physical environment and the plant community over evolutionary spans of time, but most communities suffer changes in composition, or even devastation, over much shorter time-scales (Cook, article 9). A plant may be present in a given area owing to adaptations other than the one in question, or even to an accident of history. Adaptation is the product of a long evolutionary history not necessarily related to the environment that a species currently occupies. How can one even talk about adaptations without knowing whether the unitary individual plant should be a ramet, a genet, or something in between (Hickman, article 10)?

Botanists have a history of taking these objections so seriously that zoologists often ask why zoologists always beat botanists to respectable adaptive arguments. Alternatively, why have botanists been more justifiably circumspect than zoologists in using adaptive arguments? I think that much of this difference stems from the differences in motility and taxonomy between most plants and most animals.

Motile animals are difficult to capture in the field without bias or injury. Zoologists have, therefore, a long tradition of refining the questions that they propose to answer so that minimal data from the field are required. Quantitative animal ecology has been concerned with life tables, schedules of age-specific births and deaths, and with the dynamics of populations. These data lend themselves readily to adaptive arguments and interpretations.

Since most plants stay put, the questions that immediately seem tractable are: what species are present; how many; and in what spatial distribution? Until recently, all of quantitative plant ecology has been devoted to answering questions of floral composition and spatial pattern that seem trivial to a zoologist. In fact, the questions are far from trivial, but their profundity has been exaggerated by confusing their descriptive and interpretive functions in acrimonious debate over such questions as: Are typological communities objective entities; and what is the best abstract measure of spatial pattern, local diversity, floral gradient, or similarity between communities? Communication between botanists and zoologists is hindered by the fact that many zoologists think that these debates are silly. Conversely, botanists are amused at the zoologists' debate over whether population regulation is density-dependent or density-independent. Crowding is too common and too dramatic a phenomenon in plant communities for its importance to be doubted. Nevertheless, zoologists are just discovering some of the interesting complications that spatial restrictions add to their theories, and plant ecologists are rediscovering actuarial demography. I shall say more about that later.

Most of animal taxonomy is based on characters that have ecological significance for moving about in a habitat or for seizing and processing food, for example, the bills, feet, and major feather tracts of birds. A faunal list, therefore, is almost an ecological inventory, even in taxonomically difficult and polyphyletic groups. This is in striking contrast to the situation that Raven (article 19) describes in plants, where taxonomy has traditionally been based on recondite details of reproductive organs, rather than on root, stem, branch, or leaf characters. Thus, the name of a plant's family tells a botanist much about its evolutionary pedigree, but it may leave him wondering whether the plant is a tree, vine, shrub, herb, or creeper. Adaptive patterns can become obvious to an ornithologist from a simple faunal list, but a floral list tells few ecological tales to a botanist, without local field experience, and it may even inhibit adaptive interpretations by suggesting constraints of evolutionary history. Imagine the state of ecological ornithology if simple bird-watching required intimate knowledge and close observation of cloacal morphology.

THE ARGUMENT OF EVOLUTION BY
NATURAL SELECTION

The argument of evolution by natural selection, in its axiomatic form, consists of four hypotheses, from which a fifth statement follows tautologically (after Mac Arthur and Connell, 1967). First, members of a population show variation in some features. Second, each variant has a characteristic reproductive success, measured ultimately by the number of its offspring who breed in the next generation. Third, for each variant, the characters of parent and offspring, including characteristic reproductive success, are corre-lated. Fourth, the environment remains constant. If these four hypotheses are true, then, with the passage of time, the composition of the population tends toward ever-greater dominance of the var-iant with the highest reproductive success. It is difficult to test or contest this last statement since it can easily be put in tautological form: Those variants that persist are the ones with the greatest persistence. However, the four hypotheses that lead to it are techno-logically separable and subject to empirical verification. The first and third are mainly the province of evolutionary genetics, though there are other ways besides genes to establish variation and trans-mit it from one generation to another. The second hypothesis, the relation between particular variants and reproductive success, is the target of adaptive and optimal arguments. This hypothesis can often be tested in detail without knowing the evolutionary and genetic mechanisms behind the pattern of variation. The fourth hypothesis is fatuous; the environment always changes, both periodically and capriciously. Adaptation to these changes may involve conflicts and compromises, because what is optimal at one time may be disastrous at another. In general, if the environment is changing very slowly relative to the population's adaptive response, the appropriate adap-tive response is what is traditionally called evolution by natural selection. Conversely, if environmental change is very fast, then the appropriate adaptive response is flexibility, through changes in physiology or behavior, or endurance. When environmental changes occur at about the same rate as genetic changes, a serious problem arises (Levins, 1968). In such an environment, a population is always marvelously adapted to the conditions that prevailed in the previous generation, in what Robert Mac Arthur used to call the "Barry Goldwater effect."

OPTIMALITY AND THE ADAPTIVE PATTERN

The adaptive prediction could be framed as, "Each species will have the optimal characteristics for its environment." In this form, the prediction is exceedingly hard to test. One must first ask, "Optimal relative to what?" and the answer will be something like, "Optimal relative to the characteristics that the species could have" or "Optimal relative to the environments that the species could inhabit." It would then be easy to imagine hypothetical characteristics and imaginary environments such that the real species and environment were either perfectly or badly matched. The adaptive argument would then be no more than a Panglossian just-so story that speculates on the reasons behind the wisdom of the Creator's design of a particular plant for a particular purpose. In fact, there is a respectable place for such circular and teleological thought in biology, but it is in the framing of questions rather than in the answering of them. For example, Janzen (1976) uses self-fulfilling speculations ingeniously to suggest elegant experiments that would show the mechanisms by which adaptations might have arisen.

A more tractable form of the adaptive prediction is "Species A is better adapted to environment X then species B, which is in turn better adapted to environment Y than species A." This statement is testable: It is confirmed if species A is found in environment X, and B in Y. (Some philosophers of science would say, "It is not rejected if. . . .") The optimal argument enters this prediction, not in the sense of any universal optimal behavior, but rather in deciding which species is better adapted to which environment. The prediction is not what is best in the best of all possible worlds, but what is the best available in each of many available worlds (cf. Arouet de Voltaire, 1759).

The most useful form of optimal argument is therefore not the traditional fixed optimal point explored in elementary calculus courses, but rather an exploration of the way in which this optimal point shifts with changes in some environmental parameter. For example, I have previously shown (Horn, 1975) that the photosynthetically optimal form for a tree in full sunlight is one in which leaves are scattered throughout the volume of the crown. Conversely, in the shade, an optimal tree has leaves concentrated in a continuous shell about the periphery of the crown. Each of these trees can be shown to have an optimal leaf area index (Figure 2.1). Of course, this optimal leaf area index changes with the form of the tree and with the environment that it occupies. Plotting the optimal leaf

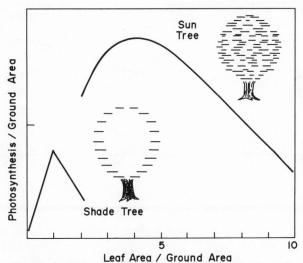

Figure 2.1. Net photosynthesis per unit of ground area is plotted as a function of leaf area per unit of ground area for shade- and sun-adapted trees (after Horn, 1971, 1975). In each case, photosynthesis rises with leaf area until some leaves are so deeply shaded by other leaves that they no longer pay the costs of their own construction. The optimal leaf area for each tree is that which results in peak net photosynthesis.

area index as a function of changing environmental parameters (Figure 2.2) gives the adaptive prediction of a correlation between leaf area index and openness of the canopy, both of which are empirically measurable parameters. The technique is similar to that used by Givnish (article 16) to generate adaptive predictions about the dimensions of individual leaves. The concept of an adaptive pattern of optimal behavior, rather than optimality alone, is also explicit in the discussions of Teeri (article 15), Miller and Stoner (article 18), Mooney and Gulmon (article 13), Solbrig (article 5), Hamrick (article 4), Levin (article 6), and Werner (article 12).

Optimality is not the only technique by which adaptive patterns can be generated. Solbrig (article 5) uses the technique of evolutionary equilibrium. Jain (article 7) uses game theory, which handles certain kinds of environmental stochasticity for which optimality is ambiguous or useless. Hamilton and May (1977) have made use of the unbeatable "evolutionarily stable strategy" of Maynard Smith and Price (1973) to show the conditions under which it would be

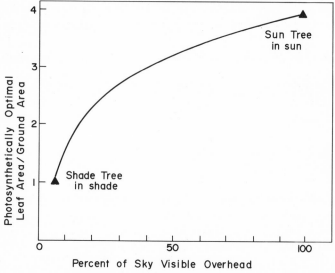

Figure 2.2. The photosynthetically optimal leaf area is inferred from a series of graphs like Figure 1.1 for optimally designed trees in different degrees of shade.

advantageous for a plant's offspring to disperse even if the plant occupies a stable and salubrious environment.

CRITERIA OF OPTIMALITY

For generating an adaptive pattern, the ultimate criterion is the number of offspring left by a plant, suitably discounted by the time that it takes them to reach breeding status. Hickman (article 10) and Jain (article 7) discuss this, and I can add little other than a cautionary note, which I reserve for later.

Ideally, any other parameter that is optimized should be some characteristic that can be translated into offspring that breed in the next generation. This characteristic may be growth and propagation in an open environment, competitive ability and survival in a crowded environment, efficiency of utilization or hoarding in a nutrient-poor environment, defense against predators if they are a threat, dispersal in a patchy or unpredictable environment, efficiency or tolerance under physiological stress, or rapid response to some environmental changes and conservative endurance of others. Optimal models involving energy fixation are especially powerful for plants since, in striking contrast to animals, much of the physical structure of plants consists of modified carbohydrates that are energetically expensive but composed of readily available nutrients. Carbohydrates can pay for structural elaboration, growth, shading of competitors, chemical warfare, dispersive appendages on seeds, photosynthetic area, root surface, and general maintenance. The addition of water economy to an energetic optimal model is a further improvement, since water is very often the crucial nutrient that limits energy fixation. This is why Miller and Stoner (article 18), Mooney and Gulmon (article 13), Givnish (article 16), and I (Horn, 1975) have had such broad success in generating adaptive patterns on the basis of light, carbon dioxide, and water alone.

There is also a more general defense of adaptive models based on a single factor or a small number of factors. The criticism that there are always other constraints can be used either constructively or unconstructively, depending mainly on the spleen of the critic. The failure to account for a long list of other factors, be they constraining or irrelevant, is often cited as an indictment against all simplified model-making. This criticism neglects the fact that all models, simple or complex, are caricatures of nature, and the sign of a good caricature is that it conveys the essence of nature with great economy of detail. The numerous constraints to a model built on a single factor can be used more constructively. As an example, by varying the constraints and examining the behavior of the optimal point of their model, Mooney and Gulmon (article 13) have generated the prediction of an adaptive pattern based on what were inconvenient constraints on the original model. There are philosophic objections to this procedure because it could allow ad hoc changes in the theory to fit initially disparate facts, but this objection is based on the notion, heard more often in parlors, cafes, and bars than in the laboratory or field, that there is only one way to do science.

From a broader perspective, what look like conflicts and compromises in terms of universal optimality are the very ingredients that generate patterns in a comparative context.

ADAPTIVE PATTERNS IN ECOLOGICAL TIME

I have already tacitly made the analogy between adaptive patterns among contemporaneous species and the evolutionary origin of adaptations within a species. It is worth making the analogy explicit because it is complete in detail and has some profound consequences.

There are four hypotheses that directly correspond to the hypotheses of the axiomatic form of the argument of evolution by natural selection. First, variants exist, namely, the species. Second, each variant has a characteristic ability to invade, exist, and propagate in a given environment. Third, each variant, i.e., each species, "breeds true" for its particular characteristics. Fourth, the environment remains constant. If these four hypotheses are true, then each environment should come to be occupied by the species best adapted to it. The first and third hypotheses are generally true. The truth of the fourth may be assured by a sufficiently circumspect choice of the environments studied. The second hypothesis is again the subject of adaptive and optimal arguments.

Note that the adaptive prediction suggests an empirical matching between species and their environments. It makes no reference either to the evolutionary history of the species or to the genetic mechanism whereby the adaptation is passed on to future generations. Therefore, unless the adaptation in question is directly related to persistence in a particular association of plants or animals, the evolutionary history of the adaptive character is irrelevant, though it may be very interesting in some other context. Similarly, as long as plants breed true for the characters in question, the genetic and developmental machinery of these characters may be interesting but irrelevant. There are, of course, adaptive arguments that require a stable association among particular species over evolutionary time, such as the coadaptation of herbivores and the herbs they devour, and there are arguments addressing the significance of genetic machinery and recombination as adaptations to particular patterns of environmental change. These problems cannot be explored without an intimate knowledge of community stability on the one hand and evolutionary genetics on the other.

Several points follow directly from the fact that adaptive arguments need not depend on evolutionary history or genetic machinery. First, it is not necessary to limit the search for adaptive patterns to climax communities or even to discover whether climax communities or stable plant associations exist as objective entities. In fact, one could argue that the clearest place to see the actual evolution of adaptations (cf. Gottleib, article 11) would be in recently disturbed communities, like those in agriculture, which have not even existed over much of the evolutionary history of the species involved. Second, for many adaptive patterns, an individual plant and its offspring need not be defined genetically. Functional stems of a grass may be defined for convenience as individuals even if the whole sward is the vegetative ramification of a single seed; independent propagules may be defined to have the same status as seedlings. The fundamental problem that vexes quantitative plant ecology—what is a plant?—need not even be answered for many adaptive questions. Third, the difficulty that Johnson (article 3) raises, the conflict between the evolution of the best individual genes and the best collection of genes in an organism, is irrelevant for a wide range of adaptive patterns, even though it is a fundamental question of evolutionary population genetics (Lewontin, 1974). Similarly, for many adaptive patterns, it is not necessary to disentangle an individual's direct fitness, counting only its own progeny, from the individual's inclusive fitness, counting the progeny of all its relatives discounted by the distance of their relation.

Finding an apparently adaptive pattern among contemporaneous species and their environments, one cannot distinguish whether an observed optimal character is an evolutionary adaptation to a long history in the current ecological role or a character that arose in a completely different context but was preadapted to the current ecological role. In a sense, this question asks, "Which came first, the evolutionary chicken or the ecological egg?" or, worse, "Who did the selection, Father Time or Mother Nature?" It cannot be answered without explicit evidence of evolutionary history. But it need not be answered for a test of the plausibility of the second hypothesis of either the argument of evolution by natural selection or the argument of origin of adaptive patterns in ecological time.

PROBLEMS AND PROSPECTS

The main lesson of this article is that many intriguing adaptive patterns can be rigorously explored without philosophical qualms about evolutionary history and genetics. The adaptive patterns can be predicted or interpreted with models based on optimality, game theory, evolutionary equilibria, and unbeatable strategies. Comparative natural history provides confirmation or inspiration.

In certain cases, optimality should be used with extreme caution. In particular, there is no guarantee that a sequence or sum of optimal choices will yield an optimal overall schedule or pattern. If each new twig on a tree originated at the optimal branching angle to fill the space near it, the tree itself would soon become a seething mass of competing branches. On a large plant, leaves that are optimal by the criteria of Mooney and Gulman (article 13) and Givnish (article 16) might deeply shade lower or older leaves. In the cases that have been explored so far, a sequence of optimal choices of age-specific reproduction has resulted in an optimal overall reproductive strategy (Schaffer, 1974a,b; Wilbur, 1976), though it is not clear why it should. The conflict between optimal genetic adaptation to the current environment and genetic flexibility to track environmental change is discussed by Solbrig (1977; article 5). Hamrick (article 4) gathers data on heterozygosity, which he suggests is an adaptation to a pattern of environmental change, but even the ideas of Levins (1968) do not explicitly explore the conflict between the optimal genotype for an individual and the optimal genetic variance for its local population.

The methods of animal demography also require caution when they are used to assess reproductive output of plants. In summarizing patterns of birth and death among animals, it makes sense to discuss schedules of age-specific mortality and fecundity because animals are motile enough for all individuals of a given age to expose themselves to a meaningful average of the environmental causes of mortality. Furthermore, all animals of a given age are likely to be of roughly the same size and vigor; at least, they are more similar than are animals of different ages. This is simply because an animal of a given age who does not maintain sufficient vigor and growth is likely to be dead before aging much further. Plants are different in both respects. Once a seed lands on a particular spot, the schedule of growth, mortality, and fecundity may be fixed and measurable but very different from the schedule at another spot. Furthermore, a plant's competitive environment is dominated by a small number of other plants that are its nearest neighbors. Thus, the effective population density that each plant experiences may be different for every individual. Hence, age-specific averages of growth, fecundity, or mortality may have much larger variance than an average of these parameters over a given size or local density or quality of site (Cook, article 9). Despite this problem, Harper and White (1974) have shown that demographic techniques and parameters are useful and illuminating in studying plant populations.

Although one can hope for a plant demography based on site-specific or size-specific rates of growth, fecundity, and mortality, such an approach would not yield a simple recipe for reproductive rate like the very powerful one discussed by May (1976). Animal ecologists face similar problems when numerical fluctuations of a population destroy the simple age specificity of mortality and fecundity.

For plants, the ecologically meaningful population may consist of an individual and its two or three nearest neighbors, whatever their species (Mack and Harper, 1977). The breeding population, or the population geneticists' "effective population size," extends further, to the borders of heavy dispersal of pollen, seeds, or other propagules. As Stebbins (personal communication, 1977) pointed out in response to Raven (article 19), the population of evolutionary interest extends ever further with the passage of time, each generation adding a concentric ring as thick as the width of the breeding population. The lack of correspondence of ecological, genetic, and evolutionary populations is both a bother and a challenge. The relations among these levels are beginning to emerge from studies that take explicit account of the spatial configuration of plant populations (see Hickman, article 10, and Cook, article 9), even in cases that are complicated by animal coevolution (Levin, article 6). I hope for more rapid progress on these problems from botanists than from zoologists because spatial patterns are easier to measure in sessile plants and marked individuals are easier to recapture (Werner, article 12).

Two other areas where botanists have advantages over zoologists are adaptive patterns of life history and the adaptive significance of sexual recombination. Because many of the costs of growth, maintenance, and reproduction can be paid for in common currency by plants, relative allocation can be measured directly and conveniently (Gottlieb, article 11; Werner, article 12). There is also some promise of uncovering further adaptive patterns relating optimal allocation of energy to the pattern of assimilation of resources, which is already better understood (Miller and Stoner, article 18; Mooney and Gulmon, article 13; Givnish, article 16). The adaptive significance of sex can be examined in plants with factorial experiments and observations using close taxonomic or ecological relatives (Solbrig and Rollins, 1977; Gottlieb, 1977a,b,c). This is in part a result of the plant taxonomists' obsession with vicarious sex, but also because plants display such a variety of sexual experience. Many procreate sexually; many shuffle their genes without explicit sex; many are asexual; and many are facultative, able to make a go of it as male, female, both, or neuter. Furthermore, developmental determination of sex is not so closely tied to chromosomal mechanisms as in most animals, so that sex determination is not confounded with other genetic problems.

It is worth reemphasizing that the solution of all of these problems depends critically on comparative observations of natural history. Although there is a wealth of information already available on all the individual species of some floras, even the most thoroughly studied flora can still yield new insights in a search for comparative adaptive patterns. Besides, as Mac Arthur (1972a, after Eddington) has noted, it is fatuous for an empiricist to go out in search of a new theory unless he has some data that he wants to prove.

3 ENZYME POLYMORPHISM: GENETIC VARIATION IN THE PHYSIOLOGICAL PHENOTYPE

George B. Johnson

STUDIES OF patterns of genetic variation within and between natural populations have always been the core of experimental population biology. Particularly with the widespread use of electrophoresis in the last decade, enzyme loci have been the subject of intensive study, both as "markers" and in their own right. Relatively little work has been carried out on natural populations of plants in this respect, however, primarily because of the technical difficulities often encountered in analyzing plant tissue (plant secondary compounds may interfere with enzyme assays or denature enzymes, polysaccharides may bind and precipitate proteins, etc.). I will not review the limited literature on enzyme polymorphism in natural populations of plants (see Hamrick, article 4). As technical solutions become more widely available, however, we can expect an increasing focus in population biology on the isozymes of plants: They constitute superb material for many questions of interest. What I wish to do in this paper is address several issues which I regard as providing particularly promising opportunities for enzyme work in plants. These issues focus around the central theme of assessing the role of enzyme polymorphism through a careful delineation of the appropriate physiological phenotypes.

SINGLE-LOCUS STUDIES

High degrees of heterozygosity at enzyme loci seem typical of animal and plant populations, with values generally exceeding 10 percent (Powell, 1975; Gottlieb, 1977a,b,c). It is difficult to explain why levels of genic variability should be so high. If variation at each of these loci affects fitness, then inbreeding depression and genetic load should be far greater than they appear to be. The high levels of gene polymorphism thus suggest that the single genetic locus may not be the usual unit of selection (Lewontin, 1974).

A variety of studies have been carried out on genic variation at single-enzyme loci. Often these studies report a cline in allele frequency at some locus along a latitudinal gradient (Vigue and Johnson, 1973; Johnson and Schaffer, 1973; Koehn et al., 1971; Merritt, 1972; Selander et al., 1971; Lakovaara and Saura, 1971) or document a correspondence of allele frequency with physical or biological characteristics of the habitat (F. Johnson et al., 1969; Rockwood-Sluss et al., 1973; Schopf and Gooch, 1971; Somero and Soule, 1974; Smouse and Kojima, 1972; Koehn et al., 1972; Levinton, 1973). Even in the absence of multilocus considerations, however, it has proven quite difficult to infer from these patterns the existence of selective interactions with the environment. Migration or other demographic factors may also produce clines in allele frequency. Also, linkage to another (unknown) locus under selection may produce spurious correlations of allele frequency with habitat or geographic patterns simply because of the low recombination fraction with the selected locus. It is for these reasons that inferring gene—environment interactions from patterns of gene frequencies requires independent estimation of the effects of migration and linkage.

Single-locus cases analyzed along the lines discussed above provide most of the evidence available to date that selection maintains polymorphism in nature. Only relatively simple systems have proven amenable to direct analysis. The rationale of such single-locus approaches has been summarized recently by Clarke (1975) and G. Johnson (1975b). The clearest examples of such single-gene polymorphisms are alcohol dehydrogenase polymorphism in *Drosophila,* an apparent adaptation to toxic levels of alcohol in fermenting fruit (Day et al., 1975, and many others), lactate dehydrogenase polymorphism in fish, an adaptation of physiological redox levels to changes in temperature (Merritt, 1972), α-glycerophosphate dehydrogenase polymorphism in insects, an adaptation of flight muscle redox levels to changes in temperature (G. Johnson, 1976a,b,c), glucose-6-phosphate dehydrogenase and hemoglobin polymorphisms in man, which confer resistance to malarial parasites (Ingram, 1957, and others), and, recently, leucine aminopeptidase in marine mussels, an apparent adaptation of ion balance to salinity stress (Koehn, personal communication, 1977). In all of these cases, polymorphic variation apparently represents a direct adaptation to changes in a specific environmental factor. The relationship is in each case a direct cause-and-effect one and may, in principle, be fully understood in terms of the difference in that enzyme's function which the polymorphism produces.

The rationale of these studies is the same as that used by Darwin in studying the evolution of finches and is ultimately unsatisfying for the same reason: Evolution cannot be understood as the simple summation of individual adaptations, with each locus interacting individually with the environment. Although the fact that between-locus interactions are also of paramount importance is now clear, this has not always been fully appreciated. Sewell Wright's conceptualization of adaptation as consisting of adaptive peaks and valleys, in which the fate of any allele could be delineated on an adaptive topography, is such a single-locus argument, and one which has had great influence (despite cautionary statements by Fisher and others).

Single-locus approaches are basically unsatisfactory for two interrelated reasons: The selection coefficient of an allele at one locus may be affected by the genotype of another locus, and the recombination fraction between the two loci is subject to change. The point here is that the target of selection is the operational phenotype, whether it be flower morphology or respiratory efficiency, and many loci interact to determine such phenotypes. In addition, because selection may involve multiple loci, the degree of linkage, and thus of coordination, between loci is itself subject to evolutionary modification. Any single-locus study which ignores such interactions and carries out a one-locus—environment comparison in an attempt to understand adaptation may misinterpret or even fail to detect key elements in the process.

MULTIPLE-LOCUS STUDIES: THE METABOLIC PHENOTYPE

For many of the loci reported to be polymorphic in natural populations, simple and direct relationships with the environment seem unlikely. Many of these enzymes are intimately involved in intermediary metabolism, and a change in the activity of one may influence the functioning of many others. Thus, a change in hexokinase, which catalyzes the reaction that generates glucose 6-phosphate, cannot help but affect the reactions of phosphoglucomutase, phosphoglucoisomerase, and glucose-6-phosphate dehydrogenase, all of which use glucose 6-phosphate as a substrate (G. Johnson, 1976). Conversely, enzymes that catalyze non-rate-limiting reactions, such as fumarase and malate dehydrogenase, would not be expected to change in concert with other enzymes such as hexokinase, since they do not determine the rate of metabolic flux through the pathway and thus do not contribute to the phenotype under selection. In this regard it is important to note that polymorphic variation among the enzyme loci of intermediary metabolism occurs predominantly at the regulatory (rate-limiting) steps in intermediary metabolism. This finding is very widespread and quite general (the matter is extensively reviewed in G. Johnson, 1974, and more recently in Powell, 1975).

The highly coordinated nature of intermediary metabolism suggests that if polymorphic alleles at regulatory enzyme loci are functionally different, then the particular allele present at one locus of a rate-determining enzyme will affect the activity not only of its own pathway but also of many other reactions. If, for a network of related regulatory enzymes, each locus maintains that allele which provides optimal activity in a particular habitat, then one may speak of a metabolic phenotype adapted to that habitat. It is important to realize that, because the particular functional variant occurring at each regulatory locus influences the overall physiological state of the individual, *selection on metabolic phenotypes implies selection on allozymic genotypes*. To understand the process, the genotypes of each of the loci contributing importantly to the phenotype must be known.

To test the hypothesis that polymorphism among the loci of metabolic enzymes reflects selection for alternative integrated metabolic phenotypes, it is necessary that at least six experimental conditions be satisfied:

1. To characterize the metabolic phenotype of individuals, the entire selected array of loci must be examined in each individual. This permits an explicit multilocus genotype to be assigned to each individual.
2. To predict (and attempt to demonstrate) selection for a difference in metabolic phenotype, the examined population(s) should encompass a broad range of environmental conditions.
3. To rule out associative overdominance (linkage to another unknown but selected locus) as an alternative explanation, it is necessary to include among the examined loci several known to be uninvolved in the postulated metabolic phenotype. Associative overdominance would predict no consistent difference between the two classes.
4. To rule out migration from an adjacent population as an alternative explanation, it must be demonstrated that migration does not occur to any significant degree. This is most easily done by studying a single genetically isolated population. Note that care must be taken to document this isolation empirically.
5. To rule out spatial subdivision within the population or differential habitat selection by individuals, it is necessary to study the patterns of migration which occur. This is most directly done by demographic characterization.
6. To rule out temporal subdivision resulting from a cyclic pattern of selection, the population should be monitored over time. The simplest situation will be one in which a single generation occurs over a short interval each year, with little variation in habitat from year to year.

Relatively few multiple-locus genotype data exist for natural populations. One study, in *Colias* (G. Johnson, 1977), describes polymorphism at 14 loci in a single population. All loci were examined in each sampled individual. The butterflies were collected at five sites along a transect through a single demographically characterized population of *C. meadii*. Because relative amounts of migration have been determined for this population from mark–release–recapture studies, the population is known to be genetically continuous. A major change in habitat occurs along this transect, which extends from alpine tundra above 12,000 ft across the timberline and down into montane forest meadows. If the different habitats exert strong differential selection on organized metabolic phenotypes, then discontinuous associations of allele frequencies might be expected at some loci; at loci not involved in these organized phenotypes, allele frequencies should vary independently of one another. Note that, if migration between subpopulations is the only significant factor contributing to observed variability, similar patterns of polymorphism should be seen at all loci. Enzyme polymorphism was studied at 14 loci, and a variety of distinct patterns of allele frequency are seen (some clinal, some uniform, some discontinuous), which provides strong evidence of selection. There is clear evidence that particular alleles at the different loci associate together preferentially at particular locations along the cline, different assemblages occurring at different locations. These appear to represent integrated metabolic phenotypes, and polymorphism may be a multilocus strategy to preserve that integration in a heterogeneous environment.

OPPORTUNITIES TO STUDY METABOLIC PHENOTYPES IN PLANTS

Plants are particularly promising subjects for the study of discrete metabolic processes which interact in a known fashion with the environment. I will discuss five physiologically distinct processes, each of which comprises a natural metabolic unit suitable for investigating hypotheses concerning metabolic phenotypes. Others could undoubtedly be chosen.

Glycolysis/Gluconeogenesis

The principal enzyme reactions of glycolysis (and its reverse, gluconeogenesis) are illustrated in Figure 3.1. For several plant tissues, relatively complete and simultaneous analyses of all the intermediates and cosubstrates have been carried out: *Rhododendron* (Bourne and Ranson, 1965), buckwheat seedlings (Effer and Ranson, 1967), peas (Barker et al., 1964), and castor bean (Kobr and Beevers, 1970). The results uniformly indicate that the regulation of glycolytic flow is at the early and late irreversible reactions catalyzed by phosphofructokinase (PFK) and pyruvate kinase (PK) (Dennis and Coultate, 1966; Kelly and Turner, 1968a,b; Lowry and Passonneau, 1964). It has been suggested that gluconeogenesis, which is important in germination, occurs in a separate intracellular region and is independently regulated at the reaction catalyzed by fructose 1,6-diphosphatase (FDPase) (Kobr and Beevers, 1970; Scala et al., 1968). These three enzymes, PFK, PK, and FDPase, thus comprise the central points of the metabolic control of anaerobic metabolism. Simultaneous survey of polymorphic variation at a variety of glycolytic enzyme loci, including both these three and others known not to be points of regulation (enzymes catalyzing the metabolism of fructose-6-phosphate, glucose-6-phosphate, triose phosphate, 3-phosphoglyceric acid, and phosphoenolpyruvate), will provide multilocus genotypes which will in turn permit an unbiased determination of the degree to which enzyme polymorphism is associated with the phenotype-determining steps and the degree to which these key enzyme loci are coordinate in their genotype expression.

Anaerobic Metabolism and Adaptation to Moisture Stress

Moisture tolerance in plants is the subject of an extensive literature, highlighted by the work of Crawford (Crawford, 1966, 1972; McManmon and Crawford, 1970) on metabolic adaptation to anaerobiosis. Although in young seedlings oxygen diffusion from the shoots is sufficient to provide an adequate supply of O_2 to submerged roots (O_2 diffusion takes place from the roots into the sur-

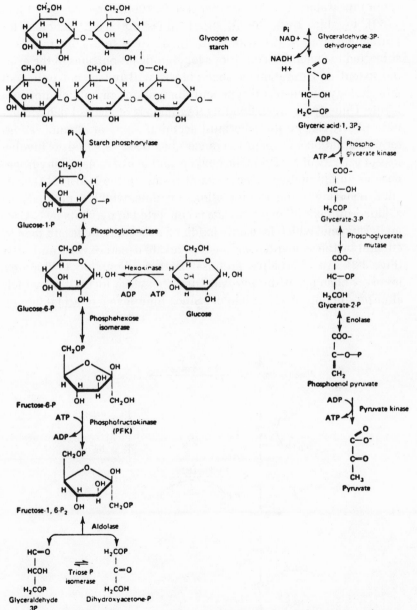

Figure 3.1. The principal enzyme reactions of glycolysis and gluconeogenesis.

rounding water; W. Armstrong, 1967, 1968; Greenwood, 1967, 1969), in older trees the O_2 supply does not seem sufficient for normal respiration (Crawford, 1972). In such trees, the roots are subjected to anaerobic conditions for significant portions of the year; the extent and duration of anaerobiosis will reflect not only root morphology but also soil type and fluctuations in habitat moisture levels. Under prolonged flooding, high levels of alcohol dehydrogenase (AdH) can produce harmful accumulations of ethanol. Long-term adaptation to "flood tolerance" has been reported to involve loss of activity of NADP-malic enzyme and use of malate dehydrogenase as an alternative mechanism of restoring the NADH/NAD balance, resulting in the accumulation of malate, which is not toxic. In a "flood-intolerant" plant, malate is converted to pyruvate by NADP-malic enzyme, which is readily oxidized to acetaldehyde and then to ethanol by the sequential action of pyruvate decarboxylase and AdH (Fig. 3.2). An alternative proposal is that adaptation to flooding involves changes at the pyruvate decarboxylase locus itself which alter production of acetaldehyde. Since high levels of acetaldehyde

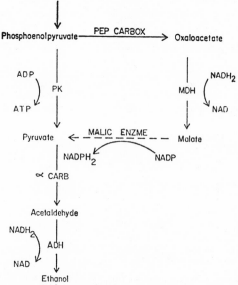

Figure 3.2. Pathways of malate and pyruvate metabolism.

or ethanol are toxic, it is easy to see why, under anaerobic conditions, malate or pyruvate accumulation is preferable.

The enzyme activity of AdH may be monitored in vivo under various levels of moisture stress by using an intact tissue assay such as that described by Mitra et al. (1978). The in vivo assay depends on the loss to water of tritium from specifically tritiated substrates and uses direct sublimation of tissue water (or surrounding buffer solution in equilibrium with tissue water) into a scintillation vial, permitting hundreds of in vivo assays per working day. It will be necessary to check malate dehydrogenase, malic enzyme, and pyruvate decarboxylase as alternative modes of response, particularly in cases of increased water potential. For each of these enzymes, in vivo assays are potentially realizable.

Thus, experimentally, the AdH system looks very promising. Growth-chamber manipulation of oxygen tension is practical, and the level of O_2 diffusion down the stem can be measured directly. The contemplated in vivo assays do not hurt the plant.

The interesting situation with respect to enzyme polymorphism concerns levels of genic variation among normally flood-intolerant plants living in habitats where they are subjected to the moisture stress of periodic flooding. In plants exposed to less chronic stress, modulation of alcohol dehydrogenase activity at low oxygen tension may be extremely important. Simultaneous survey of polymorphic variation at the seven enzyme loci PEP carboxylase, PK, malate dehydrogenase, NADP-malic enzyme, pyruvate decarboxylase, aldehyde oxidase, and AdH will permit an unbiased determination of the degree to which enzyme polymorphism is associated with moisture tolerance and, in particular, with the phenotype-determining reaction catalyzed by NADP-malic enzyme.

It is important not to lose sight of the fact that other loci may also be involved in a response to moisture stress. Certain peroxidases catalyze cell wall lignification. Their levels change with developmental stage and conditions of stress. Catalase also appears to have an obvious function in responding to oxygen stress. Its level of activity and localization change with developmental stage, and its activity is high in the microbodies of tissues in which lipid reserves are rapidly broken down. Environmental influences on lipid metabolism should be reflected in catalase levels, although the mode of this interaction is open to some discussion. The observation that catalase inhibitors promote germination has prompted Sterling Hendrix to speculate that H_2O_2 regulates dormancy in seeds, with germination actually requiring H_2O_2. The matter has been recently reviewed by Halliwell (1974), who points out that at low physiological levels of H_2O_2 (approximately 10^{-5} M), superoxide dismutase and catalase are not efficient enzymes and that H_2O_2 may have a role in formic acid and methanol oxidation.

In vivo assays of the combined activity of these enzymes in the root are practical since H_2O_2 freely diffuses out of roots. A root submerged in dilute H_2O_2 and monitored with an O_2 electrode ought to provide a direct approach. Again, it should be noted that the assay in no way hurts the plant.

Nitrogen Metabolism

Nitrogen metabolism offers many interesting avenues of investigation (Fig. 3.3). Under *increased* water potential, with metabolic conditions anaerobic, nitrate reductase inhibition is removed (and nitrite reductase inhibition is initiated). This results in high levels of nitrite, which diffuses freely out of roots into soil, and low concentrations of ammonia. If the results of Magasanik et al. (1974) are generalizable, glutamine synthetase continually removes the low

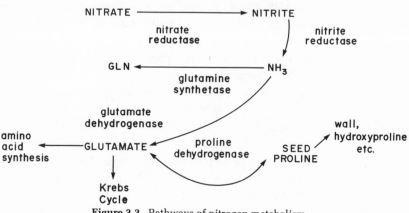

Figure 3.3. Pathways of nitrogen metabolism.

levels of ammonia into nitrogen metabolism. Thus, even under high water potential, with the Krebs cycle glutamate dehydrogenase (GdH) route unavailable, high levels of glutamine (Gln) may be accumulated in seed protein.

Under *low* water potential, a variety of other potentially interesting situations occur. Particularly in monocots, seeds contain high levels not only of Gln but also of proline. Under dry conditions, with extensive proteolysis, seeds accumulate excessive NH_3 and low nitrite. Along the lines outlined above, this would increase the rate of glutamate dehydrogenase synthesis. This enzyme, in tandem with proline dehydrogenase, would then be able to marshal the high levels of seed Gln and Pro, together with ammonia, into the Krebs cycle.

In *vivo* assays are available for the study of this system. The tritium-loss *in vivo* assay for proline dehydrogenase functions very well. It will also be necessary to examine GdH. An intact-tissue assay exists with which one can measure the sum of activity under normal or stressed conditions and compare these results to the difference seen when transaminase inhibitors such as aminooxyacetic acid are added. The differential assay does interfere with the plant's metabolism, but only transiently. Down the line, it would be desirable to monitor rates of amino acid synthesis and Krebs cycle activity in order to get a complete picture of the interacting system.

The *in vivo* assay of nitrate reductase assumes total anaerobiosis, but the assay may still be useful in studying conditions of transient oxygen stress because Magasanik's effect occurs over a time course of 4 to 5 hr, and because the turnover rate of nitrate reductase is extraordinarily rapid (3 to 5 hr; Wallace has suggested that there may be a specific protease for this enzyme). An *in vivo* assay of glutamine synthetase should be practical.

Photosynthetic Carbon Fixation

Photosynthesis is one of the most biologically important, metabolically unique, and intensively studied physiological processes in plants. Although photosynthetic carbon fixation is well suited to the type of approach suggested here, few data on genic variation at photosynthetic loci among natural populations of higher plants are available, although studies have been initiated in my laboratory (Enama, 1976). Figure 3.4 illustrates the principal reactions of C_4 photosynthesis in a NAD-malic enzyme plant. At least ten reactions are of some regulatory significance:

		EC No.
1.	Adenylate kinase	2.7.4.3
2.	Fructose diphosphatase	3.1.3.11
3.	Glutamate-oxaloacetate transaminase	2.6.1.1
4.	Glyceraldehyde-3-phosphate dehydrogenase (NADP)	1.2.1.13
5.	Malic enzyme (NAD)	1.1.1.39
6.	Phosphoenolpyruvate carboxylase	4.1.1.38
7.	Phosphoglyceric acid kinase	2.7.2.3
8.	Phosphoribulokinase	2.7.9.19
9.	Pyruvate, orthophosphate dikinase	2.7.9.1
10.	Ribulose-1,5-bisphosphate carboxylase	4.1.1.39

Regulatory aspects of photosynthetic carbon metabolism have been extensively reviewed recently (Kelly et al., 1976). The principal sites of regulation in C_4 photosynthesis are the reactions catalyzed by phosphoenolpyruvate carboxylase, ribulose bisphosphate carboxylase, and fructose diphosphatase.

An in-depth survey of enzyme polymorphism among the reactions shown in Figure 3.4 and the table above would be of major value to plant population genetics. If care is taken to characterize all loci in each individual, the resulting "multilocus genotypes" will permit a detailed analysis of the genetic basis of the photosynthetic phenotype. Genic interaction among rate-determining loci, if it occurs, will be clearly characterized by such a study.

Most importantly, the rate of photosynthetic carbon fixation may be monitored *in vivo* under various temperature, moisture, and light regimes without harming the living plant (see, for example, Berry, 1975; Mooney, 1972; Björkman et al., 1972). The opportunity exists, therefore, to assess directly the physiological significance of an allelic substitution, and to pose pointed questions about the relative significance of polymorphism among loci whose enzymes catalyze rate-determining and non-rate-determining reactions. If a thorough demographic analysis is carried out on the plants being

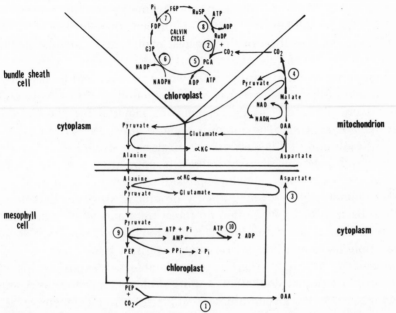

Figure 3.4. Schematic model of C_4 photosynthetic carbon fixation in a NAD–malic enzyme plant. The numbers correspond to the list of enzymes given in the text.

surveyed, the following hypotheses may be addressed, each one subject to direct experimental test:

1. Do organized genotypes constituting a photosynthetic phenotype occur among rate-determining loci?
2. Are there genotypes which are habitat specific with respect to light, temperature, moisture, etc.?
3. Does allelic substitution at a locus of an enzyme catalyzing a rate-determining reaction in photosynthesis produce a physiologically significant difference in carbon fixation *in vivo*?
4. Does physiologically significant variation translate into fitness differences as judged by demographic analysis?

By comparing rate-determining and non-rate-determining reactions in such an analytic scheme, one may approach the fundamental issue of the degree to which the occurrence of enzyme polymorphism among the loci of photosynthetic enzymes reflects the action of differential selection on the metabolic capabilities of plants. If the relationship proves significant, then further study of such isozymes will be of the greatest evolutionary interest. If the relationship is not significant, this would suggest that, from an adaptive perspective, further work with isozymes will be less than fruitful and that other avenues of investigation might be more productively pursued, leaving isozymes as material for genetic and systematic analysis. The matter is open to direct empirical test.

Resolution of this basic issue will permit pointed analyses of a variety of questions of great interest to population biology: (1) To what degree is there metabolic/physiological tracking of *spatial–environmental heterogeneity*? Very promising single-locus studies have been initiated by Mitton on forest trees and by others; in the future, a more powerful approach will require a carefully selected multiple-locus analysis. (2) Do polymorphisms track *temporal heterogeneity*? Although there is considerable evidence for this in *Drosophila* (Rockwood, 1969; Kojima et al., 1972; Dobzhansky and Ayala, 1973; Rockwood-Sluss et al., 1973; Anxolabehere et al., 1976), no detailed studies are available on plants. (3) To what degree does breeding structure act as a mechanism of *recombination modulation* during evolution? Here knowledge of the presence or absence of specific organized gene complexes affecting a given physiological phenotype will be of great value in analyzing the evolutionary significance of particular alterations in recombination fraction. (4) How does population *age structure* interact with selection? A series of pioneering investigations by Prout and Christiansen and Frydenberg (Prout, 1969, 1973; Christiansen and Frydenberg 1973, 1976, 1978) have led to systematic analyses of the stages in a life cycle at which selection operates. (Among the stages at which the action of selection may be individually characterized are gametic selection in females, female-specific selection of male gametes, differential male or female mating success, zygotic selection, and differential adult survival to reproduction.) Such studies require analysis of successive breeding seasons for a population in which the age of each analyzed individual is known, all age classes are sampled in an unbiased manner, and progeny testing may be carried out to determine parental genotypes. Plants offer a great opportunity for study since they are far more amenable to analysis than most animal systems. Again, however, the phenotype whose selection is being studied must be carefully defined.

EPISTASIS AND PHYSIOLOGICAL GENETICS

In considering metabolic phenotypes, we have discussed reaction sequences which act in a concerted way to determine the rate of a particular metabolic process such as glycolysis or photosynthetic carbon fixation. In a real physiological sense, this analysis remains incomplete since it does not take into account interacting metabolic systems. The metabolic flux through glycolysis has important influences upon pentose metabolism and aerobic respiration. Indeed, most major metabolic sequences are interrelated by a complex web of regulatory controls, mediated by allosteric interactions with cofactors such as NAD, NADP, and ATP, or effectors such as cyclic AMP. Without this coordination, an integrated metabolism would be impossible, and yet it poses an evolutionary problem of the first order (Johnson, 1975): Since different pathways almost always have different Q_{10} values with respect to habitat variables, how is integration to be maintained in a variable environment?

Phrased in a different context, this problem has been a central issue in population genetics since the initial disagreements of Wright and Fisher: In what manner may the fitness of a gene be modified by a gene substitution at another locus? Originally arising from a discussion of the evolution of dominance, the issue is just as pertinent to a discussion of the genetic characteristics of physiological systems. In a purely population genetics context, an allele may be spoken of as arising by mutation at locus 1, having a selective value s, and increasing in frequency until it reaches fixation; even though the environment is constant, the value of s may be changed during this process if other mutations occur at other loci which alter the action of selection on allele a of locus 1. The issue is the same: Analysis of the action of selection requires consideration of each component upon which selection is acting. The problem, from a physiological perspective, is to study the evolution of higher-level metabolic coordination. This issue cannot even be approached by the sort of single-locus studies available to date, no matter how carefully conducted. In the words of Paul Green, "It's like trying to understand a Beethoven symphony by studying a Stradivarius violin."

An example of epistatic interactions is provided by genic variation in binding affinities for NAD^+. Cellular redox levels (such as the ratio of NAD^+ to NADH) are set by loci such as LdH or αGPdH, which serve to regenerate NAD^+ at a variable level relative to NADH concentrations. To the extent that the functioning of such enzymes is sensitive to environmental parameters such as temperature, modulation of enzyme activity has widespread consequences through the currency of NAD^+ allosteric binding. Genetic polymorphism at a cellular redox-determining locus, as a phenotypic response to environmental variance, expresses itself by acting on a range of loci sensitive to (NAD^+). The effect is thus pleiotropic.

Some of the interactions can be quite unexpected. In *Drosophila*, the enzyme alcohol dehydrogenase catalyzes the oxidation of acetylaldehyde to ethanol, utilizing NADH as an electron donor. A variety of alleles exist, each with an apparently different conformation and binding affinity for NAD^+/NADH (G. Johnson, 1977). In the presence of high ethanol concentrations (which are toxic), there is strong selection for that allele with the lowest K_m for NAD^+, so as to maximize the rate of the (reverse) reaction. The strength of this selection will be a function of cellular (NAD^+), and thus of the genotype of redox-determining loci. Under the lower ethanol concentrations which usually obtain, the reaction utilizes aldehydes as substrates (glyceraldehyde to glycerol, etc.), and selection at this same locus will be for the allele which best modulates glyceraldehyde oxidation with respect to corresponding levels of lipid metabolism. Alcohol dehydrogenase thus has the properties of a single-locus system (ethanol catabolism), a multilocus system (regulating an entrance of carbohydrate into lipid metabolism), and an epistatic interactive system (modulation of selective coefficients by NAD^+ levels).

The key evolutionary questions concern the influence of metabolic integration on the strength of selection in a heterogeneous environment: In what sense are levels of metabolic integration a function of environmental heterogeneity, and do they act to maximize genetic fitness?

DETECTING EPISTASIS:
THE PROBLEM OF HIDDEN HETEROGENEITY

Recent studies have shown that, when different alleles are combined together into electrophoretic classes, gene associations (linkage disequilibrium) are often disguised, so that there is a much lower probability of detection (Zouros et al., 1977). Thus, in attempting to assess the degree to which genic variation at enzyme loci acts to modulate physiological processes, we need not only to phrase our questions in terms of the loci which adequately characterize the process under study but also to employ screening procedures which will detect genic variation affecting that process. Currently popular electrophoretic procedures employing starch gels are not optimal for this purpose for at least three reasons:

1. They fail to detect a considerable fraction of allozyme variation. Indeed, considerably more may go undetected than is seen, with common starch slab techniques. This occurs because variants may differ in isoelectric points but have the same realized net charge at the pH of the analysis, or because variants differ in both charge and shape and the two effects have opposite and balancing effects on electrophoretic migration.

2. They fail to provide a reasonable means of directly comparing data obtained from diverse labs or over considerable periods of time. Data obtained are generally reported as observed number of heterozygote patterns, or as frequencies of allelic classes. But this is not what is measured. What is measured in electrophoresis is the relative positions of bands on gels, and these measurements are not reported, so that the errors in their determinations are not known. Thus, there is no way in which diverse data sets may be directly compared to assess identity.

3. They fail to examine a protein property whose variation is of

adaptive significance. Variation in the net charge of enzymes (if that is what starch gel electrophoresis detects) is of no obvious physiological or adaptive importance, and its detection is not in principle different from screening for immunological variants; in both cases, variation is being detected independent of its potential influence on the process under study.

I think that modern approaches to electrophoretic analysis can to a large degree circumvent these problems.

The principal improvements in electrophoresis concern gel sieving analysis, which involves the systematic variation of gel pore size. As the pore size of a gel decreases, a protein's rate of migration in electrophoresis is progressively retarded. That this is so reflects the fact (not often appreciated in electrophoretic surveys) that the rate of migration of a protein on a gel is determined not only by protein charge but also by protein size and shape. The relationship may be expressed as

$$R_f = \frac{M_o}{u_f} e^{K_r T}$$

where R_f = mobility of the protein relative to the front

u_f = apparent mobility of a moving boundary in front of the resolving phase (a constant known for most common buffer systems)

M_o = free electrophoretic mobility of protein

K_r = the retardation (frictional/hydrodynamic) coefficient (K_R = $K_r/2.303$)

T = Percentage acrylamide (which determines pore size and is inversely proportional to it)

By examining R_f at several values of T, it is possible to estimate values of free electrophoretic mobility M_o and retardation coefficient K_R for any protein which can be discretely detected in a heterogeneous mixture. The above equation is log-linear: $\log R_f = \log (M_o/u_f) + K_R T$, which is of the form $y = b + mx$. Thus, replicate gels of differing acrylamide concentration (T) are run, R_f is determined in each case, and a linear regression of $\log R_f$ on T [a "Ferguson plot," the resulting line has a slope of K_R (a measure of protein shape) and an intercept of log M_o/u_f (a measure of charge)] is performed. This approach has been described in detail (G. Johnson, 1977, 1978). It provides a direct, reproducible, and very sensitive characterization of those physical properties of proteins which alter electrophoretic mobility. High-resolution electrophoresis requires rigorous standardization. This is particularly true in comparative studies. The simplest rigorous procedure involves running two internal-standard proteins in each gel migration path (G. Johnson, 1975). The standard should be chosen so that one resembles the experimental protein(s) of interest and the other is quite different, in both K_R and M_o. The two standards should thus respond quite differently to any alteration of experimental conditions, and the ratio of their mobilities provides a sensitive index of experimental error.

When gel sieving analyses are carried out on enzymes of natural populations, results indicate the following:

1. At least twice as many variants are detected as would have been revealed by current starch gel procedures.
2. Although crossing data are quite preliminary and variants have not yet been mapped to the structural locus, it appears that the variants detected by gel sieving analysis are genetic in nature. None are detected in inbred lines. Preliminary evidence suggests that many variants may represent genetically mediated *post-translational modifications* (Finnerty and Johnson, 1978; Johnson and Hartl, 1978).
3. Experimental error is small, so that almost all variants can be clearly and unambiguously distinguished from one another. Because the magnitudes of these errors are known explicitly from the regression analyses, explicit statements can be made of the probability that a difference would have been detected were it present.
4. Most variant types at a locus are relatively rare, whereas only a few are common.
5. Most importantly, much of the variation involves variation in conformation, K_R. Unlike net charge differences, differences in protein shape may reflect differences in how the allozymes function (Finnerty and Johnson, 1978). The screening procedure thus detects variation of potential physiological interest.

FUTURE AVENUES OF INQUIRY

A fundamental question facing those of us working with enzyme polymorpism is, "What is the role of genic variation at enzyme loci (if any) in plant adaptation?" I think that a strong approach to this question will require us to revise our thinking in several important respects:

1. We must address the proper phenotype. Little is accomplished by studying conveniently assayed but functionally unrelated enzymes or by attempting to determine "average heterozygosity" or other characterizations of the overall genetic variability. The issue is not the variability of the genome but rather that of loci affecting the phenotype under study.

2. In studying variation at enzyme loci, the appropriate phenotypes are usually metabolic and physiological ones. We must select loci so as to characterize as fully as possible a particular physiological process. The cleanest answers will come from studying small, fully understood systems. The role of polymorphism in more complex metabolic systems will almost certainly prove more difficult to analyze, although for biological reasons it is very important that we try, if only to obtain a qualitative answer.

3. We need to know the levels of physiological variation within natural populations for the phenotype under study. In studying photosynthesis, the elegant work of Mooney and others must be extended from individual "type" plants to ecologically representative samples of individuals from natural populations. The same is true for drought resistance, mineral uptake, or whatever other process is under study: If the phenotypic variation is not characterized in a biologically meaningful way, the significance of polymorphism at loci whose enzymes mediate the process is impossible to assess.

4. We must then assess, with the most powerful techniques we have, the degree to which levels of physiological variation reflect differing levels of genic variation. On the one hand, we must screen for genetically based physiological variation and then determine the degree to which the appropriate loci vary. This approach is only one of strong inference when

carried out in simple systems. It permits the phrasing of discrete rejectable hypotheses concerning the key question— the degree to which physiological adaptation is achieved by polymorphic variation at metabolic loci as opposed to alternative modes of adaptation, such as the plastic responses reviewed recently by Jain. A major effort should be made, on the other hand, to pose rejectable hypotheses which address the converse question of which loci mediating a metabolic phenotype actually vary, and to what degree this variation is or is not reflected in variation of the corresponding physiological phenotype. This is a different, although clearly related, issue and it addresses the question of the degree to which polymorphic variation at metabolic loci function genetically to achieve physiological adaptation, as opposed to more adaptively neutral variation.

5. A beginning should be made in assessing what Sokal has called "genetic inertia" at the level of enzyme polymorphism. These epistatic effects of interacting pathways are of paramount interest, despite the great experimental difficulty which their complexity presents. As a beginning we might at least start to collect genotypic data for a broad array of physiologically interrelated metabolic processes.

From the standpoint of studying the role of enzyme polymorphisms in plant populations, these are exciting times. The questions now facing us are of the broadest biological and evolutionary interest. A concerted effort approaching the problem from the perspective of a clearly delineated physiological phenotype seems to me to offer the best chance of making significant progress.

4 GENETIC VARIATION AND LONGEVITY

J. L. HAMRICK

MEASURING the amount of genetic variation in species populations has been a concern of evolutionary biologists since the time of Darwin. A variety of techniques have been used, ranging from the common garden approach of Turresson (1922, 1925) and others (e.g., Clausen, Keck, and Hiesey, 1940, 1948) to the DNA sequencing studies of modern workers. These studies, whether morphometric, physiological, developmental, or biochemical, have demonstrated that most species populations contain relatively large amounts of genetic variation. Good examples of these kinds of studies can be found in the forest genetics literature. Because of the commercial importance of forest trees, genetic variation within and between tree populations has been unusually well documented. The majority of forest tree species have considerable amounts of genetic variation within, as well as between, populations (Libby et al., 1969; Stern and Roche, 1974; J. W. Wright, 1976). A notable exception is red pine (*Pinus resinosa*), in which little genetic variation has been demonstrated (Fowler, 1964, 1965; Fowler and Lester, 1970; Fowler and Morris, 1977). This is remarkable, since red pine ranges from Wisconsin to Nova Scotia and from northern Pennsylvania to central Ontario. Other forest tree species occurring in essentially the same environments as those inhabited by red pine have been shown to have considerable genetic variation. Thus, the question arises: Why do some species have a high degree of genetic variation while others have much less?

Many hypotheses have been generated to provide answers to this question. It has been proposed that the mating system (outbreeders versus selfers), geographic range (endemics versus widespread species), species age (new versus old species), chance events (species with fluctuating or low population numbers versus species with high or stable population numbers), environmental grain (species that perceive their environments as coarse grained versus those that perceive the environment as fine grained), ecological amplitude (species in temporally variable environments versus those in temporally nonvariable environments), etc., may all have important effects on the amount of genetic variation maintained within populations. Although these hypotheses have sometimes been presented as alternatives, it is generally recognized that *all* of these factors may influence the levels of genetic variation within a species. For instance, a species may be of recent origin, have a limited range, occur in a specialized habitat, and reproduce predominately by selfing or apomixis (see Gottlieb, article 11, for an example that fits this description). This is, admittedly, an extreme example, but any attempt to explain why certain species or groups of species contain more or less variation must take all of these factors into consideration. When individual species are compared, this becomes a very difficult task unless a thorough understanding of the history and biology of each species is available. When groups of species are compared in order to test a certain hypothesis, it is assumed that the other factors, while important in producing differences between species, will not effect the differences found between groups, i.e., that the factor under consideration is not significantly confounded by any other factor. If this assumption is met, the problem then becomes one of determining whether certain factors are universally important or whether the variation maintained by each species is the result of a unique set of circumstances.

One of the most remarkable characteristics of higher plants is the diversity of longevities found among species. Plant longevities may range from a very few weeks, in the case of some annual species, to well over 4,000 years, for some perennials. Some of the variation in longevity is due to the plastic response of plants to varying environmental conditions. It is generally held, however, that the major differences in longevity among species are genetically controlled and have evolved in response to specific environmental pressures. Thus, the question may be asked, Do species which have different genetically determined longevities contain different amounts of genetic variation within their populations? In other words, do annual plants contain more or less variation than long-lived plants such as forest trees?

STUDIES OF GENETIC VARIATIONS IN THE SEED PLANTS

In the review that follows, I have divided the higher plants (angiosperms and gymnosperms) into four classes based on their longevities and life forms: annuals, biennials, herbaceous perennials, and trees and woody shrubs. To determine whether differences exist among these groups in terms of intrapopulation genetic variation, I have relied almost entirely upon a survey of the plant allozyme literature. I have included a few studies of morphological markers determined by a single gene. I have not attempted, however, to make a systematic search of this literature. Nor have I attempted to review the extensive plant literature based on surveys of morphometric variation. Readers who are interested in these topics should see the reviews of Stebbins (1950), V. Grant (1958), Allard et al. (1968), Libby et al. (1969), and Stern and Roche (1974).

There has been some discussion in the literature about the minimum number of loci needed to produce generally acceptable estimates of genetic variation. Lewontin (1974) feels that studies with fewer than 18 loci are of questionable value in this regard. Other authors (Powell, 1975) have not included studies in which less than 10 loci were used. Other criteria which are often applied concern the number of populations per species and the number of individuals per population.

If the criteria applied to the animal literature were applied to the plant literature reviewed here, the resulting data would not allow comparisons to be made between the four groups. For example, of the 150 papers reviewed, only 50 utilized 10 or more loci. Of these 50, only 36 had sampled six or more populations per species. I have therefore adopted more liberal criteria. If I was able to make a genetic interpretation of the data reported, the paper was included. Of the 150 papers reviewed, 89 met this criterion and were included in this review. The resulting data upon which this study is based are, therefore, quite heterogeneous. The number of enzyme systems, loci, and populations studied varies greatly from study to study. Furthermore, those studies which use few loci tend to depend heavily on the variable substrate enzymes such as esterase, acid phosphatase, and peroxidase. Some writers (G. Johnson, 1977a,b) feel that variation in these enzyme systems may not be representative of variation in enzyme systems with more specific functions (but see Powell, 1975).

I wish to make two points relative to the plant allozyme literature. The first is that there are few plant allozyme data available which meet the criteria usually applied to the animal literature for numbers of loci or diversity of enzyme systems. Obviously, studies which use a wide range of enzyme systems and a representative sample of populations from the range of the species are critically needed. The second point is that comparative studies of the levels of variation among plant species with different longevities are rare. Studies of genera in which a variety of life history strategies are employed would have been quite informative but do not exist.

VARIATION ANALYSIS

Data on genetic variation within and between species populations are presented in the Appendix. A summary of the data is given in Table 4.1. Three separate measures of intrapopulation variation are used: percentage of polymorphic loci per population (P), mean

TABLE 4.1.
SUMMARY OF DATA PRESENTED IN THE APPENDIX

	No. of populations	No. of loci	P (% polymorphic loci per population)	No. of alleles per locus per population	PI[a]	I[b]
Trees and woody shrubs						
Mean	8.4	9.6	75.3	2.56	0.354	0.929
Range	1–31	2–27	40–100	1.33–4.55	0.144–0.446	0.717–0.979
SE	2.58	2.40	6.50	0.267	0.029	0.031
N	12	12	11	11	9	8
Herbaceous perennials						
Mean	13.1	13.7	26.3	1.44	0.116	0.947
Range	1–63	2–27	0–94	1.00–2.82	0.000–0.336	0.853–1.000
SE	4.85	1.29	6.30	0.127	0.032	0.011
N	15	23	21	15	14	13
Biennials						
Mean	21.8	18.4	22.0	1.35	0.079	0.970
Range	3–106	7–21	0–71	1.00–2.12	0.000–0.222	0.920–1.000
SE	4.04	1.41	6.90	0.121	0.024	0.007
N	13	13	13	11	11	12
Annuals						
Mean	17.3	12.5	46.2	1.80	0.154	0.916
Range	1–149	2–32	0.0–1.00	1.00–3.75	0.000–0.414	0.648–0.992
SE	4.23	1.13	4.30	0.112	0.018	0.022
N	43	49	44	39	39	20

[a]PI, polymorphic index.
[b]I, Nei's index of genetic similarity.

number of alleles per locus, and a polymorphic index (PI). PI is calculated by:

$$PI = \frac{1}{m} \sum_{i=1}^{m} \sum_{j=1}^{n} P_{ij}(1 - p_{ij})$$

where m = number of loci analyzed, n = number of alleles at a locus, and p_{ij} = frequency of the jth allele of the ith locus. PI is equivalent to the expected heterozygote frequency, given that the assumptions of the Hardy–Weinberg equilibrium hold. For two reasons this measure was used in place of the observed heterozygosity (presented, where available, in the Appendix). First, while many studies report gene frequencies, many omit observed genotypic frequencies. Second, since PI is free of the effects of the mating system or selection, it is more indicative of the actual variation within populations. For example, most authors will agree that a population with allele frequencies of $p_1 = p_2 = 0.50$ has more genetic variation than one in which $p_1 = 0.95$ and $p_2 = 0.05$. If the former population is highly self-fertilized, however, it could have lower observed heterozygote frequencies.

Relative measures of the amount of variation between populations are supplied by Nei's index of genetic similarity (I) (Nei, 1972). I is given by

$$I = \frac{J_{xy}}{\sqrt{J_x J_y}}$$

where $J_{xy} = \Sigma x_i y_i$, $J_x = \Sigma x_i^2$ and $J_y = \Sigma y_i^2$, and x_i and y_i are frequencies of the ith alleles in population X and Y. The I value varies from 0 to 1, with I = 1 occurring when all allele frequencies are equal in the two populations and I = 0 when the two populations have no alleles in common.

GENETIC VARIATION BY PLANT GROUP

There are a number of differences between the four groups of plants (annuals, biennials, herbaceous perennials, and trees and woody shrubs) in the amount of genetic variation maintained within their populations (Table 4.1). The most notable difference is seen between the woody (mostly coniferous trees) and the herbaceous plants. The results are consistent over all three measures of variability, but the PI values are the most striking. The woody plants contain approximately three times the heterozygosity found in herbaceous plants! Examination of the Appendix indicates that, although the range of PI values of both the woody plants and the annuals is quite similar, most of the woody plant species have PI values above 0.30. The results (Table 4.1) also indicate that populations of annual plants contain more variation than either the herbaceous perennials or the biennials. These differences, however, are not statistically significant, because of the high within-group variation. The largest difference between the annuals and the other herbaceous plants is in the proportion of polymorphic loci.

The interpopulation variation, as measured by I, indicates that the life form types with higher intrapopulation variation have lower genetic similarities between populations. This is not a totally unexpected result, since species with low intrapopulation variation typically have a number of loci which are monomorphic for the same allele in all populations. Since the I values of these loci are 1.0 for all comparisons, the average genetic identities between populations are increased. Relative to their intrapopulation variation, the annual species have more interpopulation variation than any other group. When the PI and I values of each species were compared, however, nonsignificant correlation coefficients were obtained.

PLANT AND ANIMAL VARIATION

It is also interesting to compare the variation found within plant populations with that reported for animal populations. When this is done (Table 4.2), certain trends appear. First, the mean number of loci scored for animal populations was 22.5, whereas that of the plants equaled 13.1. Thus, plant studies are based on substantially fewer loci. Second, invertebrate animals have mean P and PI values of 0.469 and 0.135; vertebrates, 0.247 and 0.061; and plants, 0.416

TABLE 4.2.
COMPARISON OF GENETIC VARIATION IN ANIMALS AND PLANTS

Group[a]	No. of species or forms	Mean no. of loci per species	Proportion of loci polymorphic per population (Mean P)[b]	Proportion of loci heterozygous per individual (Mean PI)[b]
Invertebrates				
Insects				
Drosophila	28	24	0.529 (0.030)	0.150 (0.010)
Others	4	18	0.531	0.151
Haplodiploid wasps	6	15	0.243 (0.039)	0.062 (0.007)
Marine invertebrates	9	26	0.587 (0.084)	0.147 (0.019)
Snails				
Land	5	18	0.437	0.150
Marine	5	17	0.175	0.083
Total and means	57	21.8	0.469	0.135
Vertebrates				
Fish	14	21	0.306 (0.047)	0.078 (0.012)
Amphibians	11	22	0.336 (0.034)	0.082 (0.008)
Reptiles	9	21	0.231 (0.032)	0.047 (0.008)
Birds	4	19	0.145	0.042
Rodents	26	26	0.202 (0.015)	0.043 (0.005)
Large mammals	4	40	0.233	0.037
Total and means	68	24.1	0.247	0.061
Plants				
Annuals	44/39[c]	12	0.462 (0.043)	0.154 (0.018)
Biennials	13/11	18	0.220 (0.069)	0.079 (0.024)
Herbaceous perennials	21/14	14	0.263 (0.063)	0.116 (0.032)
Trees and woody shrubs	11/9	10	0.753 (0.065)	0.354 (0.029)
Total and means	89/73	13.1	0.416	0.160

[a]Animal data from Selander, 1976.
[b]Standard error of the mean, where available, is given in parentheses.
[c]Fraction represents the number of species used for P over the number used for PI.

and 0.160. Although plant populations maintain approximately the same amount of variation as the invertebrates, it should be noted that the invertebrates have higher P values but lower PI values than plants. Plants, therefore, either have more alleles per locus or the allelic frequencies are distributed more equitably. Unfortunately, the data reported for animals (Selander, 1976) are inadequate for making such distinctions. A final comparison of interest is that the standard errors of the plant estimates are higher than those of the animals. This may be an artifact of the heterogeneity of the data base or may be due to biological causes. The plant species included within each group may actually represent a wider array of adaptive strategies and resulting genetic structure than those of the animal groups.

GENETIC VARIATION, LIFE FORM, AND LONGEVITY

Two principal results have emerged from this survey of the plant allozyme literature. First, the majority of the plant species examined maintain considerable amounts of genetic variation within their populations. Second, it does not appear that longevity per se has a marked effect on the amount of genetic variation. Relatively long-lived woody species do have considerably more variation than the average, but equally long-lived perennials have somewhat less-than-average variation. Thus, the significant differences appear between woody and herbaceous species rather than between long- and short-lived species.

Unfortunately, explanations must at this time be based largely upon circumstantial evidence and speculation. The detailed comparative studies which are necessary for an understanding of the underlying causation are unavailable. Nevertheless, it may be informative to consider what appear to be the most plausible explanations.

First let us consider whether there are any underlying factors which may have produced similar levels of variation within plant and invertebrate animal species but significantly less variation in vertebrate species. It is unlikely that the average age and geographic range of plant and invertebrate species differ significantly from those of the vertebrates. Greater fluctuations in population size and increased inbreeding may be more characteristic of invertebrate and plant populations, but these considerations would be expected to decrease, not increase, levels of genetic variation. Plants and many invertebrate animals are thought to be more sensitive to microenvironmental variation than the majority of vertebrates. Since most terrestrial plants begin their lives as germinating seeds or as isolated vegetative propagules, we would expect that this assumption is certainly valid for plants (Harper et al., 1965; Harper and Benton, 1966). The assumption gains additional support from evidence that the seed to seedling stage of the plant life cycle is the most critical with regard to survival and growth (Harper and White, 1974; Cook, article 9). In addition, most plants and invertebrates are considerably less mobile than the majority of vertebrates. Thus, many plants and invertebrate species perceive their environments as coarse grained (Levins, 1968), whereas the majority of the vertebrates perceive their environments as fine grained. Levins (1968) predicted that species with coarse-grained environments would have populations consisting of mosaics of locally adapted ecotypes. As a result, relatively high levels of genetic variation would be maintained within the populations. Species with fine-grained environments would have little local adaptation and would be expected to maintain less genetic variation. Selander and Kaufman (1973a,b), Selander (1976), and Powell (1975) have suggested that these considerations may play an important role in producing differences between invertebrate and vertebrate animals in terms of the amounts of variation. It appears, therefore, that the environmental grain concept provides an explanation which is consistent with the observed results. This is not to say, however, that it is the only factor that influences genetic variation or that it is even the most important consideration in all instances.

Nor are the differences found between woody and herbaceous plants easy to explain with our current knowledge of plant populations. There may be a number of characteristics of woody plants, however, that may function to maintain high levels of genetic variation.

The majority of woody plants surveyed are gymnosperms, whereas the herbaceous plants are angiosperms. Conifer species are undoubtedly on the average more ancient and, therefore, may have had more time to accumulate genetic variation (Soulé, 1976). However, the generation time of the conifers is typically quite long, and that of the angiosperms is shorter. Therefore, if the age of the species is measured in numbers of generations since the speciation event, there may be no significant differences between the two groups. Furthermore, there is little evidence available to demonstrate that "old" species actually have more variation than relatively younger species (Selander et al., 1970; Levin and Crepet, 1973).

Conifer trees are monoecious and generally have high rates of outcrossing, whereas the reproductive modes of the angiosperms cover a spectrum from vegetative reproduction and self-fertilization to obligate outcrossing. Also, the conifers are wind pollinated while the angiosperms are more often dependent upon insect vectors. These factors could lead to more pollen dispersal, larger neighborhood sizes, and more gene flow among nearby populations in the conifers. Also, the seed of most conifers is wind dispersed. Therefore, conifers may maintain more variation within local populations. These same factors, however, retard the formation of local ecotypes and the maintenance of genetic variation by these means. Unfortunately, detailed studies of the genetic structure of forest tree populations are lacking. In contrast, since many herbaceous perennials are capable of spreading vegetatively their populations may be assemblages of many or very few clones (Wu et al., 1975).

The tree species included in this study are predominantly species of commercial importance which form large continuous stands over large geographic ranges. This may cause their populations to experience less severe fluctuations in population size than many of the herbaceous species. Large population numbers reduce the effects of drift and may allow the maintenance of more variation within the species. The two tree species in this study that lack commercial importance are Pinus longaeva and P. pungens (see the Appendix). Both are presently limited to isolated mountaintop habitats. The population sizes of P. pungens are unknown to me, but populations of *P. longaeva*, the bristlecone pine, presently contain thousands of individuals (Hiebert, 1977). Furthermore, evidence is accumulating that, in the past, *P. longaeva* had much larger population sizes. Thus, although comparative data on population sizes are not available, we might conclude that even the more isolated tree species maintain larger, more stable population numbers than do many herbaceous species.

The final consideration concerns the longevity of trees versus that of herbaceous plants. Longevity may be seen to have three effects. First, long-lived species can maintain relatively stable population numbers even though periods of establishment are infrequent or erratic. Second, if the majority of mortality occurs in the early stages of the life cycle, individuals that survive this stage will remain in the population for a long period of time. If different alleles are favored by either selection or chance events during different periods of establishment, the mature plants which result will maintain a genetic "memory" of past selection or chance events. This "memory" would retard the decay of genetic variation owing to directional selection or drift and would lead to the maintenance of genetic variation. Hiebert (1977) has argued that this may be an important mechanism in the maintenance of genetic variation in *P. longaeva*. It should be noted, however, that the seed carryover abilities of many annual and short-lived perennials also provide a similar genetic memory. A third effect of longevity would be the occurrence of significant environmental fluctuations during the lifetime of an individual. Individuals that experience such temporal heterogeneity in their environments might require more internal buffering than individuals exposed to less severe fluctuations. Although very little experimental evidence exists and the functional relationships are obscure, it is argued that individuals that are heterozygous at many loci may be better able to cope with temporal fluctuations of the environment (Lerner, 1954). Therefore, long-lived species that occur in temporally variable environments should maintain more variation than species that experience few environmental fluctuations during their lifetimes.

These considerations of the effects of longevity may provide a

satisfactory explanation for comparisons between annuals, biennials, and trees but do not explain why long-lived herbaceous perennials have less variation than woody perennials. Species age, mating
system, and population stability may be sufficient to explain these
differences, but two additional explanations which apply specifically to the temporal heterogeneity hypothesis could be offered.
Conifers are typically found in montane regions, which are some of
the most spatially and temporally variable environments in the
world, whereas many of the herbaceous perennials in this study
occur in more stable habitats. Trees growing in such variable habitats might be expected, under this hypothesis, to require more heterozygosity to provide greater physiological buffering. Comparisons
between trees and long-lived herbaceous perennials growing
together and exposed to similar macroclimatic fluctuations might
demonstrate equivalent levels of variation between the two groups.
To my knowledge no such studies are available.

A second point involves the difference in life forms (Raunkiaer,
1934) between woody and herbaceous plants. Woody plants are
phanerophytes and maintain their apical meristems above ground.
Their apical meristems are therefore exposed to environmental fluctuations at all times of the year. Herbaceous perennials are either
chamaephytes, hemicryptophytes, or cryptophytes whose apical
meristems are located at or below the soil surface during the most
stressful periods of the year. The buffering effects of snow, soil, or
litter cover may in essence reduce the selection pressures for the
internal homeostatic buffering thought to be produced by increased
heterozygosity. If this argument is valid, we would expect to find
differences in genetic variation between populations of herbaceous
perennials and trees growing in the same habitats. Furthermore,
increases in heterozygosity with increasing age of the population
should be observed.

In summary, it appears that differences between woody and
herbaceous plants may be due to a number of factors, with mating
system, population stability, and longevity playing major roles.

A number of studies have sufficiently detailed data to allow
examination in the context of the present discussion. These studies
were chosen because they illustrate that annual, herbaceous peren-

nial, and woody perennial plants may all be locally adapted to microenvironmental heterogeneity and thus may be considered to perceive their environments as coarse grained. Furthermore, these studies also have data that indicate that heterozygosity per se may be favored in temporally fluctuating environments.

Annual Plants

Populations of the slender wild oat (*Avena barbata*) occurring in the cool Mediterranean region of California (the area surrounding the San Francisco Bay) are often polymorphic at morphological loci (Jain and Marshall, 1967; Jain, 1969) and allozyme loci (Clegg and Allard, 1972). Populations occuring on well-watered, fertile sites, however, are usually monomorphic for one five-locus allozyme genotype (mesic), whereas populations on the driest, least-fertile sites are monomorphic for a second genotype (Hamrick and Allard, 1972). In the Napa Valley, there are areas where the habitat changes rather suddenly from deep, rich, mesic agricultural soils to poor, shallow, xeric hillside soils. On such a hillside (CSA), Hamrick and Allard (1972) demonstrated that frequencies of the mesic and xeric alleles were correlated with environmental changes. Figure 4.1 is a map of the CSA hillside from which 23 collections were obtained in 1971 (Hamrick and Holden, unpublished data). The numbered sites (1–7) represent the areas which were previously studied (Hamrick and Allard, 1972). These sites were sampled by collecting seed from 250 mature plants. The location of each plant was marked on a map of the site. The lettered sites (A–P) represent areas which had not been previously sampled. A subjective, a priori habitat rating, ranging from xeric to mesic, was assigned to each site. Factors used to determine these ratings included slope aspect, depth of the soil, rockiness, soil type, and plant size and density.

The results of the five-locus allozyme analyses are shown in Figure 4.1 in terms of the genetic identity of each site with a site monomorphic for the mesic genotype. Obviously, there is a great amount of genetic heterogeneity on this hillside. Furthermore, the genetic heterogeneity is correlated with the environmental variation (Fig. 4.1 and Table 4.3). Of the 23 sites examined, the genetic structure of 14 sites was predicted exactly from the a priori environmental ranking. Five additional sites were misclassified by only one class. Thus, the genetic structure of 19 of the 23 sites was predicted by their environments.

A previous study (Allard et al., 1972) demonstrated that genetic heterogeneity could be shown to exist on an even smaller scale than that shown in Figure 4.1. To determine whether this phenomenon

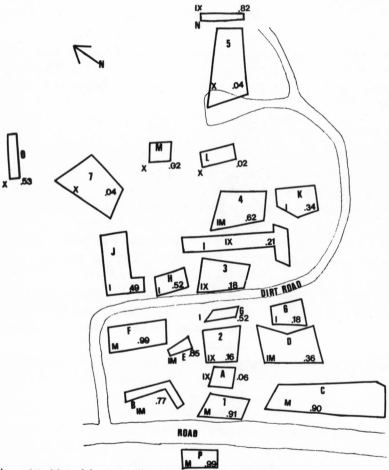

Figure 4.1. Map of the CSA hillside showing the location of 23 collection sites. Also given are the environmental ratings (X, xeric; IX, intermediate xeric; I, intermediate; IM, intermediate mesic; M, mesic) and Nei's identities (I) between the site and a population monomorphic for the mesic alleles.

TABLE 4.3.

LISTING OF A. barbata POPULATIONS FROM THE CSA HILLSIDE
RANKED BY NEI'S GENETIC IDENTITY MEASURE[a]

Identity Group				
M	IM	I	IX	X
F (0.992)	B (0.773)	O (0.532)	D (0.361)	6. (0.183)
P (0.991)	4. (0.617)	H (0.525)	K (0.343)	3. (0.177)
1. (0.914)		G (0.518)	I (0.208)	2. (0.157)
C (0.899)		J (0.490)		A (0.056)
E (0.851)				7. (0.038)
N (0.818)				5. (0.036)
				M (0.022)
				L (0.015)

[a]M, mesic; IM, intermediate mesic; I, intermediate; IX, intermediate xeric; X, xeric. Genetic identities, determined by comparison with the mesic genotype (21112), are given in parentheses.

was common throughout the hillside, each numbered site was sub-divided into six subsites, and their allozyme frequencies were deter-mined. Figure 4.2 gives the habitat rankings of each subsite within CSA4, together with the genetic identity of each. There is considera-ble variation between subsites, and it is correlated with microhabi-tat. Analyses of the other sites gave similar results. Obviously, A. barbata is sensitive to an environmental grain which is on the order

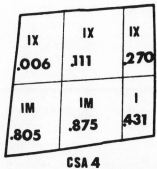

CSA 4

Figure 4.2. Map of the CSA4 collection site showing the six subdivisions, their envi-ronmental rankings, and the I values (obtained by comparison with a population monomorphic for the mesic alleles). IX, intermediate xeric; IM, intermediate mesic; I, intermediate.

of a few square feet. Furthermore, the amount of genetic heterogeneity found between microhabitats on the CSA hillside is nearly equal to that found between widely separated populations collected from throughout California. The mean genetic identity for California was 0.72 (Clegg and Allard, 1972); for CSA, 0.77; and for CSA4, 0.81 (Hamrick and Holden, 1979).

In addition to the marked microhabitat variation demonstrated for *A. barbata*, there is an excess of heterozygotes in all *A. barbata* sites that have been intensively studied (Marshall and Allard, 1970; Hamrick and Allard, 1972). Clegg and Allard (1973) have shown that heterozygotes have higher fitnesses during the viability phase (seedling to adult) of the life cycle. No differences between genotypes were found for fecundity selection. Perhaps the heterozygosity provides the individual with more physiological plasticity.

Perennial Plants

Although few plant species have been researched as intensively as *A. barbata*, studies can be found in the literature which provide similar data for perennial plants.

An example is the work of Schaal (1975) on the obligatory outcrossing, prairie perennial *Liatris cylindracea*. Schaal analyzed a single hillside population located along an environmental gradient for 27 allozyme loci. The hillside was divided into sixty 3-m² quadrats. Local genetic differentiation was marked, with allele frequencies varying by as much as 20 percent between adjacent quadrats. The majority of the variation described was not correlated with the measured edaphic factors (13 of 15 loci had nonsignificant correlation coefficients). Schaal (1975) concluded that the population structure of *Liatris* is the result of restricted gene flow, with selection playing only a minor part.

In a subsequent paper, Schaal and Levin (1976) reported the results of a demographic genetic analysis of this population. They demonstrated that average heterozygosity increases with increasing age class. There was also a positive relationship between individual heterozygosity and fecundity, longevity, and speed of development. Furthermore, the time to initial flowering was related to heterozygosity. Plants flowering after two seasons were more heterozygous than plants which did not flower. Recent papers, however, have cast doubts on the technique of aging used (Werner, 1978) and have pointed out unexplained inconsistencies in the data (Clegg, 1978).

McClure (1973) studied the population genetics of Mimulus guttatus, a riparian species of the western United States. Mimulus guttatus is a self-compatible species which also has extensive vegetative reproduction. The ten populations studied by McClure (1973) were located at intervals along the Yuba River drainage in the Sierra Nevada Mountains of California. Using 11 allozyme loci, McClure was able to demonstrate considerable intra- and interpopulation variation. The patterns of variation in M. guttatus were similar to those in Liatris in that no significant correlations could be found between the allozyme variation and such obvious environmental changes as elevation, moisture availability, or shading. Unlike Schaal (1974), however, McClure could find no relationship between genetic distance and physical distance. McClure argued that the differences were probably due to selection acting at the microhabitat level. Variation at a malate dehydrogenase locus appeared to support this conclusion.

The work of Law et al. (1977) on Poa annua and the studies of Snaydon (1970) and Snaydon and Davies (1972) on Anthoxanthum odoratum demonstrate that perennials maintain locally adapted genotypes which are correlated with subtle differences in the environment. In the Poa annua study, seed and tiller samples which were taken from disturbed habitats had shorter reproductive periods, higher seed output early in life, and shorter lives than those obtained from relatively undisturbed habitats. In the Anthoxanthum studies, seed and tiller samples were taken from a series of fertilizer plots and were grown in a common garden. The populations differed in such morphological characteristics as plant height, plant posture, plant yield, seasonal pattern of growth, reproductive strategy, and disease susceptibility. Furthermore, the patterns of morphological variation were highly correlated with differences in the environmental conditions of the plots. The authors (Snaydon and Davies, 1972) concluded that the observed differences were adaptive and had evolved over distances of approximately 30 m during a 50-year period. In both studies, the characteristics are morphometric or reproductive, and interpretations pertaining to adaptive advantages of these characteristics can be easily justified. Unfortunately, such interpretations are difficult to make for allozyme loci.

Therefore, we can conclude that herbaceous perennials also perceive their environments as coarse grained and can maintain locally adapted genotypes as well as annual plants. There is also some evidence from the *Liatris* study (Schaal and Levin, 1976) that heterozygotes have an increasing advantage as the plants become older.

Forest Trees

Bristlecone pine *(P. longaeva)* is a high montane conifer which occurs in isolated stands throughout the Great Basin region of the United States. Because of its slow growth and unusual ability to maintain life under extreme environmental conditions, bristlecone pine can live for as long as 4,500 years. Hiebert (1977) has studied the age structure and population genetics of five stands occurring in Nevada and Utah. In essence, the design of Hiebert's study is similar to that of Schaal (1975) and Schaal and Levin (1976). The results of his genetic analyses demonstrate that, despite the isolation of its populations, *P. longaeva* maintains exceptional amounts of varia-

TABLE 4.4.
INTRAPOPULATION VARIATION IN FIVE STANDS OF
P. Longaeva[a]

Population	Proportion of loci polymorphic[b]	No. of effective alleles	Mean heterozygosity	Sample size
Mammoth Creek	0.786	1.519	0.342	125
Twisted Forest	0.786	1.551	0.355	132
Wheeler Peak	0.786	1.432	0.302	114
Ward Mountain	0.786	1.506	0.336	111
Mount Hamilton	0.786	1.428	0.300	114

[a]Data are from Hiebert (1977).
[b]Fourteen loci were sampled; eleven were polymorphic in all populations.

TABLE 4.5.
NEI'S INDEX OF GENETIC IDENTITY BETWEEN
POPULATIONS OF *P. longaeva*[a]

	2	3	4	5
1[b]	.9626	.9519	.9633	.9527
2		.9814	.9896	.9824
3			.9815	.9874
4				.9887

[a]Data are from Hiebert (1977).
[b]Key: 1. Mammoth Creek, Utah.
2. Twisted Forest, Utah.
3. Wheeler Peak, Nevada.
4. Ward Mountain, Nevada.
5. Mount Hamilton, Nevada.

tion within its populations (Table 4.4). Comparisons of genetic variation between populations isolated by as much as 300 km indicate that there is a high amount of genetic similarity between the five populations (Table 4.5). Only a low elevation population at Mammoth Creek, Utah, had consistently low I values when compared with the other populations. Although a number of significant differences between populations were seen at individual loci, no consistent east–west patterns of variation were observed. When each stand was subdivided into three elevational zones, many significant differences were again found at the individual loci (Table 4.6). The mean I

TABLE 4.6.
NEI'S INDEX OF GENETIC IDENTITY WITHIN
POPULATIONS OF *P. longaeva* FROM DIFFERENT
ALTITUDES[a]

Twisted Forest[b]		Wheeler Peak[b]		Ward Mountain[b]	
A vs B	0.9882	A vs B	0.9825	A vs B	0.9620
A vs C	0.9803	A vs C	0.9678	A vs C	0.9799
B vs C	0.9749	B vs C	0.9812	B vs C	0.9784

[a]Data are from Hiebert (1977).
[b]Key: A = Upper portion of population.
B = Middle portion of population.
C = Lower portion of population.

between elevational zones within stands (0.977) was nearly identical to the mean I between populations (0.974). We can then conclude that there is as much localized genetic heterogeneity within stands of *P. longaeva* as there is between stands separated by hundreds of kilometers.

Pinus ponderosa is a montane conifer which is common at low and middle elevations of the Colorado Rocky Mountains. In a recent study of a needle peroxidase locus, Mitton et al. (1978) demonstrated consistent and often statistically significant differences between north- and south-facing slopes in the Front Range of the Colorado Rockies. In three of four comparisons, the "2" allele was in significantly lower frequencies on the south slope (Table 4.7). Low-elevation populations also had lower frequencies of this allele. In addition, all of the populations from south-facing slopes or low altitudes had an excess of heterozygotes. On the other hand, the population from the highest elevation had a significant deficiency of heterozygotes. These data were interpreted as evidence for local adaptation in *P. ponderosa*. Furthermore, it appears that heterozygotes have an advantage on the highly fluctuating temporal environments of the south-facing slopes.

To summarize, the studies discussed above provide evidence for the adaptation of plants to coarse-grained environments. Therefore, the expectation that plants should maintain high amounts of variation as a result appears to be substantiated. The case for the selective advantage of heterozygosity in temporally fluctuating environments is less well documented, but the results of the studies on oats (Clegg and Allard, 1973), *Liatris* (Schaal and Levin, 1976), and ponderosa pine (Mitton et al., 1978) are suggestive.

CONCLUSION

The results of ten years of electrophoretic studies have demonstrated that plant populations maintain as much variation as animal populations, if not more. Furthermore, there are large differences between species and between groups of species. The most striking finding of this study is the large difference in variation between

TABLE 4.7.

ALLELE AND GENOTYPIC FREQUENCIES OF A PEROXIDASE LOCUS AT SEVERAL LOCALITIES SAMPLED FOR PONDEROSA PINE (Pinus ponderosa)[a]

Sample locality	Aspect	Elevation (m)	Genotypes[b]			N	Per(2) (freq. ± SE)	X_A^2	X_B^2	Per(2) (freq. ± SE)	X_C^2
			22	23	33						
Glacier Lake		2590	135 (129.7)	44 (54.5)	11 (5.7)	190	0.826 ± 0.019	7.1**		0.826 ± 0.019	
Left Hand Canyon	North facing	≃2440	60 (60.4)	11 (10.1)	0 (0.4)	71	0.923 ± 0.022	0.5	10.0***	0.857 ± 0.021	
	South facing	≃2440	44 (45.1)	26 (23.8)	2 (3.1)	72	0.792 ± 0.034	0.6			
Coal Creek Canyon	North facing	≃2440	62 (61.8)	26 (26.4)	3 (2.8)	91	0.824 ± 0.028	0.1	0.1	0.830 ± 0.020	
	South facing	≃2440	55 (56.1)	24 (21.8)	1 (2.1)	80	0.837 ± 0.029	0.8			
Lower Sugarloaf Mountain	North facing	2130	41 (42.2)	29 (26.6)	3 (4.2)	73	0.760 ± 0.035	0.6	4.7*	0.701 ± 0.027	114.1***
	South facing	2130	23 (31.6)	52 (34.8)	1 (9.6)	76	0.645 ± 0.039	18.5***			
Boulder Canyon	North facing	1738	54 (56.5)	37 (32.0)	2 (4.5)	93	0.780 ± 0.030	2.3	3.3†	0.738 ± 0.023	
	South facing	1738	40 (45.6)	51 (39.7)	3 (8.6)	94	0.697 ± 0.034	7.6**			
Eldorado Springs		1760	26 (35.0)	64 (45.9)	6 (15.1)	96	0.604 ± 0.035	14.9***		0.604 ± 0.035	
Cheyene Lookout		1770	2 (7.5)	31 (20.0)	8 (13.0)	41	0.427 ± 0.055	12.2***		0.427 ± 0.055	

Data are from Mitton et al., 1977.

[b]Numbers in parentheses are the values expected under the assumptions of the Hardy–Weinberg model. The X_A^2 value tests the fit to Hardy–Weinberg expectations; X_B^2 tests the homogeneity of gene frequencies between slopes of different aspect; and X_C^2 tests the homogeneity of mean gene frequencies for all localities.

*P < 0.05   **P < 0.01   ***P < 0.001   †0.05 < P < 0.10

woody and herbaceous plants. The reasons for these differences, however, are obscure. Numerous theories have been proposed, but the data to support them are lacking or very sparse. Obviously, an increased emphasis must be placed on descriptive and experimental studies.

Those approaches which I feel will provide answers to the questions generated by this survey include:

1. Studies which describe levels of variation within populations of trees, perennial herbs, or annuals occurring in the same habitats. Such studies would be even more valuable if they included species of the same genus which employ different life-cycle strategies.
2. Studies in which greater attention is paid to variation at the microhabitat level. Such work is particularly important for vegetatively reproducing perennials whose populations may consist of a few genets.
3. Studies of the demographic genetics of plant populations. The pioneering work of Schaal and Levin (1976), Clegg and Allard (1973), and Clegg et al. (1978) demonstrate the increased levels of understanding which can be reached when selection is broken down into its various life-history components.
4. Studies of the relationships between variation at allozyme loci and genetic variation in quantitative traits or phenotypic variation. The studies of Jain and Marshall (1967) on quantitative variation in *Avena barbata* and *A. fatua* can be linked with later allozyme studies (Marshall and Allard, 1970; Hamrick and Allard, 1975). Also, the work of Linhart (1974) and Keeler (1975) on *Veronica peregrina* can be compared. In both situations, a positive correlation between allozyme and quantitative variation was reported. Such results, if they are general, indicate that the patterns and levels of variation as measured by allozyme studies may be indicative of variation at a majority of the loci. Furthermore, the work of Jain and Marshall (1967) on the relationship between levels of genetic and phenotypic variation in *A. barbata* and *A. fatua* has given us insights into the variety of adaptive strategies uti-

lized by even closely related species. More studies of this type are definitely needed.

In conclusion, I feel that we have barely begun to scratch the surface in understanding how plants adapt to their environments. Studies designed specifically to answer these questions are needed. Random survey work adds to our knowledge, but not at the rate needed to keep up with the questions generated by the presently rich body of theory.

APPENDIX

Species	No. of populations	No. of loci	P (% polymorphic loci per population)	No. of alleles per locus per population	PI (polymorphic index)	Observed heterozygosity	I (Nei's index of genetic similarity)	Reference
Trees and woody shrubs								
Eucalyptus obliqua	4	3	100.0	2.42	0.351	—	0.717	Brown et al., 1975
Larix decidua	1	3	66.7	2.67	0.347	—	—	Meinartowicz and Bergmann, 1975
Picea abies	20	6	96.0	2.63	0.414	—	0.945	Bergmann, 1975
P. abies	2	4	100.0	2.50	0.446	0.429	0.978	Tigerstedt, 1973
P. englemannii	3	2	50.0	1.50	0.432	—	0.943	Grant and Mitton, 1976
Pinus longaeva	5	14	78.6	2.35	0.364	0.327	0.974	Hiebert, 1977
P. pungens	3	15	40.0	1.33	0.144	—	0.943	Feret, 1974
P. ponderosa	10	27	59.0	2.00	—	—	—	Hamrick, unpublished data
P. sylvestris	31	5	—	3.80	—	—	—	Rudin, 1974
P. sylvestris	3	3	100.0	4.55	0.359	0.318	0.979	Rudin et al., 1974
Pseudotsuga menziesii	7	22	64.0	1.86	—	—	—	Hamrick, unpublished data
P. menziesii	12	11	74.2	3.17	0.332	—	0.958	Morris, unpublished data
Baptisia leucophaea	9	5	87.8	1.72	0.287	0.407	0.933	Scogin, 1969
B. nuttaliana	1	5	20.0	1.20	0.089	0.275	—	Scogin, 1969
B. sphaerocarpa	3	5	60.0	1.60	0.229	0.280	0.956	Scogin, 1969
Colobanthus quitensis	2	22	0.0	1.00	0.000	0.000	1.000	Lee and Postle, 1975

Cucurbita foetidissima	9	2	—	2.00	0.256	0.256	—	Lilley and Wall, 1972
Danthonia sericea	10	7	28.6	1.57	—	—	—	Gray et al., 1973
Elymus canadensis	63	14	9.2	1.09	0.026	0.019	0.957	Sanders and Hamrick, 1978
E. canadensis	—	13	23.0	—	—	—	—	Clegg et al., 1976
E. hystrix	—	13	7.7	—	—	—	—	Clegg et al., 1976
E. riparius	—	13	0.0	—	—	—	1.000	Clegg et al., 1976
E. virginicus	—	13	31.0	—	—	—	0.930	Clegg et al., 1976
E. wiegandii	—	13	7.7	—	—	—	—	Clegg et al., 1976
Liatris cylindracea	1	27	55.5	1.63	0.158	0.057	0.961	Schall, 1974, 1975; Schall and Levin, 1976
Lycopersicon cheesmanii	54	14	—	1.00	0.000	—	—	Rick and Fobes, 1975
L. chmielewskii	8	14	13.4	1.14	0.053	—	0.853	Rick et al., 1976
L. esculentum	7	15	2.8	1.18	0.037	—	0.981	Rick and Fobes, 1975
L. esculentum var. cerasiformae	7	15	3.3	1.16	0.018	—	0.958	Rick and Fobes, 1975
L. parviflorum	8	14	0.0	1.00	0.000	—	0.931	Rick et al., 1976
Mimulus gluttatus	10	11	94.1	2.82	0.336	—	0.930	McClure, 1973
Silene maritima	4	21	29.0	1.49	0.140	0.153	0.920	Baker et al., 1975
Tripsacum dactyloides	—	20	15.0	—	—	—	—	Senadhira, 1976
T. floridinum	—	20	20.0	—	—	—	—	Senadhira, 1976
Zea mexicana (perennial)	—	20	45.0	—	—	—	—	Senadhira, 1976

APPENDIX (continued)

Species	No. of populations	No. of loci	P (% polymorphic loci per population)	No. of alleles per locus per population	PI (polymorphic index)	Observed heterozygosity	I (Nei's index of genetic similarity)	Reference
Biennials								
Hymenopappus artemisiaefolius	12	7	71.0	2.05	0.222	0.208	0.942	Babbel and Selander, 1974
H. scabiosaeus	14	7	71.0	2.12	0.218	0.201	0.970	Babbel and Selander, 1974
Oenothera argillicola	10	20	20.0	1.25	0.075	0.080	—	Levy and Levin, 1975
Oe. biennis	106	20	30.0	1.40	0.101	0.260	0.950	Levin et al., 1972
Oe. biennis (Illinois)	44	20	6.6	1.07	0.030	0.045	0.967	Levin, 1975a
Oe. hookeri	14	20	0.0	1.00	0.000	0.000	1.000	Levy and Levin, 1975
Oe. parviflora	29	20	40.0	1.52	0.136	0.148	0.920	Levy and Levin, 1975
Oe. strigosa	29	20	25.0	1.30	0.046	0.028	0.970	Levy and Levin, 1975
Tragopogon dubius	6	21	8.9	1.09	0.028	0.001	0.983	Roose and Gottlieb, 1977
T. mirus	8	21	4.2	—	—	0.430	0.959	Roose and Gottlieb, 1977
T. miscellus	6	21	3.2	—	—	0.330	1.000	Roose and Gottlieb, 1977
T. porrifolius	3	21	6.4	1.07	0.014	0.001	0.980	Roose and Gottlieb, 1977
T. pratensis	3	21	0.0	1.00	0.000	0.000	1.000	Roose and Gottlieb, 1977

Annuals							
Arabidopsis thaliana	17	2	2.9	—	—	—	—	Grover and Byrne, 1975
Arachis hypogaea	26	27	22.0	—	—	—	—	Cherry and Ory, 1973
Avena barbata[a]	149	4	—	—	0.056	—	—	Marshall and Jain, 1969a,b
A. barbata	8	16	31.0	—	0.060	—	—	Marshall and Allard, 1970
A. barbata (Mediterranean)	73	28	—	—	0.006	—	—	Singh and Jain, 1971
A. barbata (California)	11	30	—	—	0.025	—	—	Singh and Jain, 1971
A. barbata	16	5	25.0	1.23	0.071	—	0.715	Clegg and Allard, 1972
A. barbata (single hillside)	23	5	97.0	1.97	0.290	0.037	0.771	Hamrick and Holden, 1979
A. barbata	29	32	12.8	1.20	0.041	—	—	R. Miller, 1977
A. canariensis	14	23	39.0	1.69	—	—	—	Craig et al., 1974
A. fatua[a]	101	4	—	—	0.196	—	—	Marshall and Jain, 1969a,b
A. fatua	8	13	54.0	—	0.120	—	—	Marshall and Allard, 1970
A. fatua	16	16	14.0	—	—	0.055	—	Jain and Rai, 1974
A. hirtula	23	28	—	—	0.012	—	—	Singh and Jain, 1971
Bromus mollis	10	6	92.0	1.92	0.204	—	0.991	Brown et al., 1974
Clarkia amoena	1	8	62.0	1.37	0.071	—	—	Gottlieb, 1973b
C. biloba	3	8	61.0	2.09	0.203	0.150	0.918	Gottlieb, 1974
C. dudleyana	1	8	75.0	2.38	0.250	0.160	—	Gottlieb, 1974
C. franciscana	1	8	0.0	1.00	0.000	—	—	Gottlieb, 1973b

Species	No. of populations	No. of loci	P (% polymorphic loci per population)	No. of alleles per locus per population	PI (polymorphic index)	Observed heterozygosity	I (Nei's index of genetic similarity)	Reference
C. lingulata	2	8	56.0	2.06	0.175	0.080	0.905	Gottlieb, 1974b
C. rubicunda	4	8	59.0	1.69	0.177	0.110	0.922	Gottlieb, 1973b
Gaura brachycarpa	—	12	25.0	1.25	—	0.060	—	Levin, 1975
G. demareei	2	18	22.0	1.39	0.050	0.050	0.990	Gottlieb and Pilz, 1976
G. longifolia	3	18	25.0	1.40	0.074	0.074	0.990	Gottlieb and Pilz, 1976
G. suffulta	—	12	33.0	1.42	—	0.030	—	Levin, 1975
G. triangulata	—	12	8.0	1.08	—	0.080	—	Levin, 1975
Helianthus annuus	5	2	50.0	1.50	0.162	0.176	0.992	Torres et al., 1977
Horedum jubatum	3	6	61.1	1.67	0.192	0.015	0.927	Babbel and Wain, 1977
H. vulgare	30	4	57.5	1.86	0.148	—	0.855	Kahler and Allard, 1970, and in prep.
Limnanthes alba	8	12	39.6	1.43	0.159	0.159	—	De Arroyo, 1975
L. alba	2	14	63.0	1.98	—	—	—	Jain, 1976
L. floccosa	8	12	32.3	1.37	0.130	0.130	—	De Arroyo, 1975
L. floccosa	4	14	5.0	1.04	—	—	—	Jain, 1976
Lolium multiflorum	17	15	93.3	3.50	0.350	—	—	Coleman, 1977
L. multiflorum	8	4	100.0	3.75	0.260	—	0.990	Mitton et al., 1978, and unpublished

							Reference
Lupinus nanus	1	2	50.0	1.50	0.248	0.241	—	Scogin, 1973
L. subcarnosus	8	8	88.0	1.84	0.142	0.097	0.975	Babbel and Selander, 1974
L. succulentus[a]	35	4	22.0	1.27	0.080	—	0.954	Harding and Mankinen, 1972
L. texensis	10	8	88.0	3.12	0.414	0.356	0.957	Babbel and Selander, 1974
Lythrum tribracteatum	7	18	28.0	1.39	—	—	—	Baker and Baker, 1976
Phlox cuspidata	10	16	11.0	1.11	0.044	0.013	0.950	Levin, 1975d
P. drummondii	10	16	19.0	1.22	0.075	0.040	0.974	Levin, 1975d
P. drummondii (cultivated)	16	19	11.0	1.11	0.036	0.013	0.920	Levin, 1976b
Stephanomeria exigua ssp. carotifera	11	14	57.0	2.10	0.092	0.092	0.920	Gottlieb, 1975
Zea mays	2	9	94.0	2.11	0.420	0.420	—	Brown and Allard, 1971
Z. mays	8	6	50.0	1.50	0.176	—	0.648	Brown, 1971
Z. mays (open pollinated)	—	20	60.7	2.90	0.220	—	—	Senadhira, 1976
Z. mays (commercial hybrids)	—	20	65.5	2.90	0.350	—	—	Senadhira, 1976
Z. mexicana (annual)	—	20	72.0	2.90	0.230	—	—	Senadhira, 1976

[a]Based on morphological markers

5 A COST–BENEFIT ANALYSIS OF RECOMBINATION IN PLANTS

OTTO T. SOLBRIG

THE DIVERSITY of breeding systems in plants was discovered and documented in the eighteenth and, principally, the nineteenth centuries (Darwin, 1876; 1888; Knuth, 1906–9). A comprehensive theory on breeding systems, however, was not available until C. D. Darlington's treatise, *The Evolution of Genetic Systems*, was published in 1939. In the period since this publication, a number of concepts have been clarified or developed, such as the identification and definition of the genetic system (Darlington, 1939; Stebbins, 1950); the adaptive nature of different genetic systems (Stebbins, 1958; Grant, 1958); the correlations between a species recombination system, its life-history parameters, and the environment it occupies (Stebbins, 1950, 1957; Grant, 1958, 1975); and the hypothesis that the genetic system operates so as to strike an "optimal balance between constancy and variability in reproduction" (Grant, 1958).

Nevertheless, there is still uncertainty and confusion, as well as a shortage of good experimental data, regarding the advantages and disadvantages derived by individual plants in a population from producing different degrees of variability in their offspring. In part, these problems can be traced to the fact that most theoretical work has proceeded along the lines of the statement, first introduced by Darlington (1939), that there is an evolutionary conflict between "immediate fitness" (by which is meant individual Darwinian fitness) and "long-range flexibility" (the ability to survive over a large number of generations). According to this concept, each species must find its own best compromise between immediate fitness and long-range flexibility.

It is self-evident that long-range survival of any lineage requires the ability to adjust to changing environmental conditions. But the future is always unpredictable, and there is no known mechanism that allows a plant to select a recombination mechanism that will ensure long-range survival of its lineage. Rather, long-range survival of a lineage is possible only when the recombination system that provides the maximum individual advantage (i.e., immediate fitness) is one that also generates enough variability in the population to allow it to track environmental changes. Selection is always for immediate fitness; long-range survival is an indirect result and is not under natural selection (Solbrig, 1976).

In this paper, I review the various components of the recombination system, assessing possible costs and benefits in terms of their effect on individual Darwinian fitness, and predict the probable effect of the recombination system on long-range flexibility.

The Darwinian view of the world, as it is often interpreted, carries the implication that the characteristics of organisms are evolved according to some optimality principle. The idea that nature pursues the most economical path is one of the oldest principles in science, an idea that can be traced to ancient Greek thinkers. Notions of economy or optimality are difficult to quantify in a general way since each aspect requires special consideration. Natural selection in its most extreme deterministic form indeed implies that the most optimal phenotype will be selected at each particular point in space and time. There is, however, ample evidence indicating that such a model of evolution is unrealistic. A number of internal constraints and external accidents strongly influence survival and reproductive success. The extent to which evolution is stochastic is a hotly debated subject, and at present we lack objective criteria to evaluate this question. Nowhere is this better illustrated than in the debate between advocates of the neutralist and selectionist schools of population genetics (Lewontin, 1974; Nei, 1975).

That natural selection favors the establishment and spread of characteristics that increase survival is undeniable and has been demonstrated repeatedly. What is more questionable is that every character is optimal and that changes in the phenotype of organisms always result from changes in the environment. I would like to hazard the opinion that evolutionary change due to internal changes (i.e., mutations and recombination) that create more competitive phenotypes in the absence of changes in the physical environment are taking place simultaneously with the evolution of forms that track changes in the external environment. Of course, it could be argued that the appearance of a novel phenotype represents an environmental change.

Consequently, even in an immutable physical environment, a selective advantage will accrue to the organism that maintains the ability to produce some offspring with new characteristics. An open recombination system should therefore be favored. Elsewhere (Solbrig, 1976) I have argued that, although the physical environment tends to change slowly, the biological environment, i.e., the set of competitors and predators, changes more rapidly and, furthermore, that adaptive changes that render one lineage more competitive or better able to defend itself against predators act as a selective force on these same predators and competitors, mendating the ability to produce variable offspring in order to maintain the ability to survive. Furthermore, spatial variation in the physical environment is such that it is difficult to conceive of a population occupying a uniform environment.

REGULATION OF RECOMBINATION

By now the mechanisms that produce and regulate recombination are fairly well understood. Darlington (1939, 1958) and Mather (1943) were the first to point out the integrated nature of these mechanisms, what Darlington called the "genetic system." Their ideas were further developed by Stebbins (1950, 1957, 1958), Carson (1957), and Grant (1958, 1975).

The genetic system includes a number of factors (Table 5.1) that regulate the expression of potential genetic variation of the off-

TABLE 5.1.
FACTORS WHICH REGULATE RECOMBINATION IN PLANTS
(MODIFIED FROM GRANT, 1958)

A. Not under direct selection
   1. Population size
   2. Postfertilization sterility barrier
B. Under direct selection
   a. Control operating at the stage of meiosis
      3. Chromosome number
      4. Frequency of crossing over
   b. Control operating at the stage of fertilization
      5. Breeding system
      6. Pollination system
      7. Dispersal potential
      8. Crossability barriers and external isolating mechanisms
C. Under direct selection but affected by different selective forces
      9. Length of generation
     10. Seed number and reproductive effort

spring. The operation of these factors has been discussed in detail by
Verne Grant (1958). The potential costs to the plant, however, have
so far not been assessed. These factors are now evaluated in terms of
costs and benefits to the individual plant in the population.

It is not possible at present to determine costs and benefits in a
rigorous way, because there is no known way to weigh advantages
and disadvantages using a common measuring unit. Costs are
assessed in terms of calories, benefits in terms of increased seed
production. Theoretically, calories and seed number should be con-
vertible into fitness, but that moment is still far away. Furthermore,
in most instances we lack empirical knowledge of costs and benefits
as calories and seedling number. The following discussion, therefore, will be largely subjective and speculative, with the hope that it
will stimulate more experimentation and objective assessment of the
ideas presented.

Factors Operating at Meiosis
   Chromosome number and chiasma formation. Alleles of genes
borne on separate chromosomes become combined in different ways
by the independent assortment of the chromosomes at meiosis.

Alleles of genes on the same chromosome are transmitted as a unit to the gametes in meiosis, unless a cross-over has taken place between the two genes. Since chiasmata form in the pachytene stage of meiosis, a cross-over produces two recombinant and two nonrecombinant gametes. The level of recombination between two heterozygous loci located in different chromosomes is 50 percent, whereas recombination between two heterozygous loci on the same chromosome will be 50 percent only if there is a probability of 1 of a cross-over occurring between the loci at each generation (and there are no double cross-overs). Genes in different chromosomes are subjected to more recombination than those on the same chromosome.

There is now some evidence that, within the same environment, the selective value of a gene is determined both by its direct action on the organism and by how well it interacts with its fellow genes. It is to be expected, therefore, that the architecture of the genome will reflect this fact. Genes that produce products that must interact closely during development should be closely clustered on the chromosome; other genes may be more scattered. Such is the case with bacterial operons such as the galactose (Adhya and Shapiro, 1969) or tryptophan (Wuesthoff and Bauerle, 1970) operons of E. coli and the clustered heteroloci in fungi, such as the his-4 region of yeast (Shaffer et al., 1969) and the arom region of Neurospora (Case and Giles, 1971). In higher eukaryotes, no such integrated clusters have been detected, although the "complex loci" of animals, such as some Drosophila genes (bithorax, dumpy, rudimentary, etc.), the histocompatibility loci of mammals, and the "supergenes" of plants (e.g., those genes which control heterostyly in many flowering plants), are possible candidates. On the other hand, there is ample evidence that genes governing the same trait are dispersed throughout the genetic map. Some examples are grain color in wheat (Nilsson-Ehle, 1909) and corolla length in tobacco (East, 1916). A modern example of a carefully dissected series of traits is in the green alga Chlamydomonas reinhardti, in which genes involved with thiamine and arginine synthesis and with photosynthesis are located on several different chromosomes in the complement of 15.

Changes in chromosome number occur with difficulty. Not only are there mechanical difficulties involved in the process of loss or gain of centromeres, but individuals of outcrossing species that have more or fewer chromosomes than the rest of the population may produce offspring that are totally or partially sterile owing to pairing difficulties during meiosis.

If chromosome numbers change slowly in evolution, plants must rely more on crossing-over than on chromosome number as a mechanism to regulate the combination of allelic genes in the gametes. The same argument applies to changes in the order of the genes in the genome, especially those involving exchanges of segments of different genes (translocations). Because the effect of chromosome inversions and translocations on pairing in meiosis is less drastic than differences in chromosome number, we expect changes in the order of genes to occur more often than changes in chromosome number.

The mechanism of chiasma formation and crossing-over is still shrouded in uncertainty. Since crossing-over involves breakage of a chromatid and reunion with its homolog at exactly the same spot, chiasma formation must require elaborate biochemical machinery. Repair mechanisms following UV lesions in E. coli involve no less than three separate enzymes, and we suspect that more must be involved in crossing-over. Furthermore, chiasmata appear to be formed more or less simultaneously during pachytene, and it is reasonable to assume that there are both qualitative and quantitative factors controlling chiasma formation. This machinery involves an energetic cost, although it does not seem possible at this time to assess its relative significance in terms of the overall caloric budget of the plant.

In addition to this direct energetic cost, there is an additional cost due to errors that occur during crossing-over, when breakage of the two chromatids occurs at slightly different points, so that the subsequent fusion gives rise to gametes carrying either a duplicated segment or a deficient one. Since such gametes (especially the deficient ones) are often nonfunctional, crossing-over reduces the number of viable gametes. The greater the frequency of chiasmata, the greater the number of nonviable gametes, although the relation may not be linear.

Sterility barriers. Postfertilization sterility barriers that become evident during meiosis involve both genic and chromosomal barriers, including sterility due to differences in chromosome number and chromosomal aberrations. Inasmuch as they prevent the mixing of two different genotypes, they prevent recombination. Nevertheless, they are not under direct individual selection. On the contrary, when crosses between two gene pools occur that lead to sterile or only partly fertile offspring, selection may operate to prevent hybridization (prefertilization barrier) or to eliminate the sterility barrier (Bossert, 1963; Wilson, 1965).

Factors Operating at the Stage of Fertilization

Breeding systems and pollination systems. Recombination results from the fusion of genetically different gametes. In all but cases of strictly closed recombination systems (apomixis and exclusive self-fertilization), this involves the union of gametes produced by different individuals. Sedentary organisms must rely on an outside agent to bring their gametes together. In the case of the flowering plants, pollen must be transported from the donor plant to a receptive stigma. This external agent is either a physical agent (such as wind or water) or an animal (such as an insect). Transport of pollen by wind is inefficient and wasteful of resources since stigmatic surfaces constitute a very small target. Pollen- and nectar-feeding animals serve as better agents but can also be wasteful. In Jepsonia, for example, less than 1 percent of the pollen reaches a stigma (Ornduff, 1970). In addition, animals will not visit flowers unless attracted to them by some manner of reward (normally food, but occasionally scents or nonrewards such as colors; see Dodson, 1975).

The energetic costs of these rewards can be calculated and involve such things as number of pollen grains and ovules produced, flower size, and amount of nectar. In addition to this purely energetic cost, there is also a genetic cost to outbreeding, which has been called the meiotic cost by Maynard Smith (1971; Williams and Mitton, 1973; Williams, 1975). In effect, when offspring are produced by self-fertilization or apomixis, they receive all their alleles from one parent; when offspring are produced by outcrossing, half of the alleles come from one parent, half from the other. Some of the alleles received from the two parents may be identical by descent, so that the offspring may have more than 50 percent genetic similarity with each parent. Since fitness is the proportional contribution of alleles to the next generation, each offspring produced through the sexual process transmits less of the maternal (or paternal) alleles than do offspring produced by selfing or by asexual means. For outbreeding to be of selective advantage, the fitness advantage (i.e., the increase in the differential contribution) must exceed the combined energetic and meiotic costs.

Figure 5.1 presents a graphical analysis that predicts when selection should favor outcrossing over selfing, taking into consideration only the meiotic cost and the degree of relatedness of individuals in the population. In an infinitely large population of unrelated individuals ($F = 0$) with random breeding, outcrossing individuals must produce twice as many surviving offspring as selfing individuals in order to transmit the same number of alleles to the next

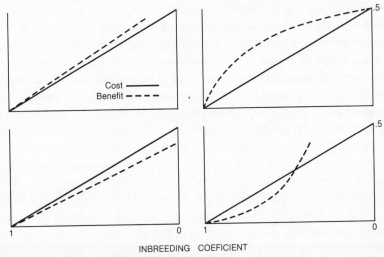

INBREEDING COEFICIENT

Figure 5.1. Prediction of breeding system and degree of relatedness in a population in four hypothetical cases assuming heterozygous individuals are heterotic, but in different degrees, and assuming that the meiotic cost is the only cost of outbreeding. *A*, Case 1: Heterosis is a linear inverse function of F; obligate outbreeder should be selected. *B*, Case 2: Heterosis decreases with decreasing F; a partially open recombination system should be selected. *C*, Case 3: Heterotic advantage does not compensate for the meiotic cost; inbreeder should be selected. *D*, Case 4: Meiotic cost greater than heterotic advantage at high levels of F, and lower at low values; unstable system, breeding system depends on initial population structure.

generation. In other words, the advantage derived by the offspring in being heterozygous must give it a probability of surviving to maturity that is at least twice as high as that of its potential homozygous half-sibs. In a population where the members are partially related (0 > F > 1), the advantage of the heterozygous over the homozygous does not need to be twice as much but will depend on the degree of relatedness according to the formula $\overline{w} = 2 - 2F$, where $\overline{w}$ is average individual inclusive fitness and F is the average degree of relatedness of two random individuals in the population. Consequently, when the fitness advantage of the outcrosser increases monotonically as F decreases in the population, outcrossing will be favored over other breeding systems. On the other hand, if the fitness advantage derived from outcrossing is a curvilinear function, selection will operate to select a breeding system such that the population maintains a certain level of relatedness (F). This could be achieved by decreasing the potential dispersal of propagules or the pollination radius. A third possibility is that at high values of F, the inclusive fitness of selfing individuals is greater than that of outcrossers, but as F decreases, a crossover point is reached, where the fitness of the outcrosser is greater. A polymorphism could be maintained in such a case or, more likely, one or the other method will be favored according to the size and structure of the population. Close inbreeding will be favored only when the inclusive fitness of the selfer is greater than that of the outbreeder over all values of F.

So far this analysis has taken into consideration only the so-called meiotic cost of outcrossing. In additon to this cost, there is the energetic cost of transporting pollen from the flower of one individual to a receptive stigma of another individual. Energetic costs have been discussed by several authors (Ornduff, 1969; Proctor and Yeo, 1972; Stebbins, 1970) and involve the production of large and attractive flowers or inflorescences and large amounts of nectar and pollen. Whole-character syndromes to attract specific pollinators have been repeatedly evolved involving flower shape, size, color, and scent. Grant and Grant (1965), Leppick (1957), Ornduff (1969), and others have documented these changes, in both general and specific terms. Although there is good documentation that outcrossing involves a positive energetic investment either to attract an animal vector or in the production of copious amounts of pollen in wind-pollinated species, there is a scarcity of good quantitative data. Cruden (1977) has recently published a quantitative estimate of the pollen cost of xenogamy (Table 5.2). Whereas the pollen/ovule ratio in cleistogamous flowers is of the order of 5/1, it climbs to 170/1 in facultative selfers and to 6000/1 in self-incompatible outcrossers. Since pollen grains are rich in protein, the difference between cleistogamous flowers and those of xenogamous species is significant not only in terms of calories but also in terms of nitrogen loss. Nectar is usually not produced in selfers but is abundant in flowers of outcrossers. Since nectar contains sugars and amino acids (Baker and Baker, 1973, 1975), nectar production is also costly, both in caloric and nitrogen terms. That these differences in investment affect fertilization has been repeatedly demonstrated (Mulligan, 1972). The net effect of these costs is to reduce the number of seeds a xenogamous plant can produce per unit of reproductive effort. Consequently, for such an investment to pay off, the seeds that are produced have to have a correspondingly higher probability of survival. Again, good quantitative data are unavailable, but, in a study of the mustard genus *Leavenworthia*, Solbrig and Rollins (1977) found that, whereas self-compatible species produced an average of 200 seeds per gram of dry plant weight, self-incompatible allogamous species produced only 100 seeds per gram of dry weight, on the average. Consequently, assuming that the lower seed number reflects the cost

TABLE 5.2.
BREEDING SYSTEMS AND MEAN POLLEN/OVULE
RATIOS[a]

Breeding system	N	Pollen/ovule ratio[b]
Cleistogamy	6	4.7 ± 0.7
Obligate autogamy	7	27.7 ± 3.1
Facultative autogamy	20	168.5 ± 22.1
Facultative xenogamy	38	796.6 ± 87.7
Xenogamy	25	5859.2 ± 936.5

[a]From Cruden, 1977.
[b]Values are means ± SE.

of outbreeding, the fitness advantage of the xenogamous species of *Leavenworthia* in the environment where they prevail over selfers must be at least two times higher on these terms alone, plus the increase in fitness necessary to overcome the meiotic cost.

Dispersal potential. Seed-dispersal mechanisms have already been surveyed (van der Pijl, 1969). The significance of dispersal in relation to the recombination system, however, has not always been appreciated.

Harper (1961, 1977; Harper, Lovell, and Moore, 1970) has pointed out that every species which does not have a niche in climax vegetation is doomed to local extinction—species differing only in the frequency with which local extinction occurs. The more frequently a species goes locally extinct, the greater is the importance of both dispersal, as a mechanism of escape, and a high intrinsic rate of natural increase. Since large seed number, small seed size, and high dispersibility go hand in hand, species occupying early phases of succession will produce many small seeds, whereas those in closed, shaded communities will tend to produce fewer and larger seeds. Species of disturbed habitats can increase seed outputs further by shifting to autogamy, which, as we saw above, would allow them to forgo the high costs of xenogamy.

It is generally accepted that differences in breeding system tend to correlate with successional stages (H. G. Baker, 1955, 1959, 1965; Stebbins, 1958, 1970; Jain, 1976). Colonizers of disturbed habitats and plants of early successional stages tend to be self-compatible or autogamous, or both, whereas those of later successional stages, both temperate and tropical, tend to be xenogamous (H. G. Baker, 1959; Bawa, 1974; Stebbins, 1958).

H. G. Baker (1965, 1974) has discussed in great detail the characteristics that increase fitness in colonizing species (in this case, with particular reference to agricultural weeds). Of the 12 characteristics cited by Baker as "ideal weed characteristics," 9 are directly related to some aspect of seed production (e.g., continuous seed production for as long as growing conditions permit, very high seed output in favorable environmental circumstances, seed adaptations for short- and long-range dispersal). Thus, it is not surprising that a large number of colonizing species are self-compatible and mostly autogamous since, as has been pointed out, xenogamy involves energetic costs that reduce the total number of seeds for a given reproductive effort (see also Ornduff, 1969).

It has also been argued (Stebbins, 1950, 1958; Stone, 1959) that disturbed habitats are more uniform and predictable than less disturbed ones. This point is not firmly established, and the truth may be just the opposite. Disturbed habitats resemble each other in only one respect—that as a result of some man-made or natural catastrophe, the natural vegetation has been totally or partly destroyed. This disturbance is, however, the cause of the unpredictability of such environments since secondary successions are characterized, especially in their early stages, by a large degree of species turnover. The probability of a species encountering the same set of competitors and of predators at different places is very low. On the other hand, in more stable environments, the opposite is true. The turnover rate slows down considerably as a community approaches the so-called climax. Harper (1961) has pointed out that selection will push a species to "climb the seral ladder" by favoring increasingly larger seeds with a correspondingly lower dispersal potential. But such a change would also necessitate changes in the recombination system, and, since such changes are difficult to evolve, it is not surprising that species replacement in succession is the rule.

Another interesting corollary of the correlation of high dispersability with early successional stages involves the genetic structure of populations. In colonizing species with a high dispersal rate, there is a higher probability that at least some of the plants are derived from seeds blown or carried in from other areas than is the case with populations of climax species, where most, if not all, seeds are locally produced. Consequently, although colonizing species are mostly autogamous and produce mostly uniform offspring, the population may be genetically polymorphic, as has repeatedly been shown to be the case with populations of apomictic dandelions (Larsen, 1959; Kappert, 1965; Solbrig and Simpson, 1974). Occasional crosses between unrelated individuals in a population of a normally autogamous species can release a great deal of genetic variability (Allard and Kannenberg, 1968). On the other hand, in populations of xenogamous species, where the influx of seeds is very restricted, individuals, although largely heterozygous, will tend to be related because of the low vagility of seeds. To reduce inbreeding under those circumstances, the species must increase its pollination radius.

In summary, colonizers must produce many small seeds to reach suitable habitats, and must multiply quickly once in them (Stebbins, 1970; Angevine and Chabot, article 8). Under these circumstances, the high cost of xenogamy probably cannot be tolerated, and a shift toward autogamy will be favored. The great vagility of the seeds will bring together diverse genotypes, and occasional crosses will release a wide array of genetic combinations. The opposite is true in stable communities. Seedling survival and establishment demand large and heavy seeds, which will have very low dispersibility. Consequently, most plants in the population will be locally produced. To reduce inbreeding, open recombination systems and gene flow via pollen are necessary, and xenogamy and self-incompatibility will thus be favored.

Sterility barriers. The relationship between prefertilization barriers and recombination has been discussed in detail by Grant (1958). It is well known that certain groups, such as many orchid genera, can be successfully crossed with species belonging to genera of different subtribes; species of oaks are well known for the lack of barriers to crossing, as are species of willows, *Rhododendron*, and *Ceanothus*. Grant (1958) has shown that there is a relationship between breeding system and interspecific sterility barriers: Xenogamous species are less well isolated than autogamous species, this being a trend rather than a fixed rule.

It is possible that the relationship between fewer interspecific sterility barriers and xenogamy is related to the open recombination system these species possess. It is more likely, however, that the patterns of speciation in xenogamous and autogamous species differ, with related species of autogamous taxa occurring sympatrically more often than xenogamous species.

In this connection, it must be remembered that autogamous species tend to be colonizers. Lack of prefertilization barriers is detrimental to the fitness of an individual only when it is expressed, i.e., when sterile or inviable zygotes are produced. This will be a rare event in species that grow in different habitats but a more common occurrence in species that colonize an area, especially if they have common pollinators (Levin, article 6). Furthermore, not only are facultative autogamous colonizing species more likely to come into contact with related taxa and hybridize, but any loss in fertility is more detrimental to a species that depends on high seed production for survival.

Factors Related to the Structure of the Population

Three more factors that affect recombination remain to be considered: population size and spatial distribution of individuals, length of generation, and seed number and reproductive effort.

It must first be mentioned that all these factors, especially population size and spatial distribution, are to be viewed more as constraints upon recombination than as adaptations to regulate the amount of recombination. In effect, population size is determined by spatial and temporal distribution of light, water, and soil type; topography; number and kind of competing species; number and kind of herbivores; and previous history of the species at that site, which determines the size of the soil—seed pool and the extent of seed flow into the area. Selection can operate on some of these components, such as competitive ability, seed longevity, or seed dispersal, but not on others, such as weather or patterns of disturbance. Although population size and spatial distribution of a species are not under direct selection, they do have an important effect on recombination.

In effect, small populations will tend to inbreed even if individuals are outcrossed. In a xenogamous species, inbreeding will reduce the genetic cost of outcrossing (F increases), but not the cost of pollinator attraction. Inbreeding will reduce the benefit of outcrossing, in that offspring will be less diverse and less heterotic. Eventually a point is reached where a switch to autogamy will increase the fitness of the individuals in the population (Fig. 5.2). Such a change has been documented in the two mustard species *Leavenworthia alabamica* and *L. crassa* (Lloyd, 1963; Solbrig and

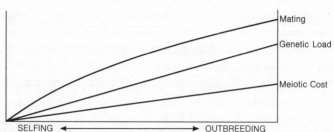

Figure 5.2. Graphical representation of relative costs and benefits of outbreeding and inbreeding systems.

Rollins, 1977), but this appears also to be the case in other groups of noncolonizers in localized habitats such as vernal pools (Jain, personal communication, 1977; Kruckeberg, personal communication, 1975) and serpentine outcrops. Subdivision of the population into smaller islands will have similar effects.

On the other hand, if a species produces seeds that lie dormant in the soil, so that generations are not synchronous, inbreeding can be delayed in small populations. This phenomenon is particularly important in annual species, which go through periods when the population is very large followed by periods when the population is very small.

According to Wright (1941), the inbreeding coefficient is equal to

$$F_n = \frac{1}{2N} + \left(1 - \frac{1}{2N}\right) F_{n-1}$$

The first term of the equation is directly attributable to inbreeding in generation n; the second term is due to inbreeding in previous generations. If the population in generation n is very small, and in generation $n + 1$ very large, F_n will have a large value because $1/2N$ will be a significant number and F_{n+1} will also be high because $(1 - 1/2N)F_n$ will be significant. If there is seed storage, however, then most plants in generation $n + 1$ will come from seeds produced in an earlier generation when the population was very large and when the F value in the population was low. Plants with seed storage can therefore avoid the bottleneck effects of bad years.

Length of generation and breeding system have already been discussed. In plants, competition for space favors longevity in stable environments (Hickman, article 10). This is the primary selective force. Long generation time is normally correlated with large standing biomass, and both these factors determine that such organisms will have a large seed output, allowing for more experimentation in recombination. Species of stable environments also tend to be subjected to lower population size fluctuation, which combines with the larger seed output in reducing considerably the ratio of successful genotypes to seed produced. Counteracting this trend is the greater competition between seedlings in stable environments and the low light levels that call for large seeds with copious nutrient reserves. It can be seen that seed number and total reproductive effort are more consequences of the ecological niche of the species than of the particular recombination system the species possesses, although seed number and generation length will affect the recombinational potential of a species.

SUMMARY AND CONCLUSIONS

The principle, first enunciated by Darlington, that the recombination system of an individual is molded by natural selection and is part of the adaptive machinery of the plant was a seminal breakthrough in our understanding of evolutionary adaptations in plants. The main benefits derived from outcrossing are the production of variable offspring and heterotic offspring with especially advantageous gene combinations, termed the "genetic elite" by Dobzhansky (Dobzhansky and Spassky, 1963). The costs are principally of two kinds. First, there is the meiotic cost of outbreeding, which increases inversely with the inbreeding coefficient of the population and lowers the fitness by $\frac{1}{2}$ at an F value of 0. Second, there is the energetic cost of petals, scents, and a food source (nectar or pollen), in the case of animal-pollinated flowers, or of additional pollen, in the case of wind-pollinated species. Although quantitative information is scarce, the energetic cost is not trivial.

Natural selection will tend to favor the recombination mechanism that maximizes fitness, i.e., benefit (in fitness) minus cost (in fitness). Because the cost is largely expressed as a loss of potential zygotes per unit of reproductive effort, whereas the benefit is in the form of a higher probability of survival, natural selection will favor outcrossing and an open system of recombination in those environments where individual survival probability and turnover rates are low, and autogamy will be preferred where individual survival probability and turnover rates are high.

6 POLLINATOR FORAGING BEHAVIOR: GENETIC IMPLICATIONS FOR PLANTS

Donald A. Levin

POLLINATOR foraging theory has been used as a tool to predict ecological phenomena, especially competitive interactions among pollinators, in addition to foraging behavior. The theory has also been used to predict the behavior of pollinators confronted with an ensemble of plant traits and to predict the manner in which pollinators alter their behavior in response to changes in the features of single plants, in population density, and in species mixtures (Levin and Anderson, 1970; Heinrich and Raven, 1972; Pyke et al., 1977). On the other hand, the many implications of foraging theory in the study of the genetic structure of plant populations have only been incompletely addressed. My purpose here is to explore the genetic implications of several components of foraging behavior, which may be best understood from the theory of optimal foraging. Foraging theory will be discussed in light of empirical data on pollinator behavior and will then be used to draw inferences about pollen and gene flow within and between populations of the same species and between populations of different species. For the most part, foraging models have not been tested in plants, but the data do suggest considerable concordance.

DEVELOPMENT AND THE REGULATION
OF PATCH SIZE

As a prelude to a discussion of pollinator–plant interactions, it is useful to consider the spatial organization of food sources, since this is an important variable to which pollinators respond. Pollinators view the environment as a set of dietary patches in a spatiotemporal arena. The composition and quality of patches in time and space is not an inherent property of a community but is pollinator-defined and must be considered in terms of the perceptions of the pollinators rather than those of the investigator (Wiens, 1976). The patch structure of a habitat will be interpreted variously by different species and types of pollinators in accordance with their own adaptations and nutritional requirements. A patch may be a stand of one species or an intermixed or juxtaposed aggregation of two or more similar species.

If a pollinator perceives a patch as a population of one species, then there will be predictable changes in patch size during succession and differences between species or communities whose populations have different agents regulating plant density. When a plant species enters a habitat, there will be a few initial nuclei which will be randomly distributed, and patch size will be small. As the species become established, the distribution of conspecifics will become more contagious or aggregated (and thus patch size will increase) owing to the tendency of seedlings or vegetative offshoots to be narrowly dispersed and to the coalescence of neighboring patches. As the community matures, patch size declines, with individuals achieving a more random distribution (Margelef, 1958; Kershaw, 1958; Laessle, 1965; Pielou, 1966; Brereton, 1971). Moreover, each stage of succession displays a patch cycle like its constituents (Yarranton and Morrison, 1974). Therefore, groups of species which may compose the same patch may emerge and decline in a roughly synchronous fashion.

Patchiness, in general, appears to decrease as communities mature along successional gradients or when going from regions of low to high abiotic and biotic environmental predictability, with an attendant increase in species diversity and spatial complexity

(Greig-Smith, 1961, 1964; D. J. Anderson, 1967; Whittaker, 1969; Williams et al., 1969; Goodall, 1970; Mac Arthur, 1972a,b; Morrison and Yarranton, 1973; Kershaw, 1973). The change in patchiness as a function of maturity is due to alteration of the factors which regulate population density. In immature communities, density regulation may be achieved principally by abiotic factors and intra- and inter-specific competition. In mature communities, regulation by herbivores and pathogens is paramount. Since the area over which these agents operate is much greater than that over which competition is effective, populations in mature communities, especially in some tropical regions, are thinned to a much greater degree than those in other areas (Janzen, 1970, 1972a,b; Burdon and Chilvers, 1975; Cromartie, 1975; Tahvanainen and Root, 1972; Root, 1973; Strandberg, 1973).

Patches exist in time as well as space. Fugitive plants provide the most ephemeral patch type. In order of increasing patch persistence, there are: (1) annuals and short-lived perennials of continuously disturbed communities, (2) long-lived perennial herbs, and (3) woody species. The latter are best represented in the nonseasonal tropics. There seems to be an inverse correlation between patch size and longevity of the constituents. Patches of fugitive species may become enormous in a few years and then become extinct before having equilibrium density imposed upon them by herbivores and pathogens. Patches of tropical plants may be small owing to devastating density-dependent control by pests but may persist at equilibrium for many generations and for hundreds of years.

Given that one or a few plant species form a dietary patch for a single pollinator species, from where do patches arise? The spatial component of the environment operates to increase species diversity and to increase patch type but decrease patch size (S. A. Levin, 1974, 1976a,b). This arises from the heterogeneity of the environment, especially with regard to edaphic and microclimatic differences and interspecific competition. Local effects of herbivores may be important. Patchiness also can arise in an initially homogeneous environment from stochastic events, such as a colonization pattern whose effects may be magnified by species interactions. In essence, patchiness is self-augmenting (Whittaker, 1969). Patchiness of the environ-

ment also is promoted by disturbance, which provides the opportunity for local differentiation through random colonization and
constantly interrupts the natural successional sequence on a local
scale.

THE FORAGING STRATEGIES OF POLLINATORS

Optimal foraging strategies for predators have been discussed
from different vantage points by various authors (Emlen, 1968; Royama, 1970; Schoener, 1969, 1971, 1974a,b; Rapport, 1971; Murdoch,
1973; Pulliam, 1974; Katz, 1974; Estabrook and Dunham, 1976;
Covich, 1976; Pyke et al., 1977). Optimal foraging strategies in
pollinators, for the most part, are expected to parallel those of
predators because the ecological economics of foraging in the two
groups are similar and pollination fits within the broad conceptual
framework of predation (Covich, 1974).

There are obvious differences in pollinator–plant and predator–
prey interactions, but not necessarily in the foraging strategy most
favorable for resources of a given quality, distribution, and ease of
handling. By definition, a pollinator is the agent of pollen transport
from flowers of one plant to those of another of the same species and
as such is essential for completing the sexual reproductive cycle in
zoophilous self-incompatible plants. From their pivotal position in
the reproductive scheme of angiosperms, pollinators may exert profound selective pressures upon floral architecture and the allure-
ment cues and reward system. A predator, on the other hand, has its
greatest impact on the immediate size of the prey population rather
than on the reproductive success of individual prey. The predator
can be the substance of population regulation, a force causing an
increase or decrease in numbers (Huffaker, 1970). Correlatively, a
predator acts as an agent favoring the evolution of avoidance mechanisms by the prey without eliciting a shift in reproductive biology.
Both pollinators and predators require energy; foraging behavior
presumably maximizes the rate of energy intake per foraging time,
within certain constraints, regardless of the effect on the food

source. The experience of the food source does not dictate foraging strategy, although the abundance of a given food source may dictate how that species is treated by pollinators or predators.

Pollinators encounter a vast array of potential food plants, which vary manifestly in their density, nutritional value, and ease of handling. Faced with this diversity, a pollinator must decide where to search, which species to feed from, and which plants to feed from and in what sequence. The first two decisions are simple. Foraging should be concentrated in habitats where expectation of yield is greatest and upon those plants whose pollen or nectar, or both, is most efficiently harvested and provides necessary nutrients as well as calories. A plant species should be exploited only if the amount of time spent in travel and extraction is more productive than the same time devoted to another species in the same habitat. The decision to visit a particular plant is more difficult, since the pollinator may have no way of knowing the quantity of nutritional reward which awaits it (Mac Arthur, 1972a,b).

Natural selection favors individuals that are genetically predisposed to forage in a manner that maximizes the net rate of energy and nutrient uptake per unit foraging time (Pyke et al., 1977). This form of foraging may be considered optimal. Selection for optimal foraging may involve short-term (a few hours or days) or long-term responses (weeks to years) and favor increased ability to choose the best plants or patches (with respect to their effects on fitness), to locate patches, to manipulate the flowers, to choose the best foraging path within each patch type, or to assess the quality of a patch relative to others in the area. In general, optimal foraging would be characterized by a dynamic response to altered resource states brought about by pollinators or the phenology of individual flowers, plants, or patches. It would also allow for exploration of surrounding patches, to the detriment of immediate foraging efficiency, so that the pollinator has information about changing spatiotemporal distributions in resources.

What kind of pollinator behavior do we expect from foraging theory? Do pollinators behave in the anticipated manner? Answers to both questions will be briefly discussed. Expectations about behavior are drawn from the predation literature and will be trans-

posed into a pollinator context. The papers cited do not discuss
pollinators; observations on pollinators are taken from the pollination biology literature.

Theory predicts that pollinators will differentiate between different plant species and form "search images" of the most favorable
species, because this behavior permits efficient localization of time
and effort and maximizes cost–benefit relationships (Tinbergen,
1960; Royama, 1970; Tullock, 1971). In fact, the tendency of individual pollinators to forage within a plant species for a period of time
rather than foraging at random among several suitable species (the
adaptation referred to as flower constancy) is a typical form of
behavior (V. Grant, 1949; Free, 1970; Baker and Hurd, 1968; Proctor
and Yeo, 1972; Heinrich, 1975). The choice of a forage species is
determined principally by the quantity and quality of the floral
reward (Butler, 1945; Linsley, 1958; Free, 1968; Mosquin, 1971;
Heinrich, 1975). The dominant factors affecting bee visits appear to
be the abundance and sugar concentration of the nectar and chemical attractants in the pollen (Martin and McGregor, 1973). Hummingbird food selection is based primarily on sugar concentration
and secondarily on ease of access to the nectar; sugar composition
has little effect on choice (Hainsworth and Wolf, 1976). In a highland Costa Rican community, the hummingbird portion of the nec-
tarivore guild seems to be organized along the foraging efficiency
gradient of bird–plant interactions (Wolf et al., 1976); each bird
species first exploits the plant species it can forage most efficiently
and then moves to the next level on the gradient.

Theory predicts that pollinators will evolve to become specialists if the density and quality of one or a few resource plants is high
and if the resources are predictable in time and space (Levins and
Mac Arthur, 1969; Schoener, 1969; Colwell, 1973; Bryant, 1971).
Pollinators will specialize only if high-quality plants are abundant
and will not choose "less preferred" plants simply because they are
most abundant (Pyke et al., 1977). In general, as the abundance of
preferred plant species in the diet increases, the number of less
preferred species declines. Pollinators whose populations are food
limited will seldom be food specialists (Emlen, 1973). In fact, pollinator specialization is most highly developed in arid regions (Hein-

rich and Raven, 1972), is well developed in the tropics (van der Pijl, 1969; Baker, 1970, 1973; Gentry, 1976), and occurs to various degrees in temperate communities (Macior, 1971, 1974; Moldenke, 1975; Heinrich, 1976a). Specialization may reach its peak in the seasonal tropics and in regions where the flowering of different species is synchronized both within and between years (Janzen, 1967, 1974; Heinrich and Raven, 1972). A few specialized pollinators depend on one or two species during an entire season, but most switch from one diet to another in a regular way with the seasons. If the reward structure were constant in time, specialists would never have to leave the one to few species upon which they concentrate. Since resources change through time, however, even specialists must occasionally visit alternate species as a necessary compromise (Oster and Heinrich, 1976a,b). This form of exploratory behavior has been well documented in individual bumblebees (Heinrich, 1976a).

Specialization and flower constancy are evolutionary and behavioral responses, respectively, to the absolute abundance of different food plants, not their relative abundance (Schoener, 1971; Pulliam, 1974; Estabrook and Dunham, 1976). For the first behavioral response to be elicited, therefore, patches must be rich by virtue of having a large number of plants or a few plants very rich in reward. The trade-off between plant numbers and reward per flower and per plant has been discussed by Heinrich and Raven (1972). If the high-quality patches are present over years, pollinators may evolve adaptations to exploit these patches better.

Theory predicts that, because of flower constancy and specialization, pollinators will avoid unfamiliar species even though they may be highly nutritious, palatable, and conspicuous (Manly, 1973). In fact, avoidance of rare hosts has been observed within species (Levin, 1972) and within species assemblages (Levin and Anderson, 1970; Macior, 1974). This behavior reduces the mating success of a novel species and retards the entry of this species into the community.

Theory predicts that pollinators will minimize their foraging space (Covich, 1976) since site specificity reduces a forager's inefficient random movements. In fact, site constancy is well documented in some bees, hummingbirds, and butterflies and may be in force for several hours to several days (Free, 1970; Levin and Kerster, 1974; Heinrich, 1975). Pollinators may spend hours foraging within an area of a few square meters and return to the same part of the population on subsequent foraging bouts. Even when artificially displaced, some pollinators will return to the initial capture site (Keller et al., 1966). Another aspect of site specificity is the tendency of some pollinator species to repeat specific long-distance foraging paths, i.e., to trapline (Heinrich, 1975). When a plant population becomes widely dispersed, relatively large energy rewards per flower and long flowering periods are required to engage the reliable service of the trapliners. Trapline pollination is best accomplished by large pollinators (Schoener, 1969; Janzen, 1971; Heinrich and Raven, 1972; Colwell, 1973). Traplining is found in some tropical bees (Janzen, 1971), bats (Heithaus et al., 1974; Heithaus et al., 1975), hummingbirds (Colwell, 1973; Stiles, 1975), and butterflies (Gilbert, 1975). Such trapline pollinators are not loyal to a single species but visit several with divergent floral adaptations, which permits some pollen to be transported between conspecific plants.

Movement patterns of foraging insects have been studied extensively. In general, insects meander until prey is encountered and then increase their rate of turning, thereby remaining in the vicinity of the encountered prey (Bänsch, 1966; Chandler, 1969; Croze,1970; Richardson and Borden, 1972). For many pollinators, the location of suitable resource patches is considerably less haphazard than that for predators, owing to the transmission of information on the locality of the patch, the stationary and renewable nature of rewards in the patch, and the ability of the pollinator to remember the locality of the patch (Heinrich, 1976; Lindauer, 1977; Gilbert and Singer, 1975). In honeybees, the route and direction of a patch are memorized from landmarks along the route; sun orientation may also be important. When a patch is approached, navigation is aided by olfactory cues and finally visual ones (Lindauer, 1977). Once within a patch, pollinators would be expected to exhibit patterns of foraging in which the frequency of path recrossing was minimized (Pyke et al., 1977). Based upon a meager body of data, it appears that pollinator flight paths within patches are not random. For example, within a field of Lythrum salicaria, honeybees and butterflies have a slight tendency to move in the same general direction in consecutive interplant flights (Levin et al., 1971).

Theory predicts that pollinator efficiency will be increased if the distances flown between plants are minimized. Although efficiency per se has not been measured, there are observations on flight distance in relation to plant distribution and density. Pollinators working colonial plants tend to move from a plant to one of its near neighbors, the exact distances being a function of plant density. As density declines, flight distance increases (Levin and Kerster, 1969; Wolf, 1969). Density-dependent foraging has been reported in several plant species and is independent of the species, floral characters notwithstanding. The response to density alterations appears to be greater in bees and hummingbirds than in butterflies.

Theory predicts that pollinators will broaden their diet and switch hosts more frequently as their environment becomes more patchy in space and time and as patch size declines relative to foraging range (Schoener, 1974a; Mac Arthur, 1972b; Emlen, 1973). Consider two habitats differing in patch size only; foraging time per plant is the same. The proportion of time spent traveling increases as patch size decreases because the distance between patches varies linearly with the linear dimension of a patch, whereas the foraging area within a patch varies as its square (Pianka, 1974). Therefore, as patch size decreases, patch selection becomes less advantageous. In fact, we observe that flower constancy and specificity decline as patch size declines (Free, 1970; Levin and Anderson, 1970; Heinrich and Raven, 1972; Heinrich 1976a) and as the habitat becomes more unpredictable (Moldenke, 1975).

Theory predicts that pollinators will broaden their diets as the differences between patches decrease (Gillespie, 1974) since the nutritional advantage of fidelity to one patch type declines as the utility of other patches increases. In fact, we have little information on the reward structure of plants within the same or different patches (Heinrich and Raven, 1972). We can only suspect that pollinators take into account the magnitude of patch differences.

Theory predicts that pollinators will leave a patch when the rate of food intake in that patch drops to the average for the habitat as a whole (Charnov et al., 1976; Parker and Stuart, 1976) and that emigration will commence when the future fitness in another patch is greater than the expected future fitness within a given patch (Gillespie, 1974). In fact, resource depletion, even though it may only be temporary, has been observed for pollinators of cultivated plants (Free, 1970) and wild plants (Heinrich, 1975; Wolf et al., 1975), but consequential pollinator emigration has not.

Theory predicts that pollinators will contract their feeding habitats, but not their diet, when faced with competition from other vectors in ecological time and that diet will remain constant or expand (Mac Arthur and Wilson, 1967; Schoener, 1974a). If competition persists over long periods of time, however, and if resources are abundant, evolution may be expected to redistribute the phenotypes of the pollinator species and reduce the level of competition (Schoener, 1974b; Sale, 1974; Roughgarden, 1976). In fact, competition for pollinator service, and ostensibly among pollinators, has been described for numerous floras in the New World (Hocking, 1968; Mosquin, 1971; Kevan, 1972; Macior, 1973, 1975; Percival, 1974; Gentry, 1976; Stiles, 1975; Heinrich, 1975, 1976b). In a study of a tropical lowland community, Heithaus (1974) showed that pollinator niche breadth decreases as the diversity of pollinators increases.

Resource partitioning by congeneric species has been demonstrated in bees (Johnson and Hubbell, 1975; Heinrich, 1976b; van der Pijl and Dodson, 1966), hummingbirds (Stiles, 1975; Gill and Wolf, 1975a,b), and butterflies (Gilbert and Singer, 1975; Schemske, 1976), and adaptations by different species to different pollinators is seen in Costa Rican species of *Cordia* (Opler et al., 1975) and in several genera of the Bigononiaceae (Gentry, 1976).

THE GENETIC CONSEQUENCES OF FORAGING BEHAVIOR

The foraging behavior of pollinators determines their effectiveness as pollen vectors for single species and their capacity for interspecific pollination. The way in which pollinators forage within a species has a profound effect on the breeding structure and organization of genetic variation within and among populations.

Foraging in response to proximal and historical factors is a prime determinant of the level and extent of interspecific hybridization, the presence of cross-compatibility barriers also being important. The models of foraging strategy upon which expected pollinator behavior is based, although somewhat simplistic, seem to have good general predictive value for many foragers, especially pollinators. Thus, using these models as a basis for inferring foraging behavior from local spatiotemporal environmental conditions seems warranted.

Foraging Patterns and Interspecific Hybridization

What conditions are most conducive to interspecific pollination, and thus hybridization? By definition, the frequency of hybridization is an inverse function of flower specialization and constancy, since both constrain the wanderings of pollinators. As noted earlier, specialization is an adaptation in evolutionary time, and constancy is an adaptation in ecological time: They are not mutually exclusive. The general relation between the two adaptations as they relate to patch switching and hybridization is depicted in Figure 6.1. We will

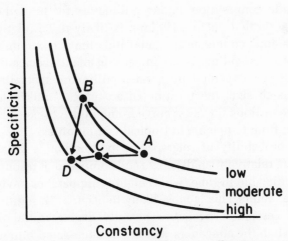

Figure 6.1. The relative levels of interspecific cross-pollination as a function of pollinator specificity and constancy. Lines of high, moderate, and low levels of cross-pollination are shown. All points along a line represent the same level of cross-pollination.

assume that related species do not occur in the same patch. The three curves represent different levels of switching, and all points along a given curve represent the same level of hybridization. Different combinations of specificity and constancy may provide the same result. A pollinator species will assume a position on the diagram as dictated by its long-term participation within a community and its present perception of patch size and quality. If the level of host specificity and constancy (in combination) is high, only a small proportion of all flights will be between patches, and the level of hybridization will be low. On the other hand, if generalized on both counts, i.e., low specificity and low constancy, the level of hybridization will be relatively high. The position of a pollinator species is fixed on the specificity axis in ecological time, but the position on the constancy axis may vary as patch quality changes in time and space. The capacity for interspecific pollination will shift accordingly.

From foraging theory we may infer that pollinators are least apt to cross-pollinate two species when patch sizes are relatively large, distinctive in composition, composed of high-quality species, and persistent over a period of several years, and where there is stringent interspecific competition among pollinators for limited resources. This suggests that patch switching is likely to be relatively infrequent in stabilized midsuccessional and climax communities, geography notwithstanding. Since increasing plant species diversity is apparently accompanied by greater pollinator specialization and smaller patch size, the position of a "typical" pollinator species should move along the outer curve from point A to point B in Figure 6.1, going from temperate to tropical communities.

The probability of interspecific pollination is greatest when patches are relatively small in size and composed of resources which are unpredictable in time, with neighboring patches having similar nutritional value. This is most likely to occur with fugitive or early to midsuccessional species—especially during initial population growth, when the congeners will not be aggregated into unispecific arrays. Hybridization also has a relatively high probability of occurring in sites subject to periodic, but nonuniform, disturbance, because this kind of perturbation diminishes patch size and brings species with different ecological requirements into closer proximity. Patch disintegration would move a pollinator species from points A to C, C to D, or A to D in Figure 6.1, because flower constancy would decline. Hybridization in late-successional temperate genera is most apt to occur along suture zones formed as a consequence of postglacial migration and range expansion (Remington, 1968). Patches of similar composition would be placed in close proximity in these zones, which would in turn reduce flower constancy, especially with regard to similar patch types. Large isolated islands offer an excellent arena for interspecific pollen exchange and hybridization because pollinators tend to be generalists and, if patch size is small, may display only moderate flower constancy. A species of continental pollinator located at point B in Figure 6.1 might shift to point D if it were transplanted to an island.

The proportion of pollinator flights between congeneric species in different patches is a crude indicator of the level of hybridization. The cross-compatibility of species may vary from one habitat or community to another, as may the extent of sites suitable for the establishment of hybrids. In spite of these factors, pollinator foraging behavior is probably a prime predictor of potential and actual hybridization and deserves more consideration than it has received. The transfer of pollen between species, pollen–pistil compatibility, and the presence of an environment suitable for establishment constitute barriers which must be overcome before reproductive hybrids can be obtained. These factors form a linear sequence: Pollen exchange must occur before the compatibility is a consideration and hybrid seed must be formed before habitat suitability is a consideration. If species are readily cross-compatible and there are abundant sites for hybrids, hybridization is still contingent upon pollinator behavior. The journey cannot be taken without the first step, and there is no assurance that the first step, pollen exchange, will ever occur.

The literature is replete with examples of closely related sympatric species with different habitat preferences. Such differences impose a degree of spatial isolation between species, which reduces the potential for interspecific pollinator movement. Moreover, the degree of spatial isolation owing to different habitat requirements is likely to be a function of the magnitude of the difference. Congeneric species distributed discontinuously along an edaphic gradient are less likely to experience interspecific pollination than are juxtaposed or intermixed species because pollination foraging areas tend to be relatively small.

The Breeding Structure of Species

The breeding structure of single populations and groups of populations is poorly understood. Studies on crop plants and some native plants have demonstrated that populations are not panmictic units and that gene exchange between populations separated by a few hundred meters is rare, regardless of the vehicle of gene transport (Levin and Kerster, 1974). The breeding structure of single populations is determined by the pollen dispersal schedule, the seed dispersal schedule, and the incidence and nature of assortative pollination. Most assortative pollination is expressed as self-pollination and pollination between plants with close spatial proximity. Both forms of assortative pollination result in inbreeding—the first for obvious reasons, the second because the relationship between plants typically is an inverse function of distance. This relationship is due to restricted seed dispersal, most seeds being deposited in the vicinity of the seed parent (Levin and Kerster, 1974). For insect-pollinated plants, then, the behavior of the pollinator emerges as the prime determinant of the breeding structure of populations.

In temperate areas, the flight distance of bees between successively visited plants is often density dependent: The greater the interplant spacing, the greater the flight distances (Levin and Kerster, 1969; Estes and Thorp, 1975; Estes et al., unpublished data; Schaal, unpublished data). Presumably , the same holds for tropical bee–plant associations when patch size is large. A different form of foraging behavior (trapline pollination) is displayed by certain tropical bees adapted to small patch size, i.e., widely dispersed conspecific plants (Janzen, 1971; Williams and Dodson, 1972; Frankie, 1976; Frankie et al., 1976).

The distances pollinators fly between plants are not an accurate descriptor of pollen dispersal distance, since some pollen may be carried past several plants before it is deposited on a stigma. The more flowers visited per plant, the smaller the pollen carryover. If there is carryover, however, its effect may not be substantial, because flight direction seems to be near-random with respect to the previous flight and resembles a drunkard's walk (Levin et al., 1971; Estes et al., 1976). With pollen carryover, mean dispersal distance equals flight mean $\times$ Σ (proportion of pollen deposited on the n_i plant $\times$ $\sqrt{n_i}$) (Levin and Kerster, 1969). Consider the effect of a liberal carryover schedule. If 50 percent of the pollen collected on a plant is deposited on the next plant, 25 percent on the second, 13 percent on the third, and 6 percent on both the fourth and fifth, the mean pollen dispersal distance would be only 30 percent greater than the mean flight distance.

Wright's (1940, 1943, 1951) neighborhood, or isolation-by-distance, model has proved valuable in thinking about breeding structure. A neighborhood is the area of a population within which mating is assumed to be random, with no mating beyond. In his models, the reference colony for panmixia is one composed of dioecious individuals, with equal numbers of both sexes, random mating, and a Poisson distribution of number of offspring per parent. A subject population is said to have a genetically effective size, N_e, if it undergoes the same rate of decay of gene frequency variance as a reference colony of size N. The effective size of a neighborhood is equivalent to the number of reproducing individuals in a circle whose radius is equivalent to twice the standard deviation of the gene dispersal distance. A circle of this type will include 86.5 percent of the parents of the individuals at its center. The plant density may be used as an estimator of effective density if the colony is stable (in age structure) and stationary (in numbers) and if the distribution per parent of offspring reaching maturity is a Poisson distribution (Kimura and Crow, 1963). Deviations from these conditions undoubtedly exist, and, accordingly, the effective density will usually be less than the standing crop (flowering plants) in the prescribed circle (Falconer, 1960). The neighborhood area is N_e/d, where d is the genetically effective density, which is approximately the density of flowering plants.

Wright (1946) applies $N_e = 12.6\,\sigma^2 d$ as the neighborhood size in a population of hermaphrodites where male and female gametes show the same amount of axial dispersion. Expanding Wright's equation so that gene dispersal is affected by pollen, p, and seeds, s, yields

$$N_e = 12.6d\left(\frac{\sigma_p^2 + \sigma_s^2}{2}\right) = 6.3d(\sigma_p^2 + \sigma_s^2) \tag{1}$$

Wright's equation is based on the assumption that populations do not move in space. Accordingly, there must be no net movement of pollen or seeds. By substituting calculated variances that have zero means in Equation (1), we obtain

$$\sigma_p^2 = \sum \frac{(p_i - p)^2}{N_p} = \sum \frac{(p_i - 0)^2}{N_p} = \sum \frac{p_i^2}{N} \tag{2}$$

and

$$\sigma_p^2 = \sum \frac{(s_i - s)^2}{N_s} = \sum \frac{(s_i - 0)^2}{N_s} = \sum \frac{s_i}{N_s} \tag{3}$$

To bring pollen (haploid) and seed (diploid) dispersal into accord, we use one-half the absolute pollen dispersal. Combining the two dispersal components in the same equation gives

$$N_e = 6.3d\left(\frac{p_i^2}{2N_p} + \frac{s_i^2}{N_s}\right) \tag{4}$$

By incorporating the proportion of outcross progeny (r) into Equation (4), we arrive at Equation (5):

$$N_e = 6.3dr\left(\sum \frac{p_i^2}{2N_p} + \sum \frac{s_i^2}{N_s}\right) \tag{5}$$

By employing the neighborhood size as an indicator of breeding structure, we are relating gene flow to the decay of genetic variance. In amphimictic plants, the narrower the area from which parents are drawn and the stronger the correlation of parental genes by descent, the smaller the neighborhood size. Self-fertilization greatly restricts neighborhood size, because it represents zero gene dispersal by pollen.

    Density-dependent pollen dispersal in bee-pollinated plants, and zoophilous plants in general, yields neighborhood sizes which are roughly constant over a range of plant densities and neighborhood areas which increase as plant density declines (Levin and Kerster, 1974). These consequences of density-dependent pollen flow have some important genetic implications. Populations are buffered against the loss of genetic variability which would accompany pronounced downward fluctuations in population density or in the proportion of plants flowering in a given year. Neighborhood size considered over generations is the harmonic mean of the genetically effective densities of each of several generations, and harmonic means are dominated by low values (Wright, 1938). In order to see the buffering effect of density-dependent pollen flow, consider the consequences of fluctuations in population density on neighborhood size over a 12-year period. Population density and the harmonic mean of neighborhood size in each year for six hypothetical populations are presented in Figure 6.2. The densities and pollinator response to such are based upon bee foraging behavior on *Liatris* as

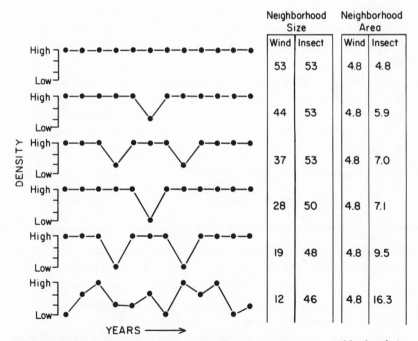

Figure 6.2. The effect of population size changes over 12 years on neighborhood size and area with density-dependent and density-independent pollination. Values are based upon pollen dispersal schedules and densities described in the text.

described by Levin and Kerster (1969). The four densities considered
are 1, 3.25, 5, and 11 plants per square meter. The variances of
pollinator flight distance, in the order of increasing plant density,
are 2.6, 1.4, 0.8, and 0.04 m², respectively. For comparative purposes, the effect of density fluctuations on a hypothetical wind-
pollinated plant whose pollen dispersal is density independent is
also presented (first and third columns in Fig. 6.2). Wind-dispersed
pollen schedules vary from species to species. For small herbs, the
schedule may be similar to that generated by bees foraging in the
dense *Liatris* population. The neighborhood size of populations of
bee-pollinated plants is largely independent of density fluctuations
of populations, whereas wind-pollinated plants experience declining neighborhood sizes as the magnitude of population contractions
increases and as the mean density decreases. The estimates of neighborhood size are based only upon pollen dispersal.

Density-dependent pollen dispersal also affects neighborhood
area. As shown in Figure 6.2, the mean neighborhood area increases
with declining density in bee-pollinated plants but remains constant
in wind-pollinated plants. Neighborhood area is defined such that
gene migration from a neighborhood to an adjacent one occurs at a
given rate per generation. Therefore, the greater the number of
neighborhoods separating two points, the greater the number of
generations required for genes to flow from one point to the other. It
follows that gene migration between two points, on the average, will
be much slower in high-density arrays than in low-density arrays
since neighborhood diameters will be smaller in arrays where high
density and close spacing prevail.

Density and spacing, through their effect upon gene migration,
will influence the ability of a population to undergo selective differentiation and subdivision. Should population dimensions remain
constant, differentiation in response to a heterogeneous environment is most likely to occur in arrays of high density and close
spacing since, under such circumstances, gene migration is
restricted and swamping effected by migration is minimized. Correlatively, high-density arrays will be able to maintain many alleles at
each locus at moderate frequency and will resemble a graded patch-
work in general structure. In contrast, low-density arrays will tend to

display a more continuous variation pattern effected by a few all-purpose alleles at each locus.

The structure and gene flow potential within high-density colonies bear a strong affinity to the population (= colony system) structure that Wright deems most conductive to rapid evolution. Wright (1948) states that, "in a large population, divided and subdivided into partially isolated local races of small size, there is a continually shifting differentiation among the latter, intensified by local differences in selection, but occurring under uniform static conditions which inevitably brings about an indefinitely, continuing, irreversible, adaptive, and much more rapid evolution of the species than in a comparatively large, random breeding population."

Levins (1964) proposed that gene flow among populations is part of the adaptive system of a species and that optimum gene flow rates depend on the statistical structure of the environment. He argues that "the optimal amount of gene flow between populations is increased by the temporal variance of the environment variable." The relationship between optimal gene flow and environmental stability permits the population to respond genetically to general, long-term environmental shifts while damping the response to local, short-term oscillations. On the basis of the arguments presented in this paper, we believe that Levins's rationale can be applied to gene flow within as well as between colonies. Furthermore, density-dependent gene flow affords an ideal means of achieving a harmonious balance between flow rates and temporal environmental fluctuations.

The breeding structure of widely spaced plants, or small clumps thereof, is somewhat deceptive. Although plants are far apart, most pollination is between neighboring plants, and seed dispersal is narrow. Thus, widely spaced neighboring plants in tropical forests may be sibs or half-sibs, as may be the case for closely spaced plants. If the feeding stations of trapline pollinators are small populations of conspecifics, the level of migration between neighboring populations will be relatively high. Whether the feeding stations are single plants or small populations, the potential for local differentiation in the absence of strong selection will be small. The neighborhood area may be thousands of square meters. The neighborhood sizes would probably be large enough to preclude random differentiation. Many tropical species whose plants are thinned to low densities have disproportionate sex ratios (Bawa, 1974; Bawa and Opler, 1975; Opler et al., 1975; Gilbert, 1975; Opler and Bawa, 1977). A deviation from a 1:1 sex ratio reduces the effective density (which in turn reduces neighborhood size) but increases neighborhood area (Wright, 1931). Thus, the potential for local differentiation in some tropical plants is further diminished. In these plants, the neighborhood area must be tens of thousands of square meters. In summary, the effect of low density in some tropical forests is the same as it is elsewhere, except that greater interplant spaces can be tolerated because of the presence of trapline pollinators. In temperate areas, the paucity of conspecifics would not be possible since they would tend to be ignored by pollinators or receive insufficient conspecific pollen to ensure adequate seed-set and plant replacement. Finally, it should be noted that foraging behavior on low-density tropical plants is an evolutionary adaptation and not an immediate behavior response to density per se.

The discussion of density-dependent foraging has focused upon bees because their response to resource distribution and quality is far better understood than that of the other groups of pollinators. The tendency of a pollinator to move from a plant to one of its near neighbors, however, is evident in hummingbirds, bats, and butterflies and presumably occurs in moths, flies, and beetles as well (Levin and Kerster, 1974).

Consider next the breeding structure of a species whose populations typically are small (less than 30 plants) but relatively dense and perceived by pollinators as a component of small patches. Under these conditions, pollinators are expected to work most of the plants in the patch in a fashion more closely approximating random mating than in large populations and then move on to another nearby patch. If the same species is present in the other patch, interpopulation pollen flow will ensue. Viewing populations as islands, this pattern of pollen and gene flow closely resembles the stepping-stone model of Kimura and Weiss (1964). According to the stepping-stone model, the gene pools comprising adjacent islands are correlated by descent to a greater degree than islands picked at random. Correlatively, the larger the number of islands lying between two given islands, the less correlated their gene pools, in spite of the fact that the absolute distance between the two islands remains the same. With small population size and considerable migration, the breeding structure of a species would encompass large numbers of plants and a large area. The opportunity for random differentiation between populations and establishment of regional variation patterns for characters free of selection is virtually nil (Kimura and Maruyama, 1971). The opportunity for selective differentiation between small populations still exists, and most likely would be manifested in terms of clinal variation rather than sharp discontinuities between neighboring populations and local mosaic patterns of differentiation (Nagylaki, 1976a,b,c; Gillespie, 1976; May et al., 1975).

The foraging behavior of pollinators can affect the pattern of variation from gene flow between adjacent populations experiencing different selection pressures. Consider a case of two large populations which grow in different environments, one abruptly giving way to the other, as discussed by Jain and Bradshaw (1966) and others. In the face of gene flow, a cline will develop whose characteristics are in part determined by the distance that migrant genes are carried beyond the population interface. The width of the cline is the ratio of the standard deviation of dispersal to the square root of the selection intensity (Slatkin, 1973). Since pollinator flight distance is density dependent, the width of clines would tend to be smaller in high-density arrays. The effective gene flow distance increases linearly with the distance pollen or migrants travel, but only as the square root of the fraction of the population which migrates (May et al., 1975). Thus, a population in which all pollen is transported 1 km each generation has about the same gene flow as one in which 99 percent of the pollen does not move any appreciable distance but 1 percent is transported 10 km. If there is a discontinuity between populations, and thus in migration rate (since population edges tend to act as reflecting barriers for pollinators; Pyke et al., 1977; Levin, unpublished data), the slope of the gene frequency would be continuous, but there would be a discontinuity in the gene frequency itself, proportional to the steepness of the slope (Nagylaki, 1976a,b,c; Slatkin, 1973; May et al., 1975). What constitutes a geographical barrier for one species or group of pollinators need not constitute a barrier for others. Accordingly, the genetic structure of discontinuous populations subjected to a given selective differential is not independent of the pollen vector.

Neighborhood size may be sufficiently small to permit population subdivision and genetic drift in a cline maintained by selection and gene flow. The primary effects of a genetic drift are a reduction in the slope of the cline and some variation in its location relative to the environmental gradient. Even when neighborhood sizes are small, however, a cline will not be greatly distorted in shape if there are high correlations in allele frequency in adjacent populations (Slatkin and Maruyama, 1975).

Models involving the interaction of gene flow with spatially varying selection also have taken into account two loci. A major finding is that a large amount of linkage disequilibrium can be generated by gene flow in subpopulations along a cline, even in the absence of epistasis (Prout, 1973; Li and Nei, 1974; Feldman and Christiansen, 1975; Slatkin, 1975). The more restricted is gene dispersal, the greater is the disequilibrium. Depending on the spatial pattern of selection on the two loci, disequilibrium can reduce a population's ability to track changes in selection, so that the locations of clines would not necessarily correspond to the environmental changes. Even in the absence of selection, gene flow restriction within populations may result in linkage disequilibrium (Nei and Li, 1973; Feldman and Christiansen, 1975).

The effect of pollinator foraging behavior on the organization of genetic variation in dense arrays is dependent upon the dimensionality of the spatial distribution. Some plant species are confined to river banks, seashores, or other such linear habitats, whereas other species occur in two-dimensional habitats. Large linear populations will usually show more differentiation of gene frequencies and lower heterozygosity than two-dimensional ones (Malecot, 1969; Maruyama, 1971, 1972; Nagylaki, 1976b).

Given that pollinator foraging behavior can result in restricted gene flow and population subdivision in large populations, it is well to consider the effect of this behavior on the genetically effective size of populations and the fate of mutant genes. According to Wright (1943), the sampling variance of a subdivided population is less than would be the case if there were random mating throughout. Accordingly, the effective size of subdivided populations is greater than that of populations without subdivision. Increasing the effective size increases both the fixation and extinction times for novel neutral alleles (Kimura and Ohta, 1971). The probability of fixing an advantageous mutant also increases as the effective size increases. If pollinator behavior results in inbreeding (as might arise from self-fertilization) not accompanied by subdivision, the sampling variance would be greater than with random mating, and the effective size of the population would decline (Wright, 1943). Then the fixation and extinction times for neutral alleles would decline, as would the probability of fixing advantageous mutants. Thus, the foraging behavior of pollinators influences the organization of genetic variation not only in space but also in time.

The foraging behavior of pollinators has profound implications for the breeding structure of populations and population systems and for the incidence of interspecific pollination, and thus the amount and organization of genetic variation within species. Pollinators should no longer be viewed simply as agents of pollen transport, and populations no longer considered simply as cross-fertilizing or partially self-fertilizing. Studies of pollinator behavior with reference to genetic implications and of population structure with reference to pollinators as causal agents, are needed for a thorough understanding of plant population genetics within the context of the community.

Acknowledgments

The author is indebted to Otto T. Solbrig and Peter Raven for their comments on an earlier version of this paper. The study was supported in part by National Science Foundation Grant DEB 76-19914.

2 LIFE-CYCLE PARAMETERS

INTRODUCTION

IN THIS SECTION we address the question of why there are so many different species of plants. As Patricia Werner (article 12) points out, the answer to this question depends to a large extent on the particular approach taken in trying to answer it. The approach taken by the various authors in this part is primarily ecological (strictly speaking, demographic), although Subodh Jain (article 7) and Leslie Gottlieb (article 11) also introduce genetic variables.

Plant ecologists have only recently become interested in demographic problems, despite the fact that half a century ago the pioneering works of Tansley (1917), Sukatchew (1928), and Clements (Clements et al., 1929) demonstrated that plants are very well suited for these kinds of studies. It has been largely through the work of the British ecologists (Bradshaw, 1965; Harper, 1967, 1977; Harper and White, 1971, 1974) that the subject has been brought into focus again. The work of an active Russian school of plant demographers (Rabotnov, 1960, 1969) went largely unnoticed in the West until recently.

The plant ecologist–demographer is basically concerned with explaining changes in plant numbers over the life cycle of the population (Figure 1). At the basic descriptive level, very little is known regarding rates of germination or seedling and adult survival and death. To be more than formal descriptions of changes in numbers over time, however, such studies must be related to the microenvironment. Given the inherent difficulties of measuring the microenvironment and its variability over time and space, a strictly correlational approach is not likely to be very productive in the long run (but see Teeri, article 15). This has led to the development of theoretical models that attempt to predict the behavior of a population or species on the basis of conflicting selective pressures. For

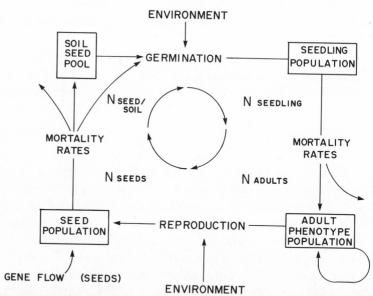

Figure 1. Generalized ecological model of the life cycle. The information of concern is the number of individuals (ramets and genets) at different stages of the life cycle. Of interest are the rates of transformation of seeds into adults, and the production of seeds by adults and the mortality rates of the seed population and the population of seedlings and adult phenotypes.

example, Solbrig (1970, 1971; Solbrig and Simpson, 1974, 1977), in a study of the distribution of different biotypes of the common dandelion *(Taraxacum officinale),* showed that there is a conflict between competitive ability and fecundity. He predicted (and subsequently confirmed) that high-fecundity biotypes are selected in disturbed sites and that good competitors are selected in more stable environments. Such models assume that the optimal solution is always selected and that the investigator possesses enough information to predict the solution. "Strategy" thinking (Harper, 1977), as this approach is often called, is analyzed by Jain (article 7), who puts in question some of the assumptions of this approach and calls for caution. Jain argues for a close connection between theoretical and empirical studies. Hickman (article 10) reviews some general plant demographic tenets; he doubts that we possess enough empirical knowledge about the biology of plants in nature to develop a general predictive theory.

It has become increasingly apparent that the characteristics of seeds and seedlings are major determinants of the success of populations of plants in nature. Angevine and Chabot (article 8) formulate a classification of different germination strategies. Robert Cook (article 9) describes his studies on the germination characteristics and fate of seedlings of species of violets in New England forest environments and reviews the patterns of selection that may be operating in *Viola* and other genera. Patricia Werner (article 12) demonstrates how resource partitioning leads to different allocations to vegetative and reproductive growth in species of *Solidago*. She also demonstrates a correlation between seed number and seed size in the goldenrod species that she investigated.

In article 11, Leslie Gottlieb describes his detailed work on the appearance of a new species, *Stephanomeria* "malheurensis," in a population of *Stephanomeria exigua* ssp. *coronaria*. In a series of studies, Gottlieb has carefully documented the genetic and morphological changes that exist between parental and derived species; in his article he suggests the possible adaptive significance and developmental correlations of the observed changes.

7 ADAPTIVE STRATEGIES: POLYMORPHISM, PLASTICITY, AND HOMEOSTASIS

SUBODH JAIN

THERE ARE at least three significant lines of development in current population biology: First, the formal descriptive parameters of how a population is born, survives, dies, or shows differing genetic structures and microevolutionary changes in space and time are now clearly recognized. Second there is a greater emphasis on the understanding of adaptive mechanisms in relation to specific morphological, physiological, or biochemical traits, and on the specific environmental factors eliciting an adaptive response. And third, a wide array of theoretical questions in evolutionary ecology are being raised on the adaptive significance of such population (supra-individual) features as the pattern of survivorship, rate of outbreeding, degree of polymorphism, or overall colonizing ability in new environments. When we ask, Why is a certain species more polymorphic than its congener? or, What are the life historical characteristics of a successful weed? we seek answers in terms of certain combinations of traits, as well as the kinds of genetic systems as a whole which we may refer to as different adaptive strategies. For example, a species may be a successful colonizer for its greater dispersal ability, combined with a high seed output, whereas its congener may have greater seed dormancy and may establish colonies through long-lived seed banks developed infrequently in time (Martins and Jain, in press).

In recent plant literature, several detailed studies of the reproductive aspects of life cycles have been discussed in relation to the concept of adaptive strategy. Harper and Ogden (1970), for instance, asked, What proportion of energy fixed during its life cycle does the plant allocate to seed production? and, Is the proportion allocated in this way fixed and characteristic of a species or group of species or is it plastic, being subject to change in response to environmental stress? They recognized the genetic versus plastic components as a range of tactics such that evolutionary changes tend to achieve an optimal strategy. Janzen (1977) restated some well-known biological information on pollination in terms of strategies, with the premises that energy is a limited and fixed resource and that the optimized quantity is some magic number of diverse parents, dioecy, monoecy, mating patterns, etc., which have evolved as optimal strategies. Orians and Solbrig (1977) concluded from a strategy analysis that "the high proportion of desert plants with mesophytic leaves is explained because their shorter amortization time and high income rates may yield higher yearly profits than xerophytic leaves." They state explicitly that cost equals calories invested in leaf construction and maintenance and that income equals calories received from photosynthesis. Likewise, Wilbur (1977) examined seed production in six *Asclepias* species, with seed output treated as investment and survival rates as payoffs, as "a suite of adaptations to the risk of herbivore damage, adult survival, colony size, and the spatial and temporal pattern of opportunities for seedling establishment." Thus, numerous approaches to evolutionary strategy questions have attempted to bring together ideas in physiology, genetics, and systematics, and it is no wonder that at least ten of the speakers at this symposium have referred to this kind of strategy analysis.

The genetic aspects of strategy building, however, have been left out in most of these writings. In order to understand the evolutionary processes, (1) genetic variation in life cycle components, mating system, etc., needs to be described, (2) genetic responses to various environmental factors should be analyzed to examine how evolution proceeds to maximize profit (income minus cost, whether it be survivorship, success in colonization, or simply photosynthetic gain per unit lifetime), and (3) it must be recognized that, since several features of genetic systems relate to the production or maintenance of genetic variation in populations, they are also subject to a strategy argument if the optimized quantities are genetic diversity, gene flow among populations, evolutionary rates, etc. Waddington, in *The Strategy of the Genes* (1957), raised some interesting questions concerning the role of genetic variation versus developmental or biochemical homeostatic mechanisms in maximizing fitness. Levins (1968) discussed the genetic issues at the levels of both genetic variation and genetic systems governing the variation. King and Anderson (1971) simulated models of r- and K-selection in specific population genetic terms, and, more recently, Giesel (1976) discussed the genetics of reproductive strategy evolution. In general, however, the literature has very few genetic studies.

As an attempt to survey our current knowledge in the area, this article has four parts: further analysis of adaptive strategy notions; a review of our own work in grassland annuals to enumerate parameters to be studied; a discussion of genetic polymorphism versus plasticity as alternative strategies; and, finally, a brief commentary on genetic system models. The different lengths of these sections is a reflection of the amount of writing in these four areas. Two main points emphasized are that comparisons of related species provide an important approach in ecological genetics, especially when alternative adaptive strategies are easier to demonstrate at this level than at the intrapopulation level. Different related species may vary greatly in their floral display, pollination mechanisms, seed production, seed germination, and seedling survival rates (e.g., Heinrich and Raven, 1972; Beattie and Lyons, 1975; Kawano and Nagai, 1975; Werner, 1977), and through a joint analysis of their variation and evolution it might be shown that certain suites of traits, taken together, represent distinct modes of adaptation, as judged by the Darwinian fitnesses to either the same environmental regime (preferably) or different ones. A bit of overenthusiasm allows the assumption that we could demonstrate the adaptive processes underlying the evolution of alternative adaptive strategies among the populations of the same species. Here, the genetic analyses of adaptive processes, or at least the genetics of different traits representing different strategic combinations, can be pursued using population genetic models.

ADAPTIVE STRATEGY AND TACTICS

To dispel any connotations of teleological design or group selection, some authors prefer to call adaptive strategies the tactics of adaptation. Stearns (1976) defines a tactic as "a set of coadapted traits designed, by natural selection, to solve particular ecological problems" (in short, a "complex adaptation"). Disregarding this minor semantic concern, it is important to recognize that the notion of an adaptive strategy is a shorthand expression for the following series of ideas: (1) Organisms living under varying environments respond (adapt) through physiologcal, biochemical, or morphological changes. (2) Individual fitnesses, as defined in terms of their relative rates of survival, persistence, or reproduction, may be based on various traits in combinations. (3) The principle of allocation of resources requires that adaptive responses in one direction, through, say, increased seed dispersal, be constrained by a loss in some other variable (e.g., seed size reduction and poorer seedling establishment) so that a cost–benefit function, $\psi = \phi(B) - \phi(C)$, must be considered; an optimal strategy then would be one that minimizes the cost $\phi(C)$ in this equation or optimizes the "net adaptive gain." (4) Analogous to the population genetic ideas on stable equilibria as multiple adaptive peaks, there might be more than one optimal strategy (equivalent peaks) represented by different combinations of evolving characteristics. (5) The reasons for a particular taxon or local population having evolved along one of these alternatives may be in developmental canalization, minimum cost of gene substitution, linkage disequilibrium in finite (small) populations, or simple chance occurrence of certain mutations or recombination events.

Thus, in simplest terms, the notion of adaptive strategy is an attempt to define adaptive responses in relation to certain character sets and to determine through comparative surveys of population and species variation if their "ways of life" somehow represent optimal outcomes. Assumptions are made about the availability of adequate (and appropriate) genetic variation, environmental challenges in the past, and, of course, sufficient time to have arrived at "high fitness peaks" through the optimization of a fitness function. I do not wish to speculate on the success of this theme in evolutionary ecology or express too much enthusiasm for current writings on this subject, but three comments are in order.

First, the metaphors of game theory are used, e.g., alternative moves, payoff function, minimax or maximax solutions (see Lewontin, 1961, for a clear discussion), but only as metaphors. Even those biologists who might "distrust" theory hopefully would agree that they too speak of certain favored character combinations and the evolution of compromises. Rosen (1967) discusses certain optimality principles in biology at length, and, as he notes, although optimal designs of organs or morphogenetic growth patterns are easier to conceive, appropriate natural selection theorems could also employ certain optimality principles in the adaptive processes.

Second, in disussing theories and modeling in population biology, Levins (1968) noted that we have a hierarchy of models in which generality, precision, and realism cannot be simultaneously maximized. Adaptive strategies at the highly aggregated levels of species pairs or communities would have to generate testable hypotheses at the lower levels of local population or a given set of genotypes. I think that the hang-up is not conceptual, but what Stearns (1976) referred to as the scholasticism versus empiricism gulf between those who develop general theory and others who are interested in explaining particular observed facts.

Third, in view of the above, analyses of adaptive strategies should proceed with the help of comparative studies of variation in several well-chosen traits, establish adaptive roles of these traits, demonstrate genetic changes under varying environments, and finally, demonstrate experimentally whether the outcomes represent optimal solutions of an adaptive challenge. Often the second and third points are ignored, so that the arguments or inferences remain somewhat speculative. The following two examples of model building for seed dormancy and life-history patterns hopefully will clarify these issues.

MODELING OPTIMAL STRATEGIES

Cohen (1967) developed a simple model for studying the role of seed dormancy in desert plants. Let us define good and bad years in terms of successful reproduction by seedlings such that a fraction of seed germinating after first rain (G) live to produce S seeds each or die. The dormant seed $(1 - G)$ is carried over to the next year. In a long sequence of T years, Cohen assumed that there will be pT good years and $(1 - p)T$ bad years. An initial batch of seeds will multiply at a rate $R = (GS + 1 - G)^{pT} (1 - G)^{(1-p)T}$, so that average increase per year will be its Tth root, $T\sqrt{R}$. In order to define optimal germination rate as the value of G that will maximize the rate of increase, in seed number, equating $dR/dG = 0$ gives $G^* = (Sp - 1)/(s - 1)$, which is usually near p, the probability of a good year. Mac Arthur (1972b) noted that here we have assumed the most likely event, exactly pT and $(1 - p)T$ bad years, to occur. But if we allow all possible combinations of good and bad years to occur, the optimal strategy is given by G^* equal to 1; i.e., all seeds germinate. He points out that now we are speaking of the average of all possible outcomes. Picture thousands of separate deserts in which numbers of good and bad years vary as a random variable with expected values p and $(1 - p)$. The prodigious seed production in a few deserts with exactly pT good years could provide seed for recolonizing some of the areas where this strategy failed. Thus, given some dispersal among a large number of populations, dormancy need not be the best strategy. This model predicts that the larger the value of G, the spottier the distribution of the desert annual and the denser it will be where found. (See Werner, article 12, for a discussion of dispersal versus dormancy in colonizing species.)

Note that both Cohen and Mac Arthur defined the optimal respose in terms of the maximized rates of seed increase. Mountford (1971) noted that the optimal germination rate will be different if we redefine optimality in terms of a minimizing of the probability of extinction, which is perhaps more relevant to the low density populations. In general, high dormancy is a better strategy in Mountford's model. Maximizing population growth rate in opportunistic environments and minimizing the risk of loss in the least favorable environments are called maximax and maximin strategies, respectively. Levins (1969) also showed that organisms with a high reproductive potential have less dormancy and spread their risk over more environments. We shall return to dormancy as an adaptive strategy later in discussing certain genetic aspects of its evolution and the role of polymorphisms.

It might be argued that since the criterion of optimality is equivocal depending on environmental heterogeneity and definitions of fitness, the notion of optimal strategy is begging the issue. But many alternative ways of looking at optimization will provide useful tools for defining "fitness"—and therein lies the value of theoretical developments and their tests in the field.

Several interrelated models of optimal life histories have been put forth in recent years. They seek to find conditions (i.e., patterns of environmental variation) that would favor annual versus perennial life cycle, semelparous vs iteroparous reproduction, and r- vs K-selection as measured in terms of resource allocation to the vegetative and reproductive organs. Schaffer and Gadgil (1975) considered life histories in plants in some detail. Let us consider the simplest model, an extension of Cole's (1954) result, in which, given the probability of seed becoming a mature plant equal to c and B_a as the seed output per plant, an annual multiplies at the rate of $R_a = cB_a$, and given p as the probability that a plant survives from one year to the next, a perennial population multiplies at the rate $R_b = cB_p + p$, where B_p is the seed output of a perennial. For an annual to outcompete the perennial $B_a > B_p + p/c$ should hold, i.e., seed output per plant in an annual should be large enough to overcome the perenniality advantage. The authors note that since $p \gg c$ is the usual case in the real world, we have to look for the special case of $p < c$ favoring the annual habit. Now, let adult survival in good and bad years be $p(1 + s)$ and $p(1 - s)$ respectively. Schaffer and Gadgil (1975) showed that for an annual to outcompete the perennial, $(cB_a)^2 > (cB_p + p)^2 - (sp)^2$ should hold. Increasing temporal fluctuations in adult survivorship, i.e., large s, would reduce the perennial's fitness. Conversely, variation in c, the seedling survival, say through large fluctuations in population density, would increase the fitness of the perennial. Note that in these models, if we use Mac Arthur's argument for taking the average of all likely proportions of good and bad years, and given $c(1 + s)$ and $c(1 - s)$ to be the relative seedling survival rates,

$$\overline{R}_p^2 = \{cR_p + p(1 + s)\} + \{cRp(1 - s)\}$$

or

$$(cR_a)^2 > cR_p(2 - s) + p(1 + s)$$

Thus, $R_a > \sqrt{R_p(2 - s) + p(1 + s)/c}$ should hold for an annual to be favored. Thus, natural selection could generally favor the annuals as predicted by Cole's model. Also, it can be easily shown that increased seed dormancy, other things being equal, would favor annuals. In general, as shown by Schaffer (1974), the so-called bet-hedging against environmental variations makes variances in parameters of life histories (c, p, etc.) more important than their means.

Following the definitions of reproductive value in relation to the size and age of growing organisms, Gadgil and Bossert (1970) and others have shown that, in a big-bang reproducer, reproductive effort rapidly increases with individual organism size to a maximum, favoring greater reproductive effort at an early age (r-selection), whereas in species with delayed reproduction the reproductive value rises slowly. Solbrig and Simpson (1974, 1977) showed in dandelion that plants growing in disturbed sites were early maturing and devoted a greater proportion of biomass to reproduction than those growing in undisturbed sites. Both genetic and experimental demographic aspects were investigated. Schaffer (1974) determined conditions favoring iteroparity versus semelparity (or their mixture, as polymorphism) in relation to the shape of the tradeoff curve between the reproductive effort and the adult survival. An intuitively obvious result was that density-dependent seedling deaths (mortal response) favored iteroparity, whereas density-dependent adult mortality favored semelparity. This dichotomy of r- and K-selection is summarized in Table 7.1 using Pianka's

TABLE 7.1.

r–SELECTION VERSUS K–SELECTION: A SHORT LIST OF VARIABLES

	r-Selection	K-Selection
Environments	unpredictable; variable	predictable; fairly certain
Mortality and survivorship	often catastrophic; density-independent; type III survivorship curve	more density dependent usually type I and II curves
Population size	variable in time; $N \ll K$ recolonization each year	fairly constant; $N \approx K$; saturated environments
Longevity	short; usually annual	usually more than 1 year
Competition	often lax; variable	usually keen
Selection for	rapid development; high r (or r_{max}); early reproduction; small body size; semelparity	reverse of r-selection traits

[a]After Pianka (1970).

(1970) list of variables. Some of the conclusions of Schaffer and Gadgil (1975) on semelparity and iteroparity are summarized in Table 7.2. Two important points should be noted: First, r- vs K-strategies might be a continuum since suites of characters are not generally found to occur in only two or three patterns. In fact, the concepts of r- and K-selection are not often clarified by the broad dichotomies of various traits used at many levels ranging from genotypes to communities. Stebbins (1974), approaching the subject from a nontheoretical viewpoint, also noted that different species show a whole range of allocation ratios, which presumably results from their unique past histories and present adaptive needs. Second, the comparisons of these strategies have been largely based on nonevolutionary analyses; i.e., dynamics of the evolution of life history strategies have not been investigated. Thus, optimal life history arguments are left incomplete and unsupported by experimental evidence.

One of the most detailed studies in this area is that of Sarukhan

TABLE 7.2.

COMPONENTS OF LIFE HISTORY FAVORING SEMELPARITY VERSUS ITEROPARITY (AFTER SCHAFFER AND GADGIL, 1975)[a]

Variable	Semelparity	Iteroparity
1. Probability of seed becoming established (c)	+	−
2. Probability of adult survival to the next season (p)	−	+
3. p/c ratio	−	+
4. B_a/B_p = ratio of seed output by annuals	+	−
5. Variation in c		
Cyclic ($\pm S_c$)	−	+
Random	±	±
6. Variation in p		
Cyclic ($\pm S_p$)	+	−
Random	±	±
7. High density-independent mortality rate (d)	+	−
8. High dispersability to transient habitats	+	−
9. Seed size and calorific value (e)	+	−
10. Sexual versus asexual reproduction	±	±
11. Seed dormancy ($1 - g$)	+	−
12. Percentage reproductive effort (E_i), V_i = reproductive value of ith age		
Concave of pV and B_i curves	+	−
Convex pV and B_i curves	−	+
13. Optimal first age of reproduction (a)	+	−
14. Density response	Complex	
15. Genetic versus nongenetic variation in life-history components	Complex	

[a] (+) favorable; (−) unfavorable.

(1974), who worked with three *Ranunculus* species which vary with respect to vegetative versus completely sexual reproduction. Vegetative reproduction allowed R. repens to persist most successfully, since this species also produced seed with high seed carryover rates. The two sexually propagated species produced large seed crops and had very little or no dormancy. Several factors apparently determine the relative allocation of reproductive resources. In fact, as suggested by Table 7.2, the list of variables is formidably large. In a

study of milkweeds *(Asclepias),* Wilbur (1976) documented the
great diversity of reproductive strategies in relation to the patterns of
flowering and seed production in adaptive responses to competition,
seed predation, seed dispersal, and overall habitat heterogeneity
factors. Table 7.3 summarizes some of his results in terms of comparison of life history components in different species. Wilbur (1976)
argued that seed number is "the product of a suite of adaptations
involving flower number, umbel number and the distribution of
pods among umbels" such that high seed predation in two of the
species is associated with large underground root storage and small
annual seed output. However, as Hickman (article 10) strongly
argues, the postselection measures (i.e., the product of evolution)
cannot give any meaningful clues to the processes and the sequence
of events involved in the development of strategies.

Harper and White (1970, 1974) summarized literature on the
dynamics of plant populations in terms of various components of
life histories. As a general principle, they stated that, "evolution has
led to an almost universal imposition of regulating mechanisms on
seed germination which will prevent the entire reproductive capacity of the species being used up in any given habitat at a given time,
except where and when the probability of survival is maximal."
Reddingius and DenBoer (1970) examined in detail the game theo-
retic approaches to the matter of spreading-the-risk. Seed germination studies suggest that numerous genetic and phenotypic polymorphisms might govern such a strategy. In most of our work on
grassland annuals, as well as wild sunflowers, almost invariably, a
few seeds are found to deviate in germination requirements from the
remainder population sample, under any combination of test environments. Harper and White (1970) reviewed the iteroparous reproductive strategies in relation to the variation among populations
living in different environments. They pointed out that perennials
have very little seed storage compared to the coexisting annuals.
Often there is little information available on the environmental
regulation of age-dependent birth and death rates of sexual vs asexual reproduction in perennials. The subject of life history strategies
in plants is clearly fascinating but so far lacks sufficient information
for any generalizations to be made.

TABLE 7.3.

ESTIMATES OF MEANS (TOP) AND VARIANCES (BOTTOM; ON LOG SCALE) OF VARIOUS COMPONENTS OF REPRODUCTION AND SURVIVORSHIP IN *Asclepias* SPP.[a]

	A. exaltata	A. incarnata	A. purpurascens	A. syriaca[b]	A. tuberosa	A. verticillata	A. viridiflora
Flowering stems per flowering plant	1.2	1.4	1.3	—	2.3	1.4	1.1
	0.115	0.164	0.172		0.786	0.210	0.068
Umbels per flowering stem	2.2	7.6	1.7	2.2	7.1	3.9	3.2
	0.289	1.036	0.325	0.258	0.621	0.282	0.331
Pods per umbel	0.6	0.5	0.5	0.8	0.2	0.6	0.2
	0.432	0.949	0	0.159	1.072	0.202	0.555
Seeds per pod	76.1	50.9	113.0	236.7	64.0	41.6	84.1
	0.025	0.035	0	0.016	0.029	0.033	0.040
Mean seed weight (mg)	5.73	6.13	3.76	4.98	5.13	2.14	5.12
	0.114	0.100	0	0.026	0.066	0.161	0.044
Flowers per flowering plant	35.3	76.0	45.5	66.2	112.6	51.3	84.3
	0.339	0.218	1.026	0.262	1.137	0.584	0.848
Pods per flowering plant	1.7	5.3	1.1	1.8	3.3	3.3	0.7
Seeds per flowering plant	113.0	270.0	125.0	417.0	209.0	136.0	59.0
Proportion of plants not flowering	0.537	0.572	0.783	—[b]	0.611	0.486	0.550
Proportion of plants visibly damaged by herbivores	0.674	0.164	0.233	0.372	0.049	0.240	0.069
Annual mortality rate of adult plants	0.131	0.226	0.235	—[b]	0.057	0.136	0.145

[a] Data from Wilbur, 1976.
[b] *A. syriaca* spreads widely by vegetative reproduction.

ALTERNATIVE STRATEGIES OF COLONIZING
GRASSLAND ANNUALS

Mac Arthur and Wilson (1967), in *The Theory of Island Bio-geography*, discussed the various features of colonizing species. Dispersal, initial establishment, values of r and K in logistic growth curve, and other fitness components were discussed in relation to the probability of founding new colonies and the relative duration of colonies. A comparative study of six grassland annuals showed that different strategies are involved in each case (Jain, 1976a; Foin and Jain, 1977). Table 7.4 gives a summary of rank-ordered estimates of various parameters related to genetic structure and colonizing ability. Rose clover has low rates of seed dispersal, but occasional long-distance dispersal with the aid of grazing animal in pastures, as well as man-made plantings, allows new colonies to be founded which, through high seed output per plant and long-term seed storage in the soil, are capable of surviving over long periods of time. Wild oats and bromes differ in their dispersal and reproductive patterns from rose clover. Seed dormancy is unimportant, but a plastic response to the environment allows their germination to occur under a wide range of conditions. Seedling survival is higher than in rose clover, and, in some situations, these species quickly build up large stands. Baker (1965, 1974) listed a number of characteristics expected to be present in an ideal weed, but it is apparent that many variables could provide even more numerous and different colonizing strategies.

A related issue is raised by the discussions of the evolution of sex. Asexual reproduction is claimed to be advantageous for its efficient reproduction as well as for genotypic constancy in stable environments. Similar arguments have been made in relation to the advantages of inbreeding. Accordingly, sexuality is favored when colonization of new environments requires dispersal and strong selection for unique gene combinations. Williams (1975) reviewed several population studies in order to illustrate this model. A review of evolutionary ideas on inbreeding in plants brought out a number of alternative hypotheses, none of which could be unconditionally supported by the existing examples (Jain, 1976b). A model for inbreeding versus outbreeding in *Leavenworthia* suggests the evolu-

TABLE 7.4.
RANK ORDERS OF ESTIMATED MEANS OF SOME PARAMETERS
DESCRIBING POPULATION STRUCTURE, VARIATION, AND
NUMBERS IN SIX GRASSLAND ANNUALS[a]

Variable Category	Annual grasses				Forbs	
	A	B	C	D	E	F
I. Distribution of species						
a. Geographical range	2	3	1	6	5	4
b. Continuity	2	3	1	5	4	6
c. Local patchiness	5	4	6	1	3	2
d. Patch stability	6	6	6	3–4	1–2	?
e. Presence in "weedy" habitats	5	2	6	4	3	1
f. Within-patch plant density	6	6	6	1	2	3
II. Gene flow and neighborhood size						
a. Range of seed dispersal	4	3	6	1	5	2
b. Range of pollen dispersal	3	4	5	6	1	2
c. Mean outcrossing rate	5	4	3	6	2	1
d. Effective neighborhood size (N_e)	3	4	2	1	5	6
III. Temporal aspects of species abundance						
a. Life cycle components						
1. Percentage seed carryover	4	3	6	5	2	1
2. Percentage survival to maturity	4	5	1	2	3	6
3. Seed output per plant	4	3	6	1	2	5
b. Overall density response						
1. Mortality type	4	3	5–6	5–6	2	1
2. Plastic type	3	4	2	1	5	6
c. Temporal stability in N_e	6	3	3	5(?)	?	?
IV. Genetic variation						
a. Between sites, within regions	4	3	2	6	5	1
b. Between neighborhoods or patches	2	4	5	6	3	1
c. Within patches						
1. Heterozygosity	5	3	4	6	2	1
2. Polymorphism index	5	4	3	6	2	1
V. Phenotypic plasticity						
a. "In nature"	2	4	1	3	5	6
b. Controlled environments	3	4	1	2	6	5
VI. Variance in fitness and indices of evolutionary potentials						
a. Heritable variation in seed output	5	5	5	5	4	?
b. "Reproductive surplus"	3	4	5	1	2	6
c. Genetic components of mortality rates	Unknown					

[a]Species: A, *Avena barbata;* B, *A. fatua;* C, *Bromus mollis,* D, *B. rubens;* E, *Trifolium hirtum;* F, *Medicago polymorpha.* The ranking of means assigns the lowest scales to the highest or greatest value for the parameter considered; the highest value is the lowest or least value. A question mark is used whenever the ranking for the species in question is too poorly known for any estimate to be made.

tion of optimal strategy as a compromise between costs of pollination mechanisms and advantages of genetic variation generated by
outbreeding (Solbrig 1976, article 5).

These strategies have evolved as "character combinations." In
heterogeneous environments, response by a population could be
described in terms of phenotypic variances and covariances. Now, a
second way of defining optimal strategies involves the study of
polymorphisms, quantitative genetive variation, plasticity, and
homeostasis. Strategies now measure phenotypic capacities to adapt
to varying environments, as superimposed on the reproductive strategies. In terms of both life history parameters and breeding systems,
the theoretical role of genetic polymorphisms and phenotypic plasticity were repeatedly alluded to; we shall turn to this topic next.

VARIATION AND ADAPTIVE RESPONSES

As outlined by Levins (1968), there are several outcomes at the
level of variation in populations living in heterogeneous environments: Genotypes may (1) produce the same phenotype (canalization) or (2) produce different phenotypes (phenotypic plasticity); (3)
distinct developmental classes (phases) may develop in response to
some threshold factor or (4) natural selection may favor different
coexisting genotypes (genetic polymorphism). Merely describing
the amount of total phenotypic variation of genotypic variation does
not allow the adaptive role of such variation to be assumed. In his
classic review, Bradshaw (1965) pointed out that homeostasis, the
tendency for certain physiological or morphological traits to display
constancy in the face of varying environments, or, conversely, specific phenotypic modifications induced by specific environments
(phenotypic plasticity) have to be shown to enhance adaptedness.
He concluded that "not all plasticity is adaptive but some is." He too
emphasized that genetic control of plasticity, demonstrated either by
genetic analyses of segregating generations or through artificial
selection experiments, was an important issue in the discussions of
its evolutionary significance. Likewise, Lerner (1954) discussed at

length the nature and evidence for genetic homeostasis as a property of polymorphic populations, particularly in relation to heterozygosis, and as measured by fitness-related traits. It is intuitively easy to see how a population of genotypes might respond to varying environments through polymorphism-based gene frequency changes (Hedrick, et al. 1976), plasticity of certain genotypes (Bradshaw, 1974; Cook and Johnson, 1968), or homeostasis of heterozygous genotypes (Lerner, 1954; Hochachka and Somero, 1973; Selander and Kaufman, 1973; Ayala, 1976). In fact, several ideas on the adaptation to heterogeneous environments were spelled out by Mather (1955) in a discussion of disruptive selection in time and space. He noted that rapid changes in environments require plastic or homeostatic responses, since genetic polymorphism involves a time lag in response through gene frequency changes, whereas spatial environmental heterogeneity produces genetic adaptations through differentiation of gene pools in different populations. The large literature on the surveys of variation among populations in the form of clines, races, or ecotypes probably bears directly on the evolution of polymorphism as an adaptive strategy. However, as shown by Hedrick et al. (1976), the roles of genetic versus plastic response might be reversed such that genetic tracking of temporally varying environments was not favored in finite populations (see also Templeton and Rothman, 1976).

Before returning to the models of optimal strategies, we need to review briefly the following questions: (1) How widespread are genetic polymorphisms in plant populations and how often are these involved in selective changes in gene frequencies? (2) How is plasticity measured and does it complement polymorphism? (3) How can we best describe the patterns of environmental variation considering organisms' phenotypic responses? and (4) Do individual and population fitness functions need to be defined separately?

POLYMORPHISMS IN PLANT POPULATIONS

Hamrick (article 4) reviews in some detail the available literature on genetic variation, in particular those studies involving electrophoretic variation. Although the available information does not seem to allow any generalizations, it appears that most populations, particularly long-lived trees, have large amounts of variation. A few studies on comparative surveys of congeners show that deriving any patterns based on range of distribution or population features might be premature. Several inbreeders and asexuals too have a rather large degree of variation both within and between populations. Furthermore, amounts of variation for morphological characters and allozymes may not be collinear.

Several classic examples of the adaptive role of genetic variation in plant populations have been reviewed by Stebbins (1950), Grant (1963), and Heslop-Harrison (1964), among others. Antonovics (1977) discussed the paradigm for studying the adaptive role of variation in plants; several detailed studies of his own provided elegant case studies. Among the highlights of this field are the presence of microgeographical differentiation in plant populations, the relative magnitudes of viability and fecundity components of selection (Clegg, 1978), and well-documented evidence on ecotypic patterns for a greater variety of species. Quantitative variation has also been widely used in drawing up the patterns among populations, ecotypes, or related species in different habitats. The adaptive significance of characters like seed size (Baker, 1972), flowering time (Grant, 1963), parallel variation in vegetative growth patterns, and response to grazing certainly provide adequate evidence for the role of genetic variation in plant populations (e.g., Stern and Roche, 1974; Wright, 1976). But adaptive strategy arguments require further evidence for character combinations and their covariation to have evolved in relation to certain fitness maximization principles. The role of environments varying in space is perhaps easier to study in the situations of strong selective forces, viz., mineral toxicity, soil pH, and topography. Temporal variation in the genetic structure of plant populations has not been sufficiently studied. Surveys of protein variation or quantitative variation have to be increasingly phased into long-term studies of both natural and experimental populations for measuring multivariate responses to certain specific environmental factors.

POLYMORPHISMS VERSUS PHENOTYPIC PLASTICITY IN RELATED SPECIES

Three pairs of congeners have been studied comparatively for patterns of variation: *Avena fatua* and *A. barbata* (Jain, 1969), *Bromus mollis* and *B. rubens* (Moraes, 1972; Wu, 1974), and *Limnanthes alba* and *L. floccosa* (Brown, 1977). Within each of these pairs (Table 7.5), the first-named species has more or less ubiquitous genetic polymorphisms, whereas the latter-named species has widespread monomorphism. Evidence from quantitative genetic studies shows that although phenotypic variance for a majority of characters in monomorphic populations is as large or larger than in polymorphic species samples, the genetic component is small, as shown by the response to selection and partitioning of variance. Thus, phenotypic plasticity appears to replace genetic variation in these species (for animal examples, see Selander and Kaufman, 1973a,b). Using common garden technique and a series of macroenvironments varying in soil type, moisture, and photoperiod regimes, environmentally induced variability, i.e., phenotypic plasticity, was indeed shown to be larger.

Adaptive role of the observed genetic polymorphisms for allozyme or morphological marker loci is not often known; nor is it clear how observed phenotypic variation in the field might be adaptive. Data on population census through seed, seedling, and adult stages do not show a consistent pattern of greater or smaller selective coefficients in the two situations. Population numbers and turnover rates observed for a short period of 5 to 7 years do not suggest that polymorphic populations were in any way more or less successful in terms of persistence. The maintenance of variation in *Avena fatua* versus *A. barbata* or in *Bromus mollis* versus *B. rubens* can only be surmised to depend on the slight differences in outcrossing rates, but the evidence for heterozygote advantage is far from convincing. The needed field studies on microniche differentiation have not been carried out.

Since it is customary, however, we might still go ahead and speculate on the apparent negative association between the genetic

TABLE 7.5.

COMPARISONS BETWEEN RELATED SPECIES FOR SEVERAL MEASURES OF PHENOTYPIC PLASTICITY AND GENETIC VARIATION

Species pair	Species with higher values						
	Total phenotypic variation in nature	Between-families variance	Within-families variance	Response to within-family selection	Induced phenotypic variation under different environmental conditions	Plastic response to density	Allozyme variation within populations
Bromus mollis vs B. rubens[a]	B. rubens	B. mollis	B. mollis	B. mollis	B. rubens	B. rubens	B. mollis
Avena fatua vs A. barbata[b]	A. barbata	A. fatua	A. fatua	A. fatua	A. barbata	A. barbata	A. fatua
Limnanthes alba vs L. floccosa[c]	L. floccosa	L. alba	same in both spp.	no data	L. floccosa	L. floccosa	L. alba

[a]Data of Wu and Jain (unpublished).
[b]Data of Marshall and Jain (1968) and Jain (1969).
[c]Data of Brown and Jain (unpublished).

and nongenetic variation components in congeneric pairs. A theoretical model by Levins (1968) suggests that patterns of environmental variation in time may be crucial. Autocorrelation between the successive environments, i.e., increased predictability, or longer sequences of one or the other "kind" of seasonal environments can be tracked through genetic variation (in fact, by a rather small amount of genetic variation). Large irregular fluctuations might favor phenotypic plasticity. Finite populations, however, may show monomorphisms under highly autocorrelated environments (Hedrick et al., 1976). Thus, several alternative explanations might be offered for the monomorphic cases: (1) a smaller outcrossing rate, allowing slight homozygote differences to fix the favored alleles; (2) rapidly changing environments in large populations; and (3) a long cyclic pattern of environmental variation in small populations. The marked patchiness observed within "polymorphic" *A. barbata* populations may be based on the last-named alternative.

Review of work on the quantitative analysis of plant growth by such noted plant physiologists as G. C. Evans, L. T. Evans, R. S. Loomis, and F. DeWit, leads to the following points: Ontogenetic factors are extremely important in relation to the observed phenotypic variances, and descriptors of phenotype might preferably be derived from growth rates, allometry, size-weight-number ratios of organs. Evans (1972) describes several integrators that tie the environmental and biological variables together and provides some extremely useful insights into the role of phenotypic plasticity in adaptation (also see Teeri, article 15; Mooney and Gulman, article 13). Moreover, resource allocation studies have always interested the physiologists; collaborative work between them and population biologists would be very useful.

HOW DOES PHENOTYPIC PLASTICITY EVOLVE?

As pointed out by Bradshaw (1965), we first need to show that plasticity is under genetic control, i.e., that there is genetic variance for varying degrees of plastic response to a given set of environments. Crop scientists often compare the relative stability of perfor-

mance, in terms of yield or its components, of varieties in different
locations and in different years. Use of the regression method developed by Finlay and Wilkinson (1963), for example, shows whether
varieties maintain their phenotype constant under different environments. The heritability of plasticity was studied in *Bromus mollis* by
the parent–progeny regression approach, using intraprogeny variance on a rank-ordered scale as a measure of plasticity. A significant

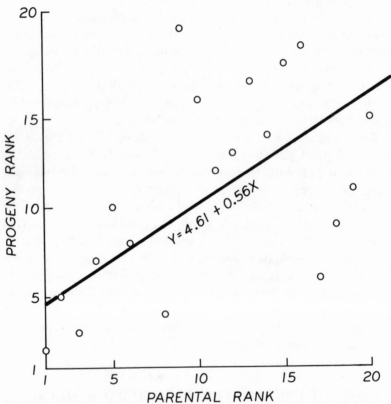

Figure 7.1. Regression of plasticity in individual plant progenies as measured by the
rank orders of the estimates of σ_w^2 = within progeny variance for flowering time plotted against plasticity of parents grown in the same environment. Parents were represented by seed collected from individuals of *Bromus mollis* in nature; progenies were
selfed seed raised from these parent individuals. Rank orders are based on the numer-

correlation between the parent and progeny rank orders (Fig. 7.1) showed shows that families of highly plastic individuals grown under similar environments are also plastic. Evolution of plasticity, like any other trait under selection, could then proceed by changes in the genotypic proportions in successive generations. Inbreeding and asexually reproducing species are particularly suitable for this area of research.

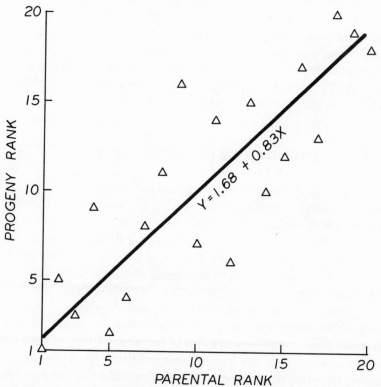

ical values of variance estimates. Circles (left) and triangles (right) represent two different populations with high and low genetic variation, respectively. Note that both lines show significant regression slopes ($P < 0.05$), but the population at right has a higher value ($P < 0.01$), i.e., a larger genetic component in the inheritance of plasticity.

EVOLUTION OF REPRODUCTIVE RATES IN
MEDICAGO POLYMORPHA

In order to develop a quantitative genetic model of reproductive strategies, an experiment was undertaken in two populations of *Medicago polymorpha*, a common annual weed of the Mediterranean region (Stebbins and Jain, in preparation). Four components of total seed output per plant, namely, number of branches, number of pods per branch, number of seeds per pod, and seed weight, were studied for the heritable component of variation and the correlations among various components. Selection to increase seed output through selection for seed number and, in another study, number of seeds per pod showed that response could not be predicted from total phenotypic variances and only weakly from heritability estimates. Evolutionary ecologists should incorporate genetic analyses of this sort in their strategy–evolution models.

A MODEL OF SEED DORMANCY

In order to examine the role of seed dormancy variation in response to varying environments, we used a Monte Carlo simulation in which the following parameters were considered: (1) a diallelic locus with three genotypes having either different means and variances of germinability in good and bad years or the same mean but different variances; (2) environments obtained using random numbers with input mean frequencies and with or without specified autocorrelation; (3) seed longevity in soil, varied between 3 and 10 years; and (4) seed output per plant. The runs were summarized for each set of input conditions in terms of the probabilities of survival of a population and the genetic composition (gene frequency changes). Specifically, we were interested in recording the number of times a population was fixed for one of the alleles before it became extinct. The results are summarized in Table 7.6.

Several conclusions can be stated from this study: (1) regular long cycles of good and bad years (high autocorrelation) combined with high rates of seed carryover favor the maintenance of genetic

TABLE 7.6.
COMPUTER-SIMULATION STUDY OF ALTERNATIVE OUTCOMES
OF SEED GERMINATION MODEL (SEE TEXT)[a]

Dormancy	Seed longevity	Constant	Pattern of environmental variation				
			Autocorrelated			No autocorrelation	
			Short cycle	Long cycle	Irregular cycles	Low	High
Low	short	GP	PP	GP	PP	GP,GH	PP
	long	GP	PP	GP	PP	GP,GH	PP
High	short	GP	GP	GP	PP	GP	PP
	long	GP	GP	GP	GP	GP	PP

[a]GP, genetic polymorphism; PP, phenotypic plasticity (allele with larger variance in germination rates is favored); GH, genetic homeostasis (either heterozygosity has lowest variance or the favored homozygote is homeostatic).

polymorphism; (2) stochastic variation (no autocorrelation) with high variance favors plastic response; and (3) in most cases, outcomes were highly indeterminate if seed output was lowered and population numbers were allowed to fluctuate widely owing to variations in the total available seed pool. It is tempting to accept results like these as proof of the existence of alternative strategies, or even for the evolutionary simplicity of achieving distinctive strategies. But numerical "proofs" from simulation studies are often artifacts of "man-made" parameter sets. Admittedly, there are many gross oversimplications in this model; for instance, polymorphism is not associated with any cost, seed dispersal is excluded, environmental patterns are simple binomial distributions, and, most important of all, fitness function is defined simply in terms of population living for specified periods of run length (up to 200 generations). Fitness functions are not likely to be simple in the real world. The results may be somewhat instructive, however, in the development of alternative strategy models. Literature survey on germination in wild populations suggests that this is a very promising area for population comparisons in relation to the notion of adaptive strategies. Genetic variation for germinability was studied in a few grassland annuals and in *Helianthus* populations. The results showed that both genetic and nongenetic responses varied among populations, so that the evolution of polymorphic plastic or homeostatic gene combinations in a particular population could be empirically described in terms of such responses over a succession of generations.

MODELING GENETIC SYSTEMS

Genetic systems are often discussed in relation to the regulation of recombination through such variables as sexual versus asexual reproduction, cross-fertilization versus self-fertilization, levels of ploidy, and other karyotypic variations. Evolution toward certain optimal levels of recombination is then discussed in relation to the following, largely unproven, statements: (1) "open" genetic systems, with higher rates of outbreeding, genetic recombination, etc., allow populations to carry a large amount of genetic variation and heterozygosity, which is adaptively useful; (2) since genetic variation also imposes "genetic load" in the sense of not allowing the most fit genotype to be fixed, conflict between immediate fitness and long-term flexibility requires genetic systems to maintain certain optimal rates of outbreeding, sex ratios, etc.; (3) most species represent such compromises in terms of several suites of variables in genetic systems. Evolution of genetic systems has been studied in relation to the models of heterostyly, sex ratios, male sterility, etc. Evolution of sex has been vigorously discussed in recent years. Whether sexuality (which is unfortunately equated to genetic recombination) evolved in relation to the colonization of highly temporally varying environments or spatial heterogeneity, or as a hitchhiking effect between linked favorable and neutral genes can only be resolved by an increased store of information on the occurrence and behavior of genes affecting variables of genetic systems (modifiers?), selection pressures, and appropriate demographic and population survival data. Plant literature so far ignored by most writings has apparently a good deal to offer in this area. Adaptive strategy analyses may be quite successful in comparative studies of genetic systems as a holistic evaluation of the adaptive role of genetic variation.

DEFINITIONS AND MEASUREMENTS OF FITNESS

The two primary questions of interest here are, Can measures of relative fitnesses derive from genetic changes in populations and life table characteristics associated with different genotypes be readily combined into a single ecological genetic measure of fitness? and Do we need to consider population fitness in relation to the arithmetic of adaptive strategies, apart from a summation of individual fitnesses? Interdeme selection, defined operationally by Wright (1969) and Levins (1970) in terms of relative successes of different subdivided populations, is not necessarily ruled out by those opposed to altruism or Wynne-Edwards' group selection. Population fitness could conceivably be measured in terms of the successful colonizing ability of local stands and their persistence in time. Population averages in dispersal rate, outbreeding rate, numerical growth rate, etc., would allow comparisons, at least in theory, between species with different adaptive strategies. This matter need not be pursued here in detail; the reader may refer to Dobzhansky (1968), Emlen (1973), and Ayala (1969) for further discussions.

OPTIMALITY, OR DO WE LIVE IN THE
BEST OF ALL POSSIBLE WORLDS?

In most discussions of optimal strategies in biology, we have a basic philosophical divergence in viewpoints. Some of us believe that organisms have evolved the best possible ways to adapt to their environments, from the molecular level to the levels of morphological and life history adaptations. The principles of fitness maximization, genetic load minimization, and ergonomically efficient ways of utilizing resources dictate that populations must reach local adaptive peaks. Optimality often invokes maximax strategy with a modification by maximin considerations in order to avoid high losses in the worst environments. Others may wish to argue, however, that, using maximin strategies, most organisms only have to be "adequately near the fitness peaks," and, in fact, most strategies as we find them in dynamic equilibrium situations are not optimal. But aside from the philosophical differences, there are also basic methodological issues involved here. Botanists often describe the common strategies of plants living together in communities by pointing to the leaf shapes of desert plants or heterophylly in aquatic habitats. Strategies in this context simply refer to the small number of ways in which groups of plants sharing the same habitats "make their living," whereas the notion of optimal strategy evolution is meaningful when we have ways to show relative payoffs of different character suites within or between different conspecific populations. Variability is again the tool for analysis in this area, and it is not sufficient to point to the characteristic life table or dispersal pattern of a species in order to invoke optimality concept. The present-day evolutionary outcomes do not demonstrate the processes of evolution but simply suggest hypotheses for further studies.

CONCLUDING REMARKS

Much of this paper may be a somewhat trivial exercise, but it may help a few like myself to learn what is meant by optimal strategy, optimal life history, and optimal genetic systems. This survey of literature may also help to overcome the awe of mathematics in this field, since analysis is often fairly general and elementary. It should also be appreciated that, although we are largely doing natural history in the classical descriptive style (leading to the beautiful conclusion that different species have different life-styles, presumably in response to different environments they live in), the strategy approach (1) makes it appear "more sophisticated," (2) draws the interest of theorists and of ardent "strong-inference supporters," and (3) promises some general conclusions in relation to the "suites of characters" (multivariate rules of evolutionary games) so appropriately emphasized by the strategy arguments. Thus, the notion of adaptive strategy in terms of both ecological parameters of life histories and genetic structure of populations living in varying environments is conceptually sound and may offer a promising theoretical framework. Several ideas, however, such as the relevant character choices, their genetic variation, definition of appropriate fitness functions, and the need to explore optimality principles through experimental ecogenetics must be carefully thought out. How environments vary and how closely populations can be shown to track the environment will be important methodological issues in this research. Should strategies emerge as discretely few combinations of characters related to an optimized fitness measure or show a continuum, this need not discourage us from pursuing the problem. We need to agree on operational definitions of cost and income variables, fitness measures, interdeme selection, etc., and to continue appropriate field studies in order to provide data for tests of a large body of theoretical ecology at hand. (Note that operationalism—i.e., first measure, then define—is well accepted in physics; the basic concepts of energy and work, and the modulus of a spring, for instance, are defined by measurable quantities.) Choices between alternative strategies may have to be generated in manipulated experimental populations and observed for evolutionary changes in order to utilize strong inference for hypothesis testing. Together, population genetics and population ecology can attack this problem effectively. Research paradigms are well known but have not often been considered, owing, I suspect, to lack of collaborative effort on our part. But there is excitement, interchange, and the search for a common meeting ground. In the meantime, the two other lines of development in population biology that I referred to in my introductory sentence, i.e., descriptive natural history studies at the level of local populations and the search for specific, simply inherited examples of adaptation through physiological and biochemical traits, should continue to accumulate the raw materials for strategy-oriented research.

8 SEED GERMINATION SYNDROMES IN HIGHER PLANTS

MARK W. ANGEVINE AND BRIAN F. CHABOT

THE STAGES of a plant's life history represent critical, distinctive episodes in the ongoing maintenance of genetic continuity for that species. Successful completion of each stage is dependent on the detection of and appropriate response to variation in the physical and biotic environment. Natural selection will shape a unique pattern of environmental sensitivity for each plant species, which is dependent on the past history of that species. Usually, however, a group of species will share a set of correlated life-history responses to characteristic environmental regimes. Such correlated responses have been referred to as a *character syndrome* (Stebbins, 1970). Insofar as we can define generalized selective regimes in the environment, we should be able to recognize a limited number of character syndromes.

This paper considers the application of the character syndrome concept to seed dormancy and germination in higher plants. We deal with germination primarily as an ecological rather than a physiological or developmental process. A scheme for classifying germination behaviors based upon general selective regimes is proposed. This classification is then used to organize the vast literature on seed dormancy and germination. We emphasize those studies which have attempted to describe and analyze germination behaviors under natural conditions. Studies that include the critical steps of observation and experimentation in the field are quite rare, yet they provide a much more satisfying and conclusive understanding of germination as an ecological phenomenon than do laboratory experiments alone.

THE ADAPTIVE NATURE OF
SEED GERMINATION

A wealth of observational and experimental data concerning the physiological processes involved in the ripening, dormancy, and subsequent germination of seeds has accumulated over many years. Several reviews have treated aspects of these subjects (Harper, 1957; Evenari, 1961; Wareing, 1966; Anderson, 1968; Heydecker, 1972; Roberts, 1972; Koller, 1972; Mayer and Poljakoff-Mayber, 1975). Studies of the physiology of seed germination have demonstrated that a remarkable diversity of mechanisms exists in plants for the purpose of recognizing particular aspects of the external environment of a seed and regulating germination in response to those aspects. In seeking an organizational framework for these various mechanisms, it is helpful to consider the ultimate objective of all of them. The germination of seeds and subsequent early growth of seedlings are not only essential phases of the life cycle of all higher plants but represent periods of maximum vulnerability to physical changes in the environment and minimal potential for homeostatic response or physiological retrenchment. The few known mortality schedules for plant populations have demonstrated that the highest mortality occurs at the seed and seedling levels (see Cook, article 9; Hickman, article 10). When the seed coat is split and epicotyl extension begins, the plant has, in a very real sense, "bet its life" on the favorability of environmental conditions until such time as the plant can successfully reproduce or accumulates sufficient photosynthetic capital to survive an unfavorable period. Consequently, selection will favor environmental cuing mechanisms that tend to decrease the probability of encountering unacceptable growth conditions in the period after germination. This is the ultimate purpose of all environmentally regulated germination responses.

There are several reasons to expect seeds to germinate as rapidly as possible after their dispersal from the parent plant: rapid germination decreases the generation time, increasing the intrinsic growth rate of the genotype without any increase in fecundity; the less time the average seed spends in the soil before germinating, the lower will be the percentage of seeds lost to mortality there; and seeds germinating quickly have a considerable competitive advantage over later-emergers germinating when seedling density is high and competition strong. Consequently, any seed that postpones germination beyond its first physiological opportunity is facing risks in these three areas and must be achieving some compensatory gains in its probability of survival. We feel that delayed and environmentally cued germination are mechanisms to improve seedling survival and overall population growth rates in the face of specific classes of environmental risk.

The nature and timing of periods of environmental risk encountered by each species will be somewhat different, as will the response pattern which has evolved in each species. But it is possible to identify broad classes of environmental risk that seem predominant in many habitats. If this is done, it is further possible to recognize sets of germination-controlling responses, correlated with these environmental risks, which are common to a large number of plant species. We may find that some classes of environmental risk have given rise to several different germination syndromes, and others to only one. But a consideration of seed germination in terms of the aspects of the environment that a species needs to avoid most will produce a grouping of germination behaviors that has ecological meaning and that can increase our understanding of the germination process in an evolutionary context.

GERMINATION IN AN ENVIRONMENTAL CONTEXT

There have been several attempts to synthesize the literature available on ecological aspects of seed germination. Environmental control of germination has been reviewed extensively by Koller (1972) and more recently by Mayer and Poljakoff-Mayber (1975). The approach of these authors is to consider each environmental factor separately and discuss its range of variation in the field and experimental evidence regarding its significance for germination. A large number of factors have been shown to affect germination. These include: average temperature and fluctuations in temperature during germination; temperature conditions during the dormant phase (chilling and heat stratification); light intensity, duration, and spectral composition; moisture availability; dissolved salts or organic chemicals in the soil solution; and the absolute and relative concentrations of atmospheric gas in the soil. For each of these germination factors, Koller attempted to relate known physiological mechanisms within seeds to laboratory germination test results and these to field germination responses and the species distributional patterns. Roberts (1972) reviewed the attributes responsible for maintaining dormancy and viability in seeds and also found a variety of factors to be acting in different species. None of these authors, however, attempted to formulate a functional classification of germination types based on these factors.

More frequently, groups of species with similar ecological relationships have been examined in an attempt to demonstrate similarities or differences in germination behavior. Several studies have shown convergence of germination behavior within defined ecological groups. Winter annuals of British chalk downs (Ratcliffe, 1961) have striking similarities in their after-ripening requirements and optimal germination temperatures. Pemadasa and Lovell (1975) examined germination in seven species of winter annuals from coastal dunes. They found three tendencies, shared by the species but varying in importance among them: strong innate dormancy, declining continuously through the first 24 weeks of storage; initially low temperature optima for the germinable seeds, with an increase in optimal temperature and tolerance range over time; and a strict requirement for adequate and continuous soil moisture at the time of seed germination. Capon and Van Asdall (1967) found a consistent tendency among nine species of desert annuals to germinate most rapidly after a brief 50°C treatment, but longer, 20°C pretreatments also produced similar germination rates and final germination percentages.

A number of studies have found several distinct germination

patterns within a given habitat. Among species common in or endemic to cedar glades in Tennessee, two patterns have been recognized: a "winter annual" type which germinates in early fall, overwinters as a rosette, and continues growth the following spring; and a "summer annual" type which overwinters as a seed and germinates in the spring (Ware and Quarterman, 1969; Baskin and Baskin, 1971a,b, 1972a, 1973). As many as five patterns of germination have been recognized in Mojave Desert species (Went, 1948). The flora of Arctic and Alpine tundras can be roughly divided into two groups, one possessing innate dormancy mechanisms and the other relying on low environmental temperatures to enforce dormancy (Bliss, 1958; Amen, 1966; McDonough, 1970; Chabot and Billings, 1972; Eurola, 1972). Among those species possessing innate dormancy, a variety of specific mechanisms are used to inhibit germination, including mechanical constraint by the seed coat, chilling, after-ripening period, and specific light requirements (Amen, 1966). This ability to arrive at the same ecological pattern of germination through several physiological mechanisms was emphasized by Harper (1957) for temperate weed species.

It is also possible to use germination behavior to explain habitat differences in groups of similar species. Harper (1957) discussed several pairs of closely related species whose strikingly different habitat distributions could be readily explained by the differences in germination behavior. Grime and Jarvis (1974) compared a number of species of herbaceous plants of grassland, woodland, and bare-ground habitats by germinating each species in a series of light conditions. Overall, there was an indication of higher light requirements for germination in species from open habitats. This was particularly strong in forb species with dark inhibition of germination; in grasses it was less pronounced. In comparing desert with alpine species from the Sierra Nevada of California, Chabot and Billings (1972) found that desert species had either a pronounced temperature optimum below 20°C or no distinct optimum temperature. By contrast, the alpine species had distinct optimum germination temperatures in excess of 20°C. A comparison of germination requirements in populations of Silene dioica, a geographically widespread species, has shown a tendency for northern populations to avoid winter conditions and for southern populations to avoid summer (Thompson, 1975).

From the existing literature on germination, we can draw three conclusions that are important in the development of our synthesis:

1. A given pattern of emergence within a habitat can be achieved by a variety of physiological mechanisms. Specific dormancy-controlling processes appear to be a proximate means for achieving an ultimate ecological goal. Thus, an emphasis on dormancy-controlling mechanisms alone will not adequately characterize the selective regime that has given rise to them.

2. Within a given habitat there may be more than one germination pattern. While it appears possible to reduce germination behaviors to a relatively small number of classes for each habitat, unique evolutionary histories for each species, competitive displacement, and selective regimes permitting alternative solutions will frustrate attempts to classify germination behaviors by habitat alone.

3. Between habitat types there are often similarities in germination behavior. As an example, the "summer annual" strategy of the cedar glades contains many elements in common with the germination of Arctic and Alpine tundra species. It is these between-habitat similarities that we will emphasize in our synthesis.

DORMANCY

Dormancy, or the lack of it, is an important factor in any germination strategy. It has at least two components: a means of maintaining metabolic torpor and a means of perceiving and predicting the state of the environment. These two components do not necessarily involve distinguishable physiological mechanisms. Several systems for classifying dormancy have been devised. A widely used scheme is that of Harper (1957), which defines three types—innate, induced, and enforced. Innate dormancy refers to the inability of seeds to germinate under favorable conditions immediately upon removal from the parent plant. The majority of plant species have an innate dormancy period, but the length of the period and the conditions favoring its cessation vary widely among species. This form of dormancy has also been termed natural (Brenchley and Warrington, 1930), inherent (Bibbey, 1948), primary (Crocker, 1916), constitutive (Sussman and Halvorson, 1966), and endogenous (Zohary, 1962; Schafer and Chilcote, 1969). After-ripening is the term generally used to describe the processes involved in eliminating innate dormancy. After a seed has passed out of innate dormancy, however, exposure to a particular set of environmental conditions may once again make germination impossible. If this dormancy remains after the unfavorable environment is removed, it is called induced, or secondary, dormancy (Crocker, 1916). If the dormancy is dependent on the maintenance of the unfavorable environmental state, it is termed enforced dormancy. Enforced dormancy is equivalent to environmental dormancy (Bibbey, 1948), or exogenous dormancy (Zohary, 1962; Sussman and Halvorson, 1966).

These classification schemes primarily describe the physiological and structural mechanisms for maintaining the dormant state. As discussed previously, they do not relate specifically to ecological patterns of germination. For our purposes, dormancy is regarded simply as delayed germination after seed maturation which allows a species to avoid certain risks. We place greater emphasis here on mechanisms for predicting periods of greater or lesser risk.

ADAPTIVE SYNDROMES OF SEED GERMINATION

We recognize two basic strategies of germination behavior: avoidance of difficult seedling growth conditions and tolerance of such conditions. Avoidance is the largest class, at least in existing studies. Avoidance behaviors involve use of environmental cues to predict favorable and unfavorable periods. Tolerance usually requires high maternal investment in seed number or size in order to overcome mortality risks. The tolerance strategy is discussed in more detail later.

AVOIDANCE STRATEGIES

Strategies of avoidance may be subdivided further according to the nature of the hazard. These are initially distinguished as physical and biotic factors. This separation is not always complete since biotic factors frequently operate through variables of the physical environment. Beyond this, predictability of favorable and unfavorable periods becomes an important selective factor. The scheme for classifying germination syndromes is outlined in Figure 8.1 and is discussed below.

Physical Factors

In most habitats, one season of the year is particularly inhospitable for the growth and survival of seedlings; considerable mortality will result from the direct effects of the seasonal climate, such as extremely high or low temperatures, drought, flooding, or low light intensities. The evolutionary responses of seed germination to these factors will depend most critically on their temporal pattern and predictability. In highly predictable climates, these mortality factors may be very strongly correlated with any one of several indicators of season, such as day length or temperature. Thus, an environmental variable which is not a direct cause of mortality may be used for

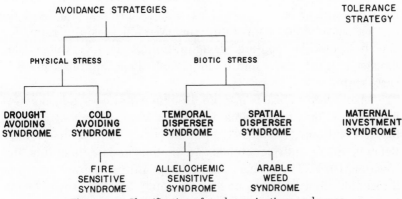

Figure 8.1. Classification of seed germination syndromes.

timing phenological events because of its greater precision. If the cessation of the inhospitable season is not consistent from year to year, however, germination must be sensitive to the primary limiting factor in the environment rather than a secondary predictor.

Drought-avoiding syndrome. Conditions of high temperatures and low moisture availability characterize the summer season of many habitats. A large number of species have responded to these conditions through development of a "drought-avoiding" germination syndrome. This syndrome has been well studied in species of cedar glades (Baskin and Baskin, 1971a,b, 1972c, 1973), Calluna heath (Newman, 1963), dunes (Janssen, 1973; Pemadasa and Lovell, 1975), chalk downs (Ratcliffe, 1961), and granite outcrops (Baskin and Baskin, 1972b). Baskin and Baskin have demonstrated, through a series of field and laboratory experiments, the suite of adaptations that this germination syndrome involves. Species utilizing this syndrome generally show strong innate dormancy when the seeds ripen in late spring or early summer, as in Viola rafinesquii (Baskin and Baskin, 1972c) and Diamorpha cymosa (Baskin and Baskin, 1972b). Some species have only limited innate dormancy but have an initially low temperature requirement for successful germination (Newman, 1963; Baskin and Baskin, 1972a). As the summer progresses, innate dormancy is relieved, but temperature ranges for germination generally stay below those present in the environment. A requirement for moisture may develop as well (Newman, 1963). By late summer, as environmental temperatures decline and the optimal germination temperature increases, successful germination becomes possible, seedlings develop through the favorable, cooler, moister autumn and winter months, and flowering takes place in early spring.

Many desert species use the drought-avoiding syndrome. Germination follows heavy rains, which occur most frequently during the late autumn (Went, 1948; Juhren et al., 1956; Zohary, 1962; Beatley, 1974). Both perennials and annuals germinate at this time (Beatley, 1974). The primary cue is the quantity of rainfall beyond a threshold, often found to be approximately 25 mm (Tevis, 1958a; Beatley, 1974). Above this threshold, germination success is related roughly to amount of rainfall. This sensitivity to moisture seems to be achieved through leachable inhibitors present in the seed or seed coat, as has been demonstrated for *Oryzopsis* (Koller and Negbi, 1959). Evenari (1961) has reviewed the evidence for the existence of soluble germination inhibitors and their importance in some desert species.

Although rainfall occurs during other periods of the year, the seasonal restriction of germination that has been observed is achieved in most species through preferential germination during periods of moderate to low temperatures (Went, 1949; Juhren et al., 1956; Tevis, 1958a; Chabot and Billings, 1972). Perennials such as *Calligonum comosum* (Koller, 1956), *Fouquieria splendens* (Freeman, 1973), and *Larrea divaricata* (Barbour, 1968) have narrow temperature optima in addition to specific moisture requirements. By germinating only upon wetting at a particular temperature range, a species can restrict its appearance to the season with the highest probability of rainfall. Some desert species germinate over a broad temperature range, apparently using moisture as the sole cue (Tevis, 1958a; Evenari et al., 1971; Chabot and Billings, 1972). A very small group of species have high temperature preferences and consequently are able to germinate after rains in the late summer or early fall (Went, 1948; Juhren et al., 1956; Tevis, 1958a). Germination and initial growth are rapid in all these species. Little appreciable growth occurs during the subsequent winter, both in desert and nondesert species exhibiting this syndrome (Beatley, 1974).

Among the few direct observations of seedling mortality are several for desert species (Tevis, 1958a,b; Beatley, 1967; Evenari et al., 1971). These support low moisture availability as being the most frequent cause of mortality. When germination has been particularly successful after above-average rainfall, mortality of seedlings occurs principally when active growth resumes in the spring. Lack of water appears to be the most likely cause (Tevis, 1958a).

Seasonal drought also occurs in many tropical communities. Recent studies of the germination behaviors of tropical species in Panama suggest that germination occurs most frequently after the winter drought period (Garwood, 1977). Species in which fruits mature at the end of the wet season possess seeds with dormancy capabilities. Fruit maturation in other species occurs during the dry season, and dispersal is timed to coincide with the onset of rain. Dormancy mechanisms are not required of these seeds.

   Cold-avoiding syndrome. Another pattern of predictable seasonal harshness is that of extremely cold winters. In these environments, the direct effects of freezing or the indirect effects of lengthy cold-enforced photosynthesis depression can account for high seedling losses during the winter. In such environments, a "vernal germination," or cold-avoiding syndrome, will be successful. Seeds that have matured at the end of the favorable growing season are prevented from immediate germination and maintained in a dormant state until the cold season has concluded. Rapid germination after the end of winter will allow the longest possible favorable season for attainment of reproductive maturity, in the case of summer annuals, or development of sufficient structure to allow a resistant, dormant phase to be achieved, in the case of biennial and perennial species. This pattern of dormancy and germination can be produced by a variety of mechanisms. In species of temperate latitudes, strong innate dormancy after dispersal is generally observed (Peters and Dunn, 1971; Baskin and Baskin, 1975; Willemsen, 1975a,b), preventing germination when the seeds mature in the summer or autumn. This innate dormancy is often broken by several weeks of cold stratification in the laboratory, which simulates winter conditions in the soil (Ware and Quarterman, 1969; Baskin and Baskin, 1971c). If germination does not occur in early spring, however, a period of induced dormancy may occur, and germination will not be possible until another period of cold stratification has passed. Another commonly reported feature of this syndrome is a light requirement for germination, even at the critical spring temperatures.

   Among tundra species, either innate dormancy or relatively high temperature preferences restrict germination to the late spring and summer after dispersal the previous summer (Bliss, 1958; Mooney and Billings, 1961; Amen and Bonde, 1964; Sayers and Ward, 1966; Chabot and Billings, 1972; Eurola, 1972). Fewer than 45 percent of the species tested in these studies possess innate dormancy. Amen (1966) speculated that late-maturing species in the Arctic may face such consistently low temperatures by the time their seeds ripen that they are effectively prevented from germination by enforced cold dormancy. Most tundra species germinate best when exposed to light, though species of disturbed habitats may do equally well in darkness (Bliss, 1971).

Biotic Stress

In addition to the potentially disastrous effects of extreme physical conditions on developing seedlings, the presence of mature plants may produce environmental conditions in which seedling maturation is extremely difficult. A dense stand of fully developed plants can outcompete small seedlings through shading, removal of moisture and nutrients from the soil, production of allelopathic substances, and physical occupation of space. These conditions would make seedling recruitment into dense, mature communities extremely difficult. In such communities, the most favorable germination sites would be in areas of disturbance where, for some period of time, the competitive advantage of the adult plants is eliminated.

The appearance of such favorable germination sites is likely to be highly variable in both space and time. Dispersal, dormancy, and longevity of seeds are important in increasing the probability that a species will encounter favorable conditions for establishment. Environmental cues used to initiate germination include a variety of physical variables that change as a result of habitat disturbance: light intensity and quality, soil temperatures, nutrient levels, and the concentration of gases in the microenvironment of the seed.

We recognize a heterogeneous group of germination syndromes as responses to biotic stresses. The major classes are the temporal disperser syndrome and spatial disperser syndrome. These are distinguished primarily according to seed longevity and dispersal characteristics. The remaining three groups are separated according to the factors that trigger germination.

Temporal disperser syndrome. One possible alternative is for the seed to lie dormant at a site, perhaps close to the parent plant, and wait for a disturbance to occur on that spot. Seeds utilizing this syndrome must have the capacity for long-term viable storage in the soil and must have a mechanism for sensing and responding to a change in the competitive situation of the site. Light-sensitive germination may provide an appropriate mechanism. King (1975) has demonstrated that germination of *Arenaria, Veronica,* and *Cerastium* is inhibited by light transmitted through a canopy of *Tilia* leaves. This inhibition of germination by canopy-filtered light has been discussed by Smith (1973; also Taylorson and Borthwick, 1969) and may involve sensitivity to the proportions of red and far-red light reaching the seed. Canopy-filtered light has highly enriched far-red:red ratios, and far-red light has often been shown to inhibit seed germination (Holdsworth, 1972).

Prunus pensylvanica has a probable longevity of 50 years in the soil (Marks, 1974). It germinates readily after disturbance of deciduous forests in eastern North America, probably in response to increased nutrient availability at the soil surface.

Limited dispersal ability may be an associated trait of this group. Werner (article 12) has demonstrated a tendency among small-seeded species of plants for seeds with limited dispersal adaptations (she calls these plants "ploppers") to also show great longevity. However, good dispersal coupled with longevity in a species such as pin cherry, which is bird dispersed, will always increase the probability of encountering a local disturbance.

Spatial disperser syndrome. An alternative response to competitively mediated mortality involves the evolution of long-distance dispersal mechanisms. Species included in this syndrome must rely on distributing their propagules over a large area and thereby encountering at least a few appropriate low-competition germination sites. Good dispersal ability will often be coupled with small seed size, shorter storage viability, and lower amounts of innate dormancy in species with this syndrome. Examples of species in this group include *Solidago canadensis* and *Asclepius tuberosa,* which show close to 100 percent germination within 10 days of seed maturation at 20 to 25°C (Chabot, unpublished data). *Populus tremuloides* possesses wind-dispersed, short-lived (less than 1 month) seeds that require exposure to light on bare mineral soil for successful germination (Schreiner, 1974; Brinkman and Roe, 1975). *Betula alleghaniensis,* an important component of northern hardwood forests, also has small, well-dispersed, short-lived seeds that germinate preferentially in areas of disturbance within the forest where light intensities are higher (Brinkman, 1974; Forcier, 1975; Trimble, 1975).

Fire-sensitivity syndrome. In communities that are regularly devastated by fires, a special variant of the temporal disperser syndrome may evolve. Here, the extreme physical conditions of a fire provide a reliable cue regarding the subsequent favorability of the site, which for some time will be rich in nutrients and poor in competitors. Jaynes (1968) has shown that dormancy in *Kalmia hirsuta*, a shrub species occupying longleaf pine savannahs in the southeastern United States, is most efficiently broken by a brief 80°C pretreatment. Similar treatments killed seeds of several other *Kalmia* species. The longleaf pine savannah community is generally maintained in this area by frequent ground fires. The frequency of serotinous cones in pine species characteristic of fire-dominated communities is another manifestation of the fire-sensitivity syndrome. These species restrict seed dispersal to the period after fires, when the closed cones open to release their mature seeds. Rapid germination of many species frequently follows fires in the chaparral communities of California (Muller et al., 1968). In some cases fire appears to volatilize allelochemical substances that repress germination (McPherson and Muller, 1969; Went et al., 1952). For other species, such as *Rhus ovata*, heat acts directly on the seed or seed coat (Stone and Juhren, 1951).

Allelochemical substance-sensitivity syndrome. Germination of a number of species has been shown to be inhibited by plant-derived chemicals. The ecological role of such a reaction has been suggested in only a few studies. In most cases, these allelochemical substances have been viewed as inhibitory agents produced by plants to suppress growth of their competitors. We feel that their absence as a result of disturbance of the community may equally well serve as a cue indicating a low-competition regime. It is possible, therefore, that some species have evolved to utilize these chemical signals as predictors of safe germination sites.

Evidence for the existence of this syndrome is largely circumstantial. Typical results demonstrate that litter on the soil will inhibit germination in species that do not otherwise have special light requirements (Wilson and Rice, 1968; Peters and Dunn, 1971;

Muller, 1974; Werner, 1975). Although the specific inhibitory chemicals have been identified in only a few instances, the litter is often from species known to produce allelochemical substances. McNaughton (1968) demonstrated that germination of *Typha latifolia* is inhibited by phenolic compounds from leaves of the parent plants, which explains the absence of *Typha* seedlings in mature stands. Similarly, extracts from the litter of *Helianthus annuus* inhibited germination in seeds of several successional species (Parenti and Rice, 1969). The fact that, in several of these studies, rapid germination occurs when the litter is removed and that all species are not equally inhibited suggests the occurrence of metabolic repression rather than toxicity.

Arable weed syndrome. Human agriculture has created a novel habitat type for wild plants to colonize—the plowed field. A large number of species have taken advantage of this ecological opportunity. Although the freshly plowed field presents an excellent site for plant growth in the absence of established competitors, it is a paradise of limited duration. As the crop matures, competition will steadily become more intense; eventually, the tremendous physical disruption of harvest and recultivation will end the development of most of the field's occupants, and the sequence will begin again.

A successful arable weed, as characterized by Chancellor (1965), appears to have several seed adaptations. Its seeds have the ability to remain dormant and viable while buried, awaiting the next cultivation of the site. Germination after cultivation is rapid in most cases, provided a seed is located in a position likely to allow successful emergence and growth. The seeds of arable weeds have the ability to germinate in any season unless burial intervenes. The seedlings usually have rapid potential growth rates, followed by early maturation and seed set, in anticipation of the next cycle of cultivation.

Studies of the germination behaviors of individual weed species are not as common as studies investigating the response of the entire buried seed pool to cultivation treatments. Several studies do indicate the importance of the suite of adaptations described above in some of the most aggressive weeds of cultivated lands, however. Two long-term studies of the viability of buried seeds have been performed (Toole and Brown, 1946; Kivilaan and Bandurski, 1973).

Both have shown that weedy species tend to outlast nonweedy species by a considerable margin. In addition, most arable weed seeds have some form of enforced burial dormancy, frequently involving a light requirement for germination. *Ambrosia artemisiifolia* has strong innate dormancy, and release is greatly facilitated by light exposure (Willemsen, 1975b). *Senecio vulgaris* and *Capsella bursa-pastoris* both have an obligate light requirement for germination, but *Capsella* also requires cold-stratification (Popay and Roberts, 1970). Dormancy is enforced by burial in *Agrostis gigantea*, and light exposure improves germination in this species under a variety of laboratory test conditions (Williams, 1973). Wesson and Wareing (1969) have demonstrated for several species that burial can enforce dormancy by apparently modifying the gaseous microenvironment of the seed and also by inducing a light requirement for germination which did not exist at the time of dispersal. The arable weed group is recently derived, composed of species which have evolved largely within successional or highly seasonal habitats. However, members of this group do possess a characteristic response to cultivation and exist in what is now a dominant habitat.

TOLERANCE STRATEGY—MATERNAL INVESTMENT SYNDROME

There appears to be a group of species that germinate under conditions avoided by many other plants. Most notable are members of closed forest communities whose seedlings face intense competition from the parent plants. In these environments there may also be a well-developed group of seed and seedling predators. Under these conditions, plants with long lifespans may be able to dispense with complex environmental cuing mechanisms for their seeds and instead produce seeds whose size or number allow them to tolerate whatever mildly adverse conditions they may encounter in the susceptible period after germination.

We have termed this the maternal investment syndrome because many of the species in this group produce relatively large seeds, and their absolute seed output, over an extended lifespan, can be huge.

Large seeds provide the seedling with sufficient energy reserves to permit establishment under conditions where the cotyledons can provide only limited nourishment. In some tropical species subjected to significant mortality from seed predators, there may be additional investment in heavy seed coat or deterrent chemicals (Janzen, 1971). Other traits of this group may be the ability to germinate under relatively low light intensities and either no dormancy or dormancy sufficient only to pass through an immediate climatic stress.

Not all climax forest species exhibit this syndrome. As mentioned, *Betula alleghaniensis* maintains its importance in the northern hardwood forest association through utilization of the spatial disperser syndrome (Forcier, 1975). The other two important trees of this association, *Acer saccharum* and *Fagus grandifolia*, germinate in nondisturbance areas under a closed canopy (Forcier, 1975). Both species possess large seeds with no light requirement and low natural longevity (Rudolf and Leak, 1974; Olson and Gabriel, 1974; Trimble, 1975). Germination of *Dacryodes excelsa* and *Palicourea riparia*, common canopy species of the rain forest of Puerto Rico, is immediate after dispersal, and germination is optimal under shaded conditions (Quarterman, 1970; Bell, 1970). These species have characteristics of the maternal investment syndrome.

CONCLUSIONS

In presenting this classification of germination behaviors we have attempted to identify the classes of mortality risk that dominate particular habitat types and to outline the suites of dormancy and germination adaptations that have evolved under the selective pressure of these mortality regimes. In doing so, we have assumed that those species which most consistently face any particular mortality regime will tend to demonstrate most completely the germination features "typical" of the adaptive syndrome they follow. Clearly, some species face a variety of mortality risks at different times and places and may as a result tend toward germination behaviors intermediate between our described syndromes. Other species may achieve the germination pattern characteristic of a particular syndrome without showing the complete set of adaptations we have described or through utilizing physiologically dissimilar means to the same ecological goal. We do not feel that these intermediate situations detract from the usefulness of this sort of classification but rather that they are logical correlates of it and help to reinforce the hypothesis that environmentally regulated germination is a means of maximizing the probability of successful maturation of seedlings. Our intention in defining a small number of germination syndromes is to identify those aspects of the physical and biotic environment which have most dramatically influenced evolution of germination characteristics.

Some of our syndromes seem to be only weakly supported by evidence from studies of the germination of particular species. Although the literature of seed germination (still scattered despite a number of excellent reviews) is a vast one, only a small fraction of these papers and books deal with the relationship between specific environmental factors and the natural germination patterns of plant species. Indeed, we are profoundly ignorant of the actual patterns of seed germination in space and time for seeds as components of field populations. Consequently, when environmental factors are chosen to be varied in laboratory germination experiments, the choices are based as often on convenience and precedent as on the probable significance of a factor in controlling the actual germination of species in nature. Experiments must be carefully designed in order to unravel the environmental control of seed germination, and as yet few investigators have properly appreciated this difficulty. The pioneering work of Went and colleagues with desert annuals and the more recent studies of the Baskins on winter annuals in Tennessee cedar glades serve as models of the kind of laboratory and field experimental approaches needed to give a better understanding of seed germination as an adaptive process in an ecological context. In laboratory studies of germination, more emphasis needs to be given to the variability of germination requirements within a species. Several reviewers (Harper, 1957; Wareing, 1966; Koller, 1972; Roberts, 1972) have pointed to the obvious significance of nonsimultaneous and polymorphic germination patterns. Yet we have found no systematic studies of variable germination behavior, and most results emphasize average behavior for the population. An additional deficiency we have found particularly troublesome in producing this review of germination syndromes is the lack of good studies of germination of woody perennials in general and tropical species in particular. Most of the existing literature emphasizes temperate zone annuals and herbaceous perennials. Beyond this, we feel that viewing seed and seedling biology in a broad, ecological context will lead to new insights and a better understanding of the evolution of plant populations.

Acknowledgments

   We wish to thank a number of people who took time to comment on the manuscript: Peter Marks, Jack Putz, Carol and Jerry Baskin, Chuck Mohler, and Otto Solbrig. Their reactions helped to sharpen our own perceptions in this area.

9

PATTERNS OF JUVENILE MORTALITY AND RECRUITMENT IN PLANTS

Robert Edward Cook

INTEREST IN the patterns of mortality and recruitment in plants comes from both ecological and evolutionary considerations. Ecologists wishing to explain the correlations between habitat and the abundance and distribution of species have stressed the critical role of seed and seedling environment in determining the subsequent dynamics of adult populations (Harper and White, 1974; Sagar and Harper, 1961). Examinations of the morphology and physiology of seedlings (Stebbins, 1974; Harper et al., 1970; Angevine and Chabot, article 8) indicate a very close relation between the biology of juvenile plants and their success in establishment in different habitats and imply that selective mortality has been important in shaping these adaptations. Biologists interested in the genetic processes of evolutionary change have also focused on the importance of juvenile mortality as a stage of natural selection (Barber, 1965). Although there has been a tendency to examine the operation of selection in extreme habitats (Bradshaw, 1971), it seems likely that the conception of very strong and stringent selection derived from these studies applies equally well to plants growing in physically stable but biologically more rigorous habitats (Bradshaw, 1972). The rapid differentiation of local plant populations implies very high selective coefficients, and much of this selection is seen to take place in the time between seed dispersal and adult reproduction. Although the ecological aspects of this phase in the biology of plants is the main focus of my review, I shall try to relate these concepts to some of the more recent work on the genetic differentiation of local populations.

Concern with the genetic and ecological aspects of juvenile mortality also derives from two recent theoretical advances. There has been considerable work on the evolution of different life-history biologies (Schaffer and Gadgil, 1975; Stearns, 1976), of which the mathematical formulation of Bell (1976) is most relevant to the question of early mortality. Following the work of Cole (1954), Lewontin (1965), Gadgil and Bossert (1970), and Charnov and Schaffer (1973), Bell constructs several very generalized expressions that indicate the selective importance of two parameters—the age at maturity and, especially, the rate of survival during the first year of life. An increase in juvenile mortality or a delay in the rate of development will favor the evolution of iteroparity, although, as pointed out by Stearns, any such change is integrally complexed with all of the other parameters in the biology of a species. Thus, the patterns of reproduction among adult plants are strongly affected by the forces determining the early survival of individuals.

A second important consideration has been concern with the evolution of sex and degrees of recombination (G. C. Williams, 1975; M. Smith, 1977). Williams argues that the adaptive significance of sexual reproduction must be explained in terms of the fitness of individual females who, through the loss of genes in meiosis, suffer considerable fitness disadvantage relative to females who reproduce asexually. Through the construction of a number of models, Williams suggests that the major compensating advantage of sexual reproduction is the generation of a diversity of genotypes in the progeny of a female facing highly unpredictable future environments. This notion, the uncertainty facing females, must be evaluated relative to the biology of the individuals concerned—clearly, what to a beech tree appears very predictable could prove disastrous to a buttercup. The biological world may be more uncertain for many organisms than biologists generally believe. This uncertainty could be due to the unpredictable presence of physical stress or to the effects of intense biological competition, particularly among siblings. Maynard Smith feels that the degree of uncertainty required to

favor the evolution of increased recombination does not really exist
in nature, and he sees purely genetic reasons for the evolution of
recombination. Two corollaries of Williams's hypothesis are that
sexuality requires high fecundity (greater than 10^6 per female) and
that the intensity of selection is positively related to fecundity. He
argues that much of mortality involves the elimination of low-fitness
genotypes, even in populations where little cumulative genetic
change takes place. Thus, "high fecundity, in providing larger numbers of genetically different individuals, allows for more genetically
selective deaths" (G. C. Williams, 1975, p. 67). Williams reviews
what little evidence exists on viability variation among genotypes; it
is clear that further study of fitness variation in natural populations
is greatly needed, and the differential survivorship of juveniles may
be the most important component of such variation (Gottlieb,
1977a).

FIELD STUDIES

Annuals and Perennials in Extreme Habitats
    As might have been expected, many investigators of mortality in
plants have chosen species growing in extreme habitats whose life
cycle is short enough to accommodate the patience of the field
worker. In a study of winter annuals of the Mojave Desert, Beatley
(1967) found that, despite earlier claims to the contrary, most desert
annuals germinating in the fall do not survive to maturity, with
mortality taking from 37 to 90 percent of the individuals. Even
higher frequencies of death before maturity were found by Steenberg
and Lowe (1969), in an excellent study of sources of mortality for
perennial saguaro cactus. Concerned with dwindling numbers of
these plants in parts of the Southwest, these investigators observed
the death of nearly all naturally germinated seedlings, primarily
from the effects of erosion, drought, frost, and grazing rodents and
insects. The study found clear interactions between different habitats (rocks, flats, hills), season of germination, and the various
sources of mortality. In a related study (Turner et al., 1969) involving
transplanted seedlings, larger plants were better able to withstand

potential factors of mortality, although again nearly all seedlings eventually succumbed. These studies, as well as a number of others, have stressed the importance of microhabitat features, particularly in relation to shade and the loss of soil moisture. Thus, a positive association is frequently found between the survival of seedlings and the presence of established vegetation (Turner et al., 1966). It appears that two of the major sources of mortality for seedlings are lack of soil moisture (Sharitz and McCormick, 1973; Mack, 1976) and presence of herbivores (Christensen and Muller, 1975). Nearly every study of seedling recruitment cites the importance for young plants of drought, whether due to lack of precipitation, excessive exposure to drying conditions, or competition with established vegetation. The greatest susceptibility occurs shortly after germination when the seedling switches from internal reserves to external sources of support (Harper, 1965). Although these studies generally correlate mortality with decreasing soil moisture, the difficulty in measuring the microenvironment perceived by an individual seedling has prevented a precise knowledge of why one plant survives while another perishes.

One critical factor is the precise temporal relations between germination and early growth of individuals and the sequence of seasonal factors that determine the availability of resources. The papers by Baskin and Baskin (1972d, 1973a, 1974b, 1976b) on the germination and survival of winter annuals of cedar glades and disturbed limestone soils demonstrated the balance between early germination in late summer (low probability of survival, high individual fecundity) and later, autumn germination (high survival, low fecundity). By marking individual seedlings that germinated at different times and censusing mortality through the fall and winter, they found that seedlings of most species appearing before September 1 were decimated by summer drought. However, those individuals of Leavenworthia that did survive (7.6 percent) had a very high fecundity (45 seeds/plant) relative to early October germinants (78.4 percent survival but only 5.8 seeds/plant). Individuals appearing after October 15 suffered relatively high mortality (27.2 percent survival) and had somewhat higher fecundity (8.2 seeds/plant). A

similar advantage to early germination, with its attendant risks of
high mortality, was found by Naylor (1972) for *Alopecurus*, a weedy
annual grass. Unfortunately, these studies do not give any information on the distribution of fecundity among individual survivors.

The importance of disturbance in the recruitment of new genotypes into populations has received considerable attention. In most
studies, the concept of a disturbance remains undefined, although
disturbance is usually coincident with a major catastrophe (fire,
erosion, hurricane). Clearly, however, the relation between disturbance and recruitment depends upon the biology of the species
under consideration. Two components frequently observed are the
removal of the inhibiting effect of existing vegetation (competition
for light and water) and the disruption of the soil, the latter because
many individuals are recruited from light-requiring seeds lying
dormant in the soil (Cook, 1978). Two recent studies (Platt, 1976;
King, 1977c) have examined the recruitment of individuals on
mounds of soil created by biotic activity (ants, badgers) in the
middle of dense, perennial vegetation. Both studies found that seedlings of many species colonizing such disturbances were unable to
establish themselves in the surrounding vegetation owing to competition for resources, particularly soil moisture. King reports a number of detailed observations on recruitment in ten species of annuals
and perennials colonizing ant hills. Many deaths occurring from
rabbit scraping, soil heaping by ants, erosion, and desiccation were
independent of density, whereas some density-dependent mortality
resulted from self-thinning and fungal infection. King suggested
that selection had favored high fecundity, efficient dispersal, and
rapid flowering among annuals, and forms of growth resistant to
burial among perennials (vigorous axillary branching and rhizome
growth). Similar studies by Miles (1973) on the fate of marked
seedlings of 19 perennial heath species showed that experimentally
sown seed was far more likely to lead to the establishment of plants
on disturbed ground than in the adjacent mature vegetation; only
Calluna became established equally well under itself. In naturally
sown seed of these species, Miles (1972) found that summer establishment was ten times more likely in bare areas caused by natural

disturbance than in the surrounding vegetation. Interestingly, however, winter mortality was relatively less among the seedlings in closed vegetation.

Long-Lived Woody Forest Perennials

Implicit in the notion of heavy selection at the seedling stage is the idea that the genetic composition, abundance,and distribution of mature individuals may reflect a strong historical dependence upon the factors that determined seedling survival. Sagar and Harper (1961) suggest that "correlations between the distribution of mature plants and existing environmental factors may be spurious, since the crucial factors determining the present position of individuals were those operating in seedling stages." Because mortality and competition in plants are usually size dependent, established individuals may prevent the establishment of competitors for long periods of time. It might be suspected, therefore, that evidence concerning such historical inertia would be found among the largest members of the vegetation.

There have been two sources of data concerning mortality and recruitment in trees: the historical reconstruction of past patterns from the ages of living and dead stems present, and silvicultural studies on the patterns of early seedling recruitment in economically important forest species. Two examples of the former (Yarrington and Yarrington, 1975; Henry and Swan, 1974) describe the critical role of major disturbance in determining forest composition. The Yarringtons used the age of dead trees to study the demography of a jack pine *(Pinus banksiana)* stand established after a fire in 1915. Rates of mortality coincided with periods of drought stress, though the increasingly regular distribution of surviving individuals suggested that competition intensified the effects of low precipitation. Henry and Swan reconstructed the history of a 0.1-acre plot of virgin forest in southwestern New Hampshire to 1665 and found little evidence that autogenic succession contributed to compositional changes. Fire, wind storms, and hurricanes were far more significant, and there were dramatic differences between species in the frequency and location of recruitment. Black birch *(Betula lenta)* was most often found colonizing tree throwmounds created by major destructive storms, whereas a beech tree *(Fagus grandifolia)* was recruited, on the average, once every 10 years after 1815.

Both biotic and physical factors strongly affect the probability of establishment of relatively large-seeded woody species. The classical studies of Watt (1919, 1923), Wood (1938), and their modern tropical counterparts (Connell, 1970; Janzen, 1970) indicate the critical role of herbivores (rabbits, rodents, insects) in the destruction of seeds and seedlings. It would appear that successful establishment of seedlings can only occur in years of extremely high seed production (Janzen, 1976). Further successful recruitment into the reproductive population may require the kind of disturbance that creates gaps in the canopy. Many species have displayed remarkable "resistance to inanition" (Chippindale, 1948) while waiting for such unpredictable opportunities. Merz and Boyce (1956), for instance, found, in the shaded forest floor, seedling oak shoots that were aged to 6 years by ring counts, growing from roots that were as much as 31 years older.

Silviculturalists interested in forest regeneration have also stressed the importance of the precise conditions of the seedbed that promote seedling establishment. A good, in addition to being classical, example is the study of Smith (1951) on the regeneration of white pine. Through experimental sowings and observation of natural germination and mortality, Smith demonstrated that a bed of litter, pine slash, and *Polytrichum* moss substantially reduced the risk of death from heat injury or drought. The deep shade of denser vegetation, however, increased the likelihood of fungal infection. Some losses were due to rodents and pine locusts. Thus, the microclimatic and topographic conditions clearly affect the probability of survival.

Demographic Studies

A number of recent demographic studies (reviewed by Harper and White, 1974) have been concerned with quantifying the mortality of seedling and juvenile plants. If censuses are made frequently, in order to quantify seedling mortality with accuracy, concave survivorship curves for cohorts of individuals are usually the rule, indicating a period of high risk early in the life of the plant. In the annuals *Minuartia* and *Sedum*, studied by Sharitz and McCormick (1973), mean life expectancies were 2.6 months and 4.4 months, respectively, with most of the mortality occurring before the mean. Likewise, Hett and Loucks (1971) found that a power-function model, implying increasing probability of survival with age, fit data from sugar maple *(Acer saccharum)* better than a negative exponential model and that mortality was probably related to the stresses imposed by the canopy trees on established seedlings. Hett and Loucks noted that increased survivorship with age is biologically sound since each year seedlings gain in height, develop deeper roots, and become less dependent upon surface moisture input. Harper and White (1974) also note that the length of the juvenile period appears to depend upon the size of the individual, though this may be quite independent of age.

Two detailed demographic studies (Hawthorn and Cavers, 1976; Sarukhán and Harper, 1973) found that seedlings suffered a very high risk of mortality relative to mature plants, and this decreased with age, reaching a point where the death risk was relatively constant each year, though showing a seasonal oscillation. Hawthorn and Cavers studied sown and naturally dispersed seedlings of two *Plantago* species and found that survivorship differed between species and habitats and displayed a seasonal variation. Although moisture conditions were not thought to influence mortality, in the wetter summer the death rates were higher, coincident with the greater growth of survivors, which suggests competitive interaction for some resource. Sarukhán found that survivorship increased with age for all three species of *Ranunculus* established from seed and that the major mortality period occurred during the seasons of active growth, which again suggests competition, though the effects of cattle grazing at this time might have been significant. Neither of these studies provides much real information on precise sources of mortality or the relationship of mortality to size of the individuals.

One other generalization to emerge from studies of well-established herbaceous vegetation is the relative infrequency of recruitment from seed among species capable of vegetative reproduction (Tamm, 1972; Harberd, 1961; Putwain et al., 1968; Thomas and Dale, 1975). Vegetative propagules have considerably greater probabilities of recruitment; thus, in *Ranunculus repens* (Sarukhán and Harper, 1973), the average life expectancy of ramets was 1.2 to 2.1 months, whereas for seeds it was 0.2 to 0.6 months. In a study of *Hieraceum floribundum*, Thomas and Dale found that nine times as many individuals were established from stolon shoots as from sexually produced seed. Harberd concluded that fields of *Festuca* and other grasses may be dominated by a few clones of very great age that are the competitive survivors among the many genotypes that became established coincident with some major disturbance in the past. In some populations, therefore, recruitment of new genotypes may be exceedingly rare, occurring when rather exceptional environmental conditions coincide with an abundant production of seeds.

EXPERIMENTAL STUDIES AND THE "SAFE-SITE" CONCEPT

A second approach that has been taken by John Harper and his students is the manipulation of natural populations in order to quantify the importance of different parameters in the regulation of the number of plants in populations. In addition, specific experiments have been designed to elucidate those biological attributes that allow the coexistence of closely related plant species in the same habitat, with attention particularly focused on seedling germination and early growth. Out of this work has come the concept that the abundance of each species may be largely determined by the number of species-specific "safe-sites" available in a habitat (Harper et al., 1961). By a "safe-site" Harper means a set of conditions in the immediate environment of the seed (i.e., soil microtopography, moisture status, light conditions, temperature) that make germination and establishment of the seedling highly probable. The existence and nature of such safe-sites have been examined in two ways.

The classic test for the existence of safe-sites as a regulator of plant populations is the experimental sowing of seeds in natural communities at different densities to determine whether the proportion of seeds failing to establish themselves increases with seeding density (Sagar and Harper, 1960; Cavers and Harper, 1967; Putwain et al., 1968; Putwain and Harper, 1970; Hawthorn and Cavers, 1976). Often in these experiments, phytometers, control-grown seedlings with one or two pairs of true leaves, are transplanted into the natural community to be compared with the naturally germinated seedlings.

The findings of such studies have been generally consistent. When seeds of a particular species are sown into mature vegetation with adults already present, very few seedlings become established, so that increasing the density of sowing increases the rate of failure; establishment usually occurs only with some form of disturbance to the soil. This implies that the establishment of the living members of the population occurred under conditions very different from those presently found. In addition, the conditions required for supporting seedling growth are considerably more restricted than those required by adults. Where a species is not found in a habitat, this is often due to failure of the seedling to grow after germination. That this is the ontogenetic stage of singular importance is shown by the relative success of transplanted phytometers in such habitats; once a seedling has formed several leaves, it shows considerable resistance to inanition and may successfully produce seed. Thus, concludes one paper, "favorable conditions for seedling establishment may occur for very short periods of time, at infrequent intervals or in locally disturbed microsites" (Putwain and Harper, 1970).

A second series of experiments was conducted by Harper and associates (Harper et al., 1965; Harper and Benton, 1966; Sheldon, 1974) to determine the species specificity of safe-sites. This work grew out of early agricultural investigations examining the microenvironmental conditions favoring the germination and establishment of maize (see Ludwig and Harper, 1958, for earlier references to this series). In these earlier experiments, Harper clearly demonstrated the importance of such soil characteristics as color and aspect on the fluctuations of soil temperature around planted maize seeds germinating in early spring. These temperatures in turn interacted with soil humidity to inhibit or promote the attack of soil pathogens, which ultimately determined the success of germination and establishment of maize seedlings.

The experimental approach was extended to the comparative response of species to variation in soil characteristics, particularly microtopography. In the first of this series, Harper et al. (1965)

demonstrated that sown seeds of three species of *Plantago* responded quite differently to the microenvironments created by glass plates, small boxes, and holes pressed into the soil surface. *Plantago media* exhibited increased germination beneath sheets of glass, whereas *P. major* displayed germination independent of the presence of many objects. The authors concluded not only that the frequency of such microtopographic disturbances could determine the abundance of a species but that different species might well be independently controlled. Further work under precisely controlled laboratory conditions (Harper and Benton, 1966) demonstrated the importance of the water relations of the germinating seed, particularly the ratio of surface area of water loss to the atmosphere and surface area in contact with the substrate. Finally, it was argued (Sheldon, 1974; Harper et al., 1970) that apparently inconsequential aspects of seedling morphology may have great adaptive significance in relation to safe-sites for successful establishment. Thus Sheldon demonstrated that the precise angle of contact between an achene-pappus unit and the surface of the soil could influence germination patterns. Implicit throughout is the notion that much mortality, and therefore selection, in the seedling stage shapes the response of individual species to different safe-sites and allows the cohabitation of species in floristically rich communities.

GENERAL CONCLUSIONS

This review of previous work leads to a number of generalizations:

1. Rates of mortality among juvenile plants are very high. In some trivial sense, if populations are not continuing to increase or if they do not decrease to extinction, juvenile mortality must approach the reciprocal of average female fecundity. More important are the observations that mortality decreases with age and that the seedling shortly after germination is the most susceptible phase in the ontogeny of the individual.

2. The probability of death decreases as the individual increases in size. Thus, larger individuals not only withstand

stress for longer periods but also display greater competitive abilities. In addition, the length of the juvenile period appears to depend upon size more than age.

3. Two sources of mortality appear particularly important—grazing herbivores and pathogens, and drought stress. Many aspects of the biology of seeds and seedlings can be interpreted in terms of adaptations to appropriate soil–water relations.

4. The abundance of individuals in a population may be determined by the number of safe-sites available for successful seedling germination and establishment. Such safe-sites may be temporally dynamic and involve soil characteristics such as surface microtopograhy, temperature, and light. Individual species may be adapted to different safe-sites in a particular habitat; therefore, their populations are independently regulated, and cohabitation is possible. Thus, fecundity and dispersal do not appear to limit the distribution of species between habitats.

5. The recruitment of new genotypes into populations is frequently coincident with some form of disturbance. This may involve destruction of the canopy, soil disruption, or the creation of microclimatic conditions appropriate to new safe-sites. Consequently, species may differ dramatically in their response to a source of disturbance, and the floristic composition of vegetation may be dependent upon the frequency and quality of disturbance.

6. In established vegetation, recruitment from sexual seed may be very infrequent, with most new shoots arising asexually. Thus, present environments may give no indication of the conditions which determined the recruitment of existing genotypes.

SOME PATTERNS OF MORTALITY IN *VIOLA*

Many of the above generalizations can be seen in work which I have begun in conjunction with Otto T. Solbrig on three species of the genus *Viola* growing in white pine–black birch–red oak forests north of Concord, Massachusetts. Seedlings of each species germinate primarily in May, and entire cohorts can be followed throughout their life to determine growth, fecundity, and sources of mortality. During the spring of 1976, a number of populations were observed by marking all new seedlings with numbered plastic flags and censusing individuals at approximately 10-day intervals from May through September. I shall recount three examples of some preliminary results emerging from this work.

Example 1. *Viola sororia* is a blue-flowered, stemless violet found growing in dry, second-growth woodlands. When mature, it has a thick underground rhizome but seldom reproduces asexually; it does produce cleistogamous and chasmogamous seed. Three hundred and eighty individuals were marked in May and censused through the growing season, with one additional census at the end of April, 1977. The survivorship curve of the cohort is shown in Figure 9.1, as is rainfall through the season. Also shown are the proportions of the populations that died because of drought (*Dr*, leaves wilted, dried, and disintegrated) or after having all their leaves grazed off (*Gr*). None of these individuals grew large enough in the first year to produce any seed.

The first observation is that mortality is high, with only 21 percent of the cohort surviving into the second year. The rate of mortality does appear to decline with time, but this may reflect the seasonality of the sources of mortality. Considerable mortality occurs during winter, and individuals failing to make much growth or grazed in the previous season do not appear in spring. The other two major sources of mortality are grazing and drought, the intensity of the latter coinciding very well with the period of low rainfall in June. The cause of winter mortality is unclear, but it is generally confined, as expected, to smaller individuals.

Example 2. *Viola blanda* is a white-flowered, stemless violet that forms asexual ramets from axillary stolons, as well as cleistogamous and chasmogamous seed. It is found growing in habitats ranging from dry woodland patches to wet, shady hollows and streambeds. The population in this example was growing in accumulated litter, overlying granite rock, at the top of a woodland slope. On May 15, I marked a cohort of 93 seedlings and, upon returning to census 15 days later, discovered a vast amount of additional germi-

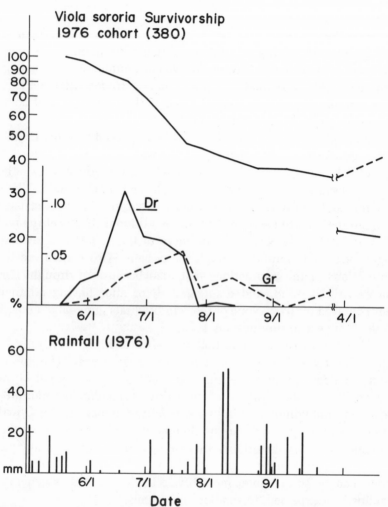

Figure 9.1. Survivorship of a cohort of *Viola sororia* seedlings during 1976. *Dr* is the proportion of individuals that had all leaves dry out and that died. *Gr* is the proportion of individuals that had more than 70 percent of their leaf area grazed and that died.

nation. These individuals were marked (318 plants), and the observations presented will distinguish between early and late germinants. At each census, the size of an individual was observed by
measuring the length of the midrib of the lamina of each leaf, and the
size distributions represent the sum of these measurements for all of
the leaves produced by that plant. The survivorship of this cohort is
shown in Figure 9.2, and its growth in size at different times can be
seen in Figures 9.3, 9.4, and 9.5.

   Mortality was not as severe in V. *blanda* as in V. *sororia*, but it
is clear that those individuals germinating later suffered greater
rates of death. The differences between the two groups are even
more striking in the size distributions. On July 3, early germinants
were considerably larger than late germinants, and this advantage
continued throughout the growing season. It can be seen that mortality was confined to the very smallest individuals. By the second
year, most of the small individuals in the population were late
germinants and were suffering greater mortality. One individual of
the early germinants had grown large enough in 1976 to produce

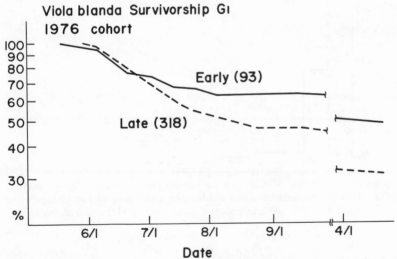

Figure 9.2. Survivorship of a May 15 cohort (early) and a May 30 cohort (late) of
Viola blanda seedlings during 1976.

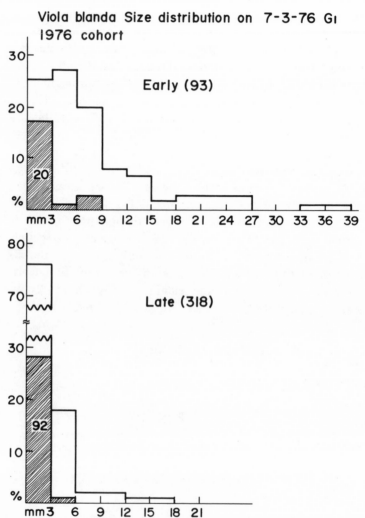

Figure 9.3. Size distribution of the early and late cohorts of *Viola blanda* seedlings on July 3, 1976. Dead individuals are shaded. Size was assessed by summing the mid-rib lengths of all leaves produced on each individual up to the date of census.

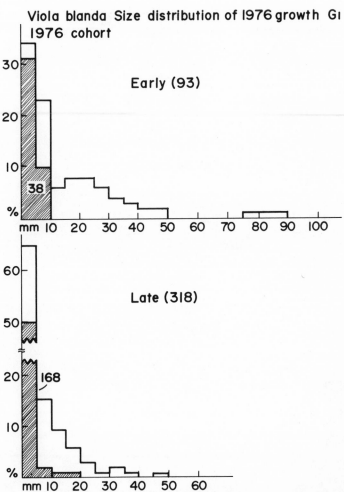

Figure 9.4. Size distribution of the early and late cohorts of *Viola blanda* seedlings for total growth during 1976. Dead individuals are shaded. Size assessed as in Fig. 9.3.

two stoloniferous ramets. Both of these daughter shoots and the adult survived the winter and occupied the very largest size class (Figure 9.5). With three physiologically independent shoots, the risk of death to this genotype must be considerably lower than that to any other genotype.

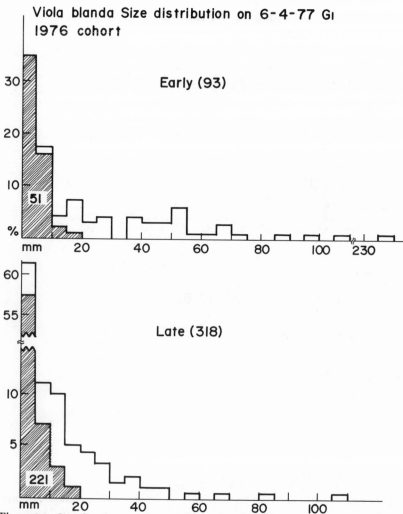

Figure 9.5. Size distribution of the early and late cohorts of *Viola blanda* seedlings for growth up to June 4, 1977. Dead individuals are shaded. Size assessed as in Fig. 9.3.

Example 3. *Viola fimbriatula,* a blue, stemless violet closely related to *V. sororia,* is usually found growing in old fields, pathways, and locations showing recent signs of disturbance. It reproduces primarily by cleistogamous and chasmogamous seed, though occasionally the rhizome may split. The population which I began to follow in June 1975 was found growing in the pit of raw soil created by the recent blowdown of a 54-year-old white pine in the same location as the previous two species. Cohorts of this species germinated from seeds lying dormant in the soil in the spring of 1975 and 1976, and their survivorship, along with rainfall data, is shown in Figure 9.6.

The mortality rate of the 1975 cohort was fairly constant throughout, declining somewhat during the second winter. Mortality was primarily due to the effects of grazing, which defoliated many plants in 1975. The 1976 cohort suffered considerably greater mortality, mostly from drought stress, and it is clear that the older and larger individuals of 1975 were better able to withstand the lack of precipitation in June of 1976. Just as occurred in the *Viola blanda* population, a single individual from the 1975 cohort grew large enough in its first year of life to produce two cleistogamous seed capsules. This individual grew larger in its second year and produced four additional capsules. In 1977 it produced the first chasmogamous flower of any violet under study and has contributed four times as many seeds to future generations as any other genotype in the population.

Several points can be made from these observations on *Viola.* First, the probability of death depends upon the size of the individual and declines with growth. A corollary is that asexual reproduction further reduces this risk for the genotype by creating physiologically independent units capable of suffering independent mortality. Second, early differences in timing during the ontogeny of individuals may be decisive. Small differences in germination time lead to compounding differences in growth and survival. Finally, from any given cohort, one or two unusual genotypes may begin to dominate and contribute disproportionately to the fitness of the population. These may well turn out to be the Sisyphean individuals of which Williams wrote.

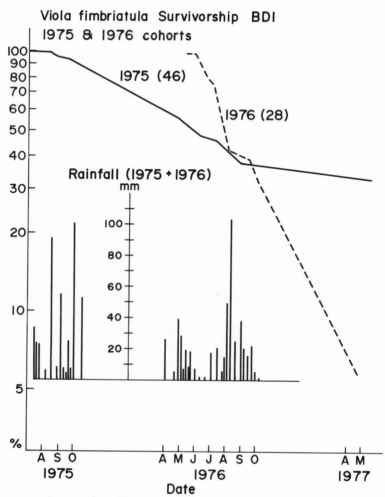

Figure 9.6. Survivorship of the 1975 and the 1976 cohorts of *Viola fimbriatula* seedlings germinating in a tree blowdown pit.

DISCUSSION

Perhaps the single most important characteristic of juvenile plants that influences their fitness is their size at the time an environmental stress is imposed upon them. All of my observations on

Viola, as well as much of the research reported in the literature, indicate that the effects of sources of mortality are strongly size dependent, with smaller individuals suffering greater hazards. For many plants, the total length of the juvenile period, as well as their fecundity as adults, will also be a function of size. It might be concluded, therefore, that selection should minimize the time spent in the highly susceptible seedling stage by maximizing the potential growth rate of juvenile plants. Yet, from the recent work by Grime and Hunt (1975), it is clear that there is great variation in the intrinsic relative growth rate of seedlings of different species. Woody perennials, for instance, have very low potential growth rates, whereas weedy herbs may grow ten times as fast under identically optimal conditions. Even the growth rate of species in the same genus (Poa, Senecio) may differ by a factor of two. Why would it not always benefit an individual to have the potential to grow as fast as possible in order to escape the severest effects of many environmental stresses and enhance fecundity?

There may be intrinsic physiological and developmental factors that prevent individuals from displaying both adaptations for particular kinds of environments and high potential growth rates. For instance, species normally found in stressful habitats have low intrinsic relative growth rates, even under optimal conditions when light, water, and nutrients are not limiting. Grime and Hunt suggest that there is a critical link between high growth rates and a consequent high and inflexible respiration rate that would prove detrimental in habitats where low resources require long-term persistence and resistance to inanition. A second connection may exist between the growth process itself (cell division in apical meristems) and the morphological differentiation of the organs so produced. Selection for anatomical or physiological adaptations that increase resistance to sources of mortality, particularly in seedlings, may necessarily result in a decrease in the potential for a high rate of growth. In the face of these limitations to selection, species may adapt to sources of juvenile mortality either by minimizing the time at risk through a high intrinsic rate of growth or by developing functional adaptations that lower the probability of death over a prolonged juvenile period.

Yet the magnitude of juvenile mortality for every case studied remains great. It seems clear, for instance, that predators, pathogens, and the lack of available soil moisture are continually removing a large number of the seedlings from the populations of most species. Although minimizing the time at risk may be a successful adaptation to predator and pathogen attack, a high intrinsic growth rate is clearly dependent upon the availability of soil water and would be a poor adaptation to drought stress. It may be asked why seedlings of species suffering heavy drought-related early mortality have not evolved xerophytic traits that would lower the death rate.

One possible answer to this question is that there may be no genetic variation in the population for drought-related adaptations, and much of early mortality is selectively random with respect to such genes. Although this is difficult to disprove without specific selection experiments, the whole thrust of recent work on the genetics of populations has revealed a vast amount of variability in most populations of sexual reproducers (Lewontin, 1974; papers in this volume). The second possibility is that much variation for drought resistance exists in the population but that it is maintained, despite a great amount of juvenile mortality, by some form of balancing selection. Of particular significance may be a form of selection which Ford (1975) has referred to as endocyclic selection on traits whose selective value reverses during the course of the life cycle of the individual (see also Christiansen and Frydenberg, 1973). This concept suggests that selection affects different genotypes at different times during ontogeny. If the expression of two characters is negatively correlated during development, then selection later in the life cycle may negate the effects of earlier selection. Thus, drought resistance may be strongly linked to traits which diminish competitive ability or fecundity, and early selective mortality which increases resistance in seedlings may be rapidly reversed by differential fecundity in adult plants. At the end of each generation, the net effects of such a series of events would appear to be random with respect to the characters in question, but, clearly, selection has been occurring. Only when the epigenetic linkage between two characters is broken can selection achieve a rapid and permanent change in gene frequencies.

In recent years, views concerning selection in natural populations of plants have been changing, largely owing to the work of A. D. Bradshaw (1971, 1972) and his students. The major contribution of this work has been the demonstration of very rapid microevolution over very short distances in populations found growing at the boundary of two habitat patches. Such a result could only occur with very high coefficients of selection. Particular interest has been shown in the rapid evolution of heavy-metal tolerance in populations occupying the soil wastes from mines (Antonovics et al., 1971). Here individuals that are not tolerant suffer very high mortality on contaminated soil, and tolerant individuals appear to be unable to compete successfully and establish plants in nontolerant populations. Thus, there is strong, divergent selection at the boundary, leading to the differentiation of the populations in such characters as growth form, response to soil nutrients, and flowering time. Comparison of the several tolerant grass species and the seeds produced by plants either on or off mine soil clearly indicates that strong selection is counteracting the effects of gene flow from nontolerant pasture populations in the vicinity. It also indicates that much of the selection is in the form of differential viability of seedlings.

Examination of seedlings germinated from seed of normal pasture populations of tolerant species has demonstrated that variation for metal tolerance is maintained in such populations (Gartside and McNeilly, 1974) despite the strong selective pressures against tolerant individuals (s, 0.68–0.99) when grown from tillers in competition (Hickey and McNeilly, 1975). Gene flow can explain some of the variation occurring in populations in the vicinity of mines, but even seed from populations growing on noncontaminated soil contains a few tolerant individuals (Wu et al., 1975). In this latter study, Wu et al. found that these rare tolerant individuals were able to establish a continuous cover of *Agrostis stolonifera* through clonal growth, which indicates very strong selection at the seedling stage.

That such rapid population differentiation over very short distances has not been confined to species growing in extreme habitats can be seen in the work on *Anthoxanthum odoratum* in the Park Grass Experiment (Snaydon and Davies, 1976). Genetically distinct populations of the grass were found to have evolved in less than 60 years across a 10-cm boundary separating two plots receiving contrasting fertilizer treatments. Significant heritable differences in height, yield, flowering time, and other morphological characters had developed despite considerable gene flow across the boundary. Since this species maintains its population through frequent recruitment from seed, selection pressures on seedling stages must have been very intense—selection coefficients of 0.15 to 0.65 were found for established tillers (Davies and Snaydon, 1976), and those for seedlings would be even higher. The major selective force is probably the hazards accompanying seedling establishment in the face of competition from mature plants, and slight differences in growth rate could lead to strongly cumulative effects on fitness (Hickey and McNeilly, 1975). Despite the inherent difficulties, a demographic study of these seedlings would be extremely valuable.

The major problem with identifying the patterns of selection in the life cycle of plants is the disjunction in the kinds of studies that have been done. On the one hand, the careful demographic studies of mortality and recruitment have ignored genotypic differences within a species and concentrated only on comparative differences between species; the action of selection is missed because the scale of resolution is not refined enough. On the other hand, the detailed genetic studies of microevolution in contrasting habitats have not been concerned with the differential fecundity and mortality of genotypes but have concentrated upon comparative differences of already established populations. Similarly, although the concept of the species-specific safe-site implies the action of past selection molding differences in the biology of species, it does not shed much light on the mechanism of selection that leads to population differentiation. Even though it seems clear that the most important parameter of the safe-site is soil moisture availability, at this point safe-sites can only be identified by the successful germination and establishment of seedlings; it is unclear what an empty safe-site would look like for any species. For future work on patterns of recruitment in plants, it would seem better to concentrate on a rigorously quantifiable parameter affecting seedling mortality, such as growth rate in relation to soil moisture availability, and to focus such a study on the demographic behavior of different genotypes within natural populations of a species undergoing divergent selection. Without the conjunction of such approaches, we will continue to see the action of selection through its products rather than its process.

10 THE BASIC BIOLOGY OF PLANT NUMBERS

James C. Hickman

IN THIS ARTICLE I wish to call attention to several topics that need more or different kinds of work than they have received in the recent past. In part, I aim to provide a perspective on the biology of plant numbers that will complement, extend, and perhaps challenge some others developed in this volume. I shall refer repeatedly to theory and shall provide some new variations on existing theoretical themes, but a major objective is to emphasize the *biology*, as opposed to the theory, of plant numbers, with the hope of helping to stimulate some redirection of both theoretical and experimental efforts. I shall occasionally present as yet unpublished data in support of particular points and make some bold assertions concerning emerging ideas that require more data. There are two themes in this article. The first is that an understanding of the basic biology of plant populations will require more comprehensive study than is now standard. Plant strategies involve compromises among a diversity of selective factors. Determining the relative importance of these factors under different conditions is one of the most important goals of plant population biology. Most population studies have been short-term, however, and have addressed few hypotheses or, in extreme cases, a single prediction or assumption of a mathematical model. The results of such narrow studies are difficult to interpret correctly and will not readily produce the basis for a comprehensive and coherent theory of plant populations. Remedy of this deficiency will require focusing attention on careful perception and quantification of the biologies of plants and on broad and continued study of exemplary systems. We need to know our plants as well as Mac-Arthur knew his warblers.

The second theme of this paper is that annual plant populations are excellent subjects for the kind of comprehensive study discussed above. Although annuals have been used extensively in evolutionary studies, recent and ongoing demographic work has emphasized perennials. This is primarily because of the simplicity of turnover patterns in annual populations, with the result that extrapolations often cannot be made to perennials, which dominate vegetation in terms of numbers of species, biomass, and importance in ecosystem processes. Nevertheless, under some circumstances, the simplicity and rapidity of the annual cycle are great experimental boons. Furthermore, annuals show widely divergent variations within their overall demographic limitations; occur frequently in pure natural populations or few-species communities at a wide range of densities; span nearly the entire range of reproductive allocations (3–77 percent net reproductive biomass; Hickman, 1977, and unpublished data); show immediately interpretable reactions to stresses; are easy to harvest completely without affecting the next generation, because underground parts are quite limited (cf. Antonovics, 1976b); and can be analyzed in large numbers. For many kinds of questions that will be raised in later sections, these advantages override the disadvantages of generally greater plasticity of size and the normal impossibility of cloning.

BASIC NUMEROLOGY OF PLANTS

Most population theory has been devised for animal populations. There is increasing recognition that these theories are not well suited to plants (see, e.g., Bradshaw, 1965, 1972; Harper, 1967; Antonovics, 1976; Schaffer and Leigh, 1976), but the problem of developing a theory suitable for highly plastic, sessile organisms that respond markedly to spatial heterogeneity is only now being recognized, and few attempts have been made at solutions [Roughgarden (1974) is an exception]. Indeed, Schaffer and Leigh (1976) suggest that the prospects for ever adequately incorporating spatial heterogeneity into mathematical population models are gloomy. The impasse appears to be similar to that reached in multilocus population genetics. It may happen, however, that complex mathematical models incorporating spatial heterogeneity can be done without. Heterogeneity in space can be analyzed in a variety of ways, and the data derived thus far (e.g., Yeaton and Cody, 1976; Mack and Harper, 1977; A. P. Smith, unpublished data) indicate that an understanding both broad and precise can accrue from studies based only on verbal or functional models (Solbrig, 1976b).

Phenotypic Plasticity

The problem of phenotypic plasticity, especially of plant size but also of the ecological tactics that make up a plant's genetically determined strategy, may prove to be more severe. On the one hand, plant populations readily differentiate genetically over both short and long geographical distances (Ehrlich and Raven, 1969; Antonovics and Bradshaw, 1970). This means that intraspecific studies of plant populations cannot assume genetic uniformity, as has been the prevailing habit.

On the other hand, plants are notoriously plastic (Bradshaw, 1965; H. G. Baker, 1965; Cook and Johnson, 1968; Hickman, 1975; Zimmerman, 1976; Roos and Quinn, 1977; Soule and Werner, unpublished data). There results a strong temptation to assume, without firm evidence, either genetic or plastic mechanisms for population-level differences, according to the experience of each investigator. Electrophoretic analysis can determine the probability that populations are genetically uniform, but, without carefully developing at least possible adaptive significances for the differing kinetics of isozymes (e.g., G. Johnson, 1975b), electrophoretic evidence cannot be used to resolve the genetic/plastic dilemma at a causal level. Habitat manipulations, transplant experiments, and progeny tests can help supply solutions, but all have inherent pitfalls and must be used and interpreted carefully.

The ubiquity and severity of plastic responses to differing environmental cues in plants play havoc with population studies in a very basic way. Although size, rather than age, is a good predictor of fate in some perennial (especially monocarpic) species (Hartshorn, 1975; Werner, 1975a; Inouye and Taylor, unpublished data), other species—particularly, but not solely, annuals—have size differences of at least 2.5 orders of magnitude at reproduction and seed production differences of at least 4.5 orders of magnitude (Harper, 1967).

Both numbers (N) and carrying capacity (K) are basic concepts in population biology, whether or not they are interpreted in strictly logistic ways. High variance in size at reproductive maturity makes both these parameters difficult to quantify such that comparisons either within or between populations are meaningful. Strong arguments have been made for a closer fusion of population ecology and population genetics (e.g., Antonovics, 1976; Roughgarden, 1977). Accordingly, appropriate plant population theory must be capable of considering population growth and genetic change simultaneously. At the same time, theory must allow for a reasonable treatment of both the similarities and differences between populations that have, for example, high versus low genet (= genetic individual) density and small versus large genet size (possibly interpreted as different numbers of ramets or vegetative offsets per genet), but equal foliage and reproduction densities.

The only approach to counting plants used widely in ecology is that of measuring N as genet number. This confounds comparisons of sparse populations of large individuals with dense populations of small individuals. Both may have equal foliage or reproduction densities, but genetic variance should differ because most recruits to the sparse population are likely to be progeny of a few of the most robust individuals.

Two other possibilities present themselves. Measurement of N (or K) as foliage density, cover, reproductive density, or the like, permits reasonable estimates of the biological effects of the population but allows no assessment of phenomena at the individual level, such as selection.

Measurement of N (or K) as ramet number has received some attention (e.g., Harper and White, 1974), but, for some kinds of plants (stoloniferous or rhizomatous herbs, self-layering or root-sprouting trees), ramets and genets are difficult to differentiate.

Doing so would allow adequate population size comparisons when ramets are functionally equivalent units but not when ramets are structurally separable portions of a genet that are of varying size, such as stems of *Solidago* or *Populus,* branches of *Polygonum,* or tillers of *Agropyron.* The most serious problem with the ramet approach is that it does not allow simultaneous consideration of population size and population genetics.

The most reasonable approach to N and K in plant populations is to adopt a hierarchical system, even though this will make theory more cumbersome. The minimal set of components of N is: (1) number of genets and (2) number of functionally equivalent ramets per genet. For some kinds of studies, it may be important to recognize more than one kind of ramet on each genet (vegetative or reproductive) or to add further, possibly multistate, categories to the hierarchy, such as number of sexual (or vegetative) propagules per ramet. Some advantages to this kind of analysis in comparative plant population biology are considered later.

This hierarchical approach requires two kinds of additions to current thinking. First, the concept of ramet must be enlarged, as has been suggested by Harper and White (1974). Even nonvegetatively reproducing plants such as some trees and annuals are susceptible to an extension of the ramet concept, in that they are composed of functional aboveground units (stems with leaves and buds). In many instances, these structural units can be considered functionally equivalent across genets of widely varying size. Where they cannot, a third hierarchical component of N must be utilized: functional size of ramet. Equivalence of ramets (or ramet size, if equivalence fails) might be estimated from biomass, leaf surface area, net photosynthetic rate, or propagule production.

Second, attention to ramets as functional units of plants will require much closer consideration of the patterns of plant growth and development than has been customary. There is a rich literature in this area that has not been exploited adequately by population biologists. A few divergent examples of attempts to model growth and development in ways that could be useful in the study of plant populations are the papers of Erickson and Michelini (1957), Kawano (1970, 1975), Ogden (1974), Kawano and Nagai (1975), and Yokoi (1976a,b).

A number of studies have found that increasing environmental severity reduces both plant stature and overall plant mortality (Raynal and Bazzaz, 1975; Hickman, 1975; earlier papers reviewed by Harper, 1967; White and Harper, 1970; Harper and White, 1974). Attempts to incorporate this effect of plasticity into models have been limited in number but appear to have broad applicability. Particularly important is the so-called "3/2 power law," which was empirically documented and theoretically justified by Yoda et al. (1963). Harper (1967) and White and Harper (1970) reviewed both data (mostly from crop plants) and theory and found remarkable evidence of generality for Yoda's relationship $w = Cp^{-3/2}$, where w is mean plant weight, C is a constant, and p is plant density. This model was derived for pure self-thinning stands, and the classic experiments that support it have followed the mean weights and densities of replicated populations sown at very high density and harvested at intervals after establishment. In this kind of experiment, density drops through time from self-thinning as some individuals capture increasing proportions of the resources. The thinning exponents calculated by White and Harper range around -1.5 (from -1.39 to -2.18), and I know of no data that contradict the law.

Two aspects of this relationship, however, will bear further thought and investigation. Its derivation assumes that when plant cover (which is proportional to the square of a linear plant dimension) exceeds 100 percent, density-dependent mortality of smaller plants will occur, reducing density and allowing further growth of survivors that will restore (and again exceed) 100 percent cover (White and Harper, 1970). The emphasis is on full utilization of resources. If the assumption is made that natural populations of plants, starting with different resource levels at different seed densities, grow so as to utilize fully the available resources, the end states of different populations should also show the proposed relationship. Yoda's data for *Erigeron canadensis* show that, within a species, a single regression line predicts behavior, even for populations with quite different resource pools. Likewise, no assumption need be made concerning whether any thinning mortality has occurred. Patchy populations of annual plants can be used particularly readily to test this extension of the 3/2 relationship. A preliminary test is cited below.

Five small, pure populations of *Polygonum kelloggii* Wats. were
harvested from a uniform mountain slope in Oregon after growth
had totally ceased. Densities ranged from 22,000 to 58,000
plants· m^{-2}. This ephemeral species has so short a life cycle that no
leaves (or even cotyledons) are lost during its lifespan and most
seeds are retained. Numbers of seeds lost from larger plants can be
determined readily, and corrections can be made for their absence.
Biomass as total net production was determined. Population means
ranged from 2 to 6.5 mg. Only six of the 370 plants in the harvested
populations died before setting seed, so self-thinning was negligible. The data fit a power-curve regression extremely well (r^2, 0.98).
The exponent is -1.22, which is nearly within the 5° envelope used
for comparative purposes by White and Harper but is lower than all
exponents they cite.

Harper (1967) suggests that where mortality does not occur, a
linear reciprocal yield model of the form $1/w = a + bp$ may be
preferable to the 3/2 power model. Without mortality, all density
stress as plants progressively impinge upon neighbors is absorbed
plastically by diminshed growth. This model has been useful in
agricultural studies but seems no more clearly tied to mortality
assumptions than does the 3/2 model. Moreover, the *P. kelloggii* data
do not fit the reciprocal yield model as well (r^2, 0.72; r is not
significant).

The question remains of why the exponent is so low. The
answer may lie in the nature of limiting resources for growth of *P.
kelloggii* and the assumptions of the 3/2 model. The denominator of
the exponent was derived assuming that the limiting resource is
distributed in proportion to the square of a plant dimension. The
emphasis by White and Harper on "cover" reinforces this assumption, which is not unreasonable because many crop plants have been
shown to compete intraspecifically for light rather than for other
resources (Jennings and Aquino, 1968). However, not all competition among plants is for light. If limiting factors are distributed in a
three-dimensional soil matrix rather than in a two-dimensional
plane (as for incoming radiation), a log slope fluctuating about -1.0
($-3/3$) would be expected.

In part to test this idea, one of my classes at Swarthmore set up a replicated pot experiment in which soil volume was severely limited, so that competition would occur for nutrients before light became limiting. Supplementary incandescent lighting was provided, and plants were watered regularly. We used wheat (Triticum aestivum L.), white mustard (Brassica hirta Moench), and alfalfa (Medicago sativa L.) at four planting densities corresponding to 226 to 4520 plants·m^{-2} (1, 5, 10, and 20 plants per 7.5-cm-diameter pot containing about 230 cm^3 of soil). Preestablishment mortality somewhat increased the number of density treatments, and where there were two or more replicates of an unexpected treatment, data were included in the regressions. Essentially no mortality occurred between establishment and termination of the experiment. Statistics for both power and reciprocal yield models are given in Table 10.1.

The outcome was surprising in that there is equally good fit to both models and the absolute values of the exponents are unusually and consistently low. Several tentative conclusions are possible. First, these data also fail to follow Harper's caveat for experiments with no mortality. Regressions with more points will help decide this issue. Second, the pot-produced stress on the plants was the probable cause of the low exponents, if only partly for the reasons we expected. Our prediction was for exponents close to −1.0. The actual values, less than half that, may be explained by the fact that the experimental conditions put all plants under obvious and serious stress soon after establishment. We may have been so overzeal-

TABLE 10.1.
REGRESSION STATISTICS USING TWO MODELS FOR RELATING
DRY WEIGHT AND DENSITY FOR THREE SPECIES

	n	3/2 Power model[1] Coefficient of determination	Exponent	P	Reciprocal yield model[2] Coefficient of determination	P
Alfalfa	4	0.95	−0.44	<0.05	0.94	<0.05
Mustard	6	0.86	−0.44	<0.01	0.96	<0.01
Wheat	5	0.996	−0.42	≪0.01	0.95	<0.01

[1]Mean weight = a(density)^b, where a and b are constants.
[2](1/Mean weight) = $a + b$(density), where a and b are constants.

ous in trying to ensure that competition was for soil factors rather than light that even plants growing alone were unable to utilize resources fully. In other words, much of the stress may have been density independent, masking the expected negative-log relationship.

Whatever the reasons for the low exponents in the pot experiment, these preliminary attempts to discover the limits of the 3/2 law deserve to be followed up. The two most intriguing aspects of the data are, in the field study, the slight lowering of the exponent but excellent fit and, in culture, the consistency of the aberrant exponents for the three species grown in a uniform and deficient soil.

STUDY OF LIFE HISTORIES

Cole's seminal work (1954) focused attention on fecundity patterns in organisms having different life histories. His result on the level of fecundity that must be maintained by iteroparous, seasonally breeding organisms to maintain reproductive output equal to that of semelparous (annual) organisms might be succinctly stated by the question: Why are there any perennials? This "paradox" has produced a shower of papers reworking Cole's mathematics and assumptions (Murdoch, 1966; Harper, 1967; Murphy, 1968; Gadgil and Bossert, 1970; Bryant, 1971; Cody, 1971; Charnov and Schaffer, 1973; Goodman, 1974; Schaffer and Gadgil, 1975; Bell, 1976; Hart, 1977; see also the lucid review of Stearns, 1976). Murdoch, Cody, Charnov and Schaffer, Bell, and Hart incorporated the concept of age-specific mortality, which has received considerable attention relative to life-history evolution (e.g., Emlen, 1970; Schaffer, 1974a,b; Pianka and Parker, 1975).

The inclusion of mortality goes partway toward filling the ecological vacuum of Cole's result, which was, I believe, first recognized as such by Murdoch (1966). Some of the more precise analyses (e.g., Charnov and Schaffer, 1973; Schaffer and Gadgil, 1975; Hart, 1977) specify that juvenile mortality (which is equal for annuals and perennials) is independent of either environment or community structure and does not vary with time. "Adult" mortality is construed as any mortality after the first year, regardless of time of reproduction, and selection is considered to be "competition." Thus, the vacuum is only partly relieved, but, because of these theoretical attempts, there will be more emphasis on determining the age distribution of mortality in plant populations and, it is to be hoped, the causes and community correlates of mortality patterns as well.

Recognition of age specificity in mortality patterns is basic, but we also need to examine the dynamism of survivorship probabilities in a broader context. Particularly important are the roles of competition (which is not a function of fecundity nor synonymous with selection) and of predictability and variability of environments. While sometimes verbally recognized by recent theorists as being important, these phenomena have so far been given short shrift in mathematical life-history theory.

It is helpful to consider the basic biological differences between annuals and perennials. One of the basic tenets of succession is that, in the presence of surviving perennials, the "fitness" of annual seeds, regardless of their number, will decline year by year as the perennials capture an increasing proportion of available resources. This phenomenon obtains in all sites that are intrinsically suitable for both annuals and perennials (e.g., core habitats of Terborgh, 1973). If annuals are able to persist at all, it is because environments are sufficiently variable or unpredictable that perennials are unable to monopolize all resources and "interstitial" habitats remain that can be exploited periodically. The controlling variable here is neither innate fecundity pattern nor static mortality pattern but ability to compete for resources. Differences in true competitive ability and the pattern of resource distribution in time and space provide the primary explanation for the distributions of annual and perennial plants.

It is in sites that are marginal for perennials because of continuous high mortality of "yearling" plants (the high "adult mortality" of Charnov and Schaffer) that the more secondary questions of static mortality pattern and relative fecundity gain importance. It is to these finer trade-offs in marginal areas that most life-history theory is addressed. Unpredictable yearling mortality can be added to continuously high yearling mortality as a phenomenon of interest in this kind of life-history theory, but there is not yet much work exploring fine trade-offs of this sort [see, however, Hart's (1977) exploration of biennial strategy].

Considering first-year mortality to be "juvenile" and second-year or later mortality to be "adult" produces some disturbing effects in life-history theory. All annuals then die as juveniles whether or not they have set seed, and second-year and older perennials die as adults, even if they have set no seed. A biologically more realistic way of differentiating juveniles from adults is to consider all pre-reproductive individuals to be juveniles and all reproductive and postreproductive individuals to be adults. This convention requires somewhat different conclusions concerning the relationship between life histories and age-specific mortality patterns, as is elaborated below.

These ideas can be brought together and developed by means of a simple graphical model in which resource availability (or environmental quality, the inverse of environmental stress) is plotted through time. The plot is similar to that used by Stearns (1976) in his summary of theoretical work on several kinds of fluctuating environments. The particular line graphed in Figure 10.1 is hypothetical, but its details are suggested by seasonal trends of water availability in several water-limited habitats in the Oregon Cascades (Hickman, 1970, and unpublished data).

Although resource limitation is a complex process that cannot be fully represented graphically, plants do have minimum resource levels below which they become dormant or die. Lines A through C in Figure 9.1 are simplified representations of this minimum level relative to the resource line only, and without reference to any single scale on the vertical axes. Thus, line C represents the most benign environment because resources, although variable, are always in adequate supply for growth (the resource line is always above the limiting level). Such an environment should support perennials predominantly or exclusively. The environment represented by line B is harsher because resources occasionally fall below the limiting level. Such a habitat should support a mixed community of annuals, biennials, and those perennials that have unusual tolerance to low

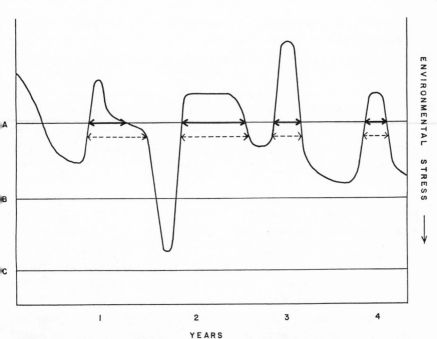

YEARS

Figure 10.1. Resource availability (here considered equivalent to environmental quality and inversely related to environmental stress) as a function of time in a hypothetical environment. Lines A through C are simplified representations of minimum tolerable resource levels relative only to the resource line, and without reference to any single scale on the vertical axes.

resource levels, access to "outside" resources such as deep water supplies, the ability to maintain dormancy in a vegetative state, or the ability to recover from occasional vegetative die-back. The environment represented by line A is physically the harshest and should support a diverse assemblage of annuals because resources annually fall below limiting levels. Perennials in this environment must either have access to "outside" resources or be adapted to an annual dormancy period. In either case, they will not be able to monopolize the seasonal influx of resources and so will be competitively unable to eliminate annuals. Many desert regions exemplify this type of resource fluctuation relative to plant requirements.

This approach suggests that resource variability (measured by the amplitude of the curve) is important in restricting the growth of perennials to the point that annuals can utilize temporally and spatially interstitial resources. That is, for any mean resource level, the less variation in resource level, the more likely that perennials will be able to survive the lowest levels. Thus, it is expected that the greatest abundance of annuals would be found in those extreme environments where resource variability is high and critical levels of resources are frequently exceeded. Perennials should predominate in environments where resources are always adequate (even if variable) or where resource variation is minimal (if levels are low relative to requirements).

In each of the 4 years graphed, the length of growing season for the environment indicated by line A is represented by a solid arrow. Because rate of production should be roughly proportional to resource levels above the minimum, the hatched areas under the resource curve represent the production possible each year. Production in years 2 and 3 can be approximately equal, despite the difference in growing season length.

Coefficients of variation for growing season length, maximum growth quality, and growth area are all measures of environmental unpredictability (see Colwell, 1974, for a more detailed approach). Unpredictability of environments on a time scale comparable with lifespan should select for plastic growth and reproductive responses that are cued by environmental parameters. The mortality costs of genetically tracking temporal change would be immense in such an environment (Hickman, 1975). Compare this with Baker's all-purpose genotype (Baker, 1965). Alternatively, as discussed below, plants may evolve mixed but genetically fixed reproductive strategies that satisfy different environmental demands simultaneously or sequentially.

Years like year 4 should select for rapid growth under suboptimal conditions and also for precocious seed set to replenish deteriorating seed pools (Cohen, 1966; Mulroy and Rundel, 1977), year 3 should select for ability to grow rapidly given good conditions, and year 2 should select for both high production ability at moderate stress and delayed switching to reproductive growth (Cohen, 1971). These patterns are not wholly incompatible, although compromises will be necessary depending on the recurrence rates of the different kinds of seasons. *Polygonum douglasii* Greene, for example, is susceptible to early-season mortality from water stress in dry years but normally survives much longer. In sparse populations, where any but catastrophic mortality is rare, plants set one or two seeds very early and then continue vegetative growth. If the season is sufficiently long, many more seeds will be set late in life, which maximizes total production, but if the season ends abruptly, the plant has at least ensured a genetic contribution to the next generation (Hickman, 1977).

By rearrangement of the graph, the four growth curves can also be considered to be averages for different annual environments, perhaps in a spatial mosaic in one region (see Roughgarden, 1974, for an analysis of population tracking of such a mosaic). Consider the relative effectiveness of selection for increased physiological tolerance of low resource levels. Assume that it is genetically possible to increase tolerance in all populations to the level indicated by the dotted arrows and that a uniform cost is levied on each unit of tolerance added. We see that such selection should have differential effects on the four populations. For the specified increase in tolerance, the potential increases in production in populations 1 through 4 are about 120, 50, 25, and 70 percent, respectively, but the costs of these increases are uniform. Further, we see that this is due not only to the area under the preselection growth curve but also to the slope of the environmental deterioration curve. The maximum productive effect of increased tolerance (hence maximum selection for increased tolerance) should be on growth-limited populations where environments deteriorate slowly (1). The minimum effect will be on populations for which growing conditions are favorable but the season is short, ameliorating and deteriorating rapidly (3). Here selection should favor rapid growth and early seed set.

The model points to the importance of actually measuring environmental quality in comparative studies of plant life histories. The best measures are integrated ones that include the response of the plant rather than just a determination of isolated attributes of soil or radiation load to which no organism may be responding. The simplest integrated measure is, no doubt, change in the rate of biomass accumulation through time. Sophisticated approaches to net photosynthesis or monitoring water potential are also available and may sometimes be necessary for determining comparative physiological responses.

The model also bears on a confusion found in much of the life-history literature. The early assumption was that short life cycles in semelparous species give them the crucial advantage of more generations per time period. Even Harper (1967) puzzled over why annuals would evolve rapid life cycles and high reproductive output only to become dormant and lose much of their fecundity advantage over perennials. The model makes it clear that high and early seed set in annuals need not be related to a fecundity race with perennials but rather to environmental limitation through periodic resource depletion. There is no advantage to immediate reproduction in the environment represented by line A in Figure 10.1. Rather, seed-dormancy mechanisms may be required to keep losses from the seed pool at a minimum. These ideas compare favorably with the work of Went (Went, 1949; Went and Westergaard, 1949) and Cohen (1966, 1967, 1968).

It is now possible to consider more fully the consequences of adopting the biologically more realistic convention mentioned above for differentiating juvenile from adult mortality. Consider the effects on life-history evolution of resource deterioration in an environment that formerly had abundant but variable resources. (That is, consider the resource curve in Figure 10.1 to drop toward line C until it intersects it annually, like line A.) Species in the original perennial community may be selected in a variety of ways as resources become more limiting. Tolerance or dormancy mechanisms may evolve that preserve the perennial habit. If there is variation in time to reproduction, however, plants with shorter life cycles may be selected, which would eventually produce annual plants. This happens through mortality of plants that delay reproduction (that is, that die as juveniles), whereas plants that achieve reproductive maturity (adulthood) earliest contribute disproportionately to the next generation. Thus, it is an increase in the rate of *juvenile* mortality that brings about a shift from perennial to annual life histories. Conversely, with increases in resource availability, annuals may give rise to perennials through a decrease in juvenile mortality rates that allows slow-maturing individuals to survive longer and capture more resources. Theory utilizing age-specific mortality with the convention that first-year mortality is "juvenile" and later-year mortality is "adult" has consistently concluded that the higher the *adult* mortality, the greater the advantage to annuals. The two approaches to mortality patterns as they affect life histories are not fully comparable, but this example points to the importance for theory of apparently small changes in the biological realism of assumptions. The most important difference between the assumptions of the two models is the conceptual departure here from true age specificity of mortality (across organisms of variable longevity) to stage specificity, in which age also is considered in relative terms that allow biological comparisons.

The model presented here, of course, can also be faulted as deficient in realism, precision, generality, or all three factors. Nonetheless, the model has considerable explanatory power and can be used to generate a host of falsifiable hypotheses that can be tested under natural conditions. Furthermore, the general form can be modified readily to fit (and help describe) a wide variety of environments.

A different approach to the study of environments supporting annuals is being taken by Platt and his colleagues in their studies of colonization of prairie badger mounds that are spatially heterogeneous and temporally undependable (Platt, 1975, 1976; Werner and Platt, 1976; Platt and Weis, 1977). These environments primarily support fugitive species and can be studied using concepts of island biogeography. Many annuals are well adapted to colonize and reproduce in areas of temporary disturbance before succumbing to competition from perennials, but other annual populations are able to persist indefinitely (as in deserts and other seasonally arid habitats). This numerically important latter group cannot be understood if it is assumed that all annuals are fugitive species, and alternative approaches, such as described above, must be taken. Results obtained from radically different systems will help in understanding the array of compromises involved in life-history strategies.

r-Selection and K-Selection

Since the original formulation by Dobzhansky (1950) of what is now known as r- and K-selection, and its later development by MacArthur and Wilson (1967), the theory has had a turbulent history. It has been expanded and developed theoretically (Roughgarden, 1971; King and Anderson, 1971; Gadgil and Solbrig, 1972; Pianka, 1972; Southwood et al., 1974; Armstrong and Gilpin, 1977); purportedly supportive data have been gathered, particularly from plants (Pianka, 1970; Solbrig, 1971; Gadgil and Solbrig, 1972; Abrahamson and Gadgil, 1973; Gaines et al., 1974; Solbrig and Simpson, 1974, 1977; McNaughton, 1975; Tramer, 1975), and arguments of conceptual or theoretical deficiency and ostensibly conflicting data have been presented (Hairston et al., 1970; Wilbur et al., 1974; Demetrius, 1975; Hickman, 1975, 1977; Hirschfield and Tinkle, 1975; Nichols et al., 1976; Wilbur, 1977).

Armstrong and Gilpin (1977) pointed out that no two-parameter trade-off scheme is likely to permit an inclusive classification of life histories. This has been the major focal point of critics. It is obvious that life-history trends in nature are generally the sort predicted by r and K theory, but these trends have a multiplicity of possible causes. Because all of the experimental work performed thus far has failed to address the basic tenets of the theory, however, confirmation cannot be assumed.

The rigorous theoretical foundation of r- and K-selection rests on the efforts of Roughgarden (1971), King and Anderson (1971), and Armstrong and Gilpin (1977). These and other approaches to the r–K continuum depend on the importance of density-independent mortality (DIM) as opposed to density-dependent mortality (DDM) in selecting for phenotypes with high r. It has been uniformly assumed that increased stress increases DIM and total mortality. A further assumption is that there exists a necessary trade-off between survival potential and reproductive potential (McNaughton, 1975). Although both the trade-off assumption and the distinction between DIM and DDM seem basic and obvious (a DIM/DDM distinction is implied by the model just presented), some aspects of plant plasticity and some of the data I have collected on Polygonum species make me wonder whether the assumpion need be true and whether the DIM/DDM distinction can be made operational for postselectional studies with plants.

The difficulty with the assumption of a trade-off between survival and reproduction stems from the widely documented phenomenon that increased stress normally lowers plant mortality by keeping individual size small and thereby minimizing DD thinning. Thus, in many plants, probability of survival is greatest in populations exposed to the greatest stress—precisely those populations that r and K theory assumes to have the lowest probability of survival. In *P. cascadense* Baker, the populations that have the largest reproductive allocation also have the greatest stress, greatest environmental unpredictability and variability, highest survival rates, and (except for one aberrantly sparse population) the highest K (as genet number) and highest production densities of biomass and seeds. At least for this species, there is now no trade-off between survival potential and reproductive potential, even though several characteristics of the sequence of populations follow the predictions of r and K theory (Hickman, 1975).

A further suggestion from this approach is that DIM may normally constitute quite a small proportion of postestablishment but juvenile plant mortalities, even for those species or populations thought to grow in DI environments.

Two caveats are necessary here. The most important is that postselection mortality patterns cannot provide critical evidence either for or against the selective effects of mortality patterns. To argue evolutionary hypotheses on the basis of postselection data is to fall into the trap of *post hoc, ergo propter hoc* reasoning. Stearns (1976) argues, correctly, that testing of evolutionary hypotheses requires selection experiments, preferably in the field. Such experiments have not yet been performed and should not be undertaken lightly, but a well-devised attempt, with adequate controls, would be most enlightening. Surveys of the results of selection can suffice only when the selective factors that have operated in the past can be known with precision. Thus, the question of whether plants that should be highly r-selected have primarily DD mortality after establishment may be irrelevant to r- and K-selection theory. Yet postselection data are normally all we have to bring to bear on evolutionary hypotheses. They must be considered, but their limitations must also be recognized.

The second caveat is really a corollary of the first: DIM and DDM may both be of crucial importance in a given species at different points in its history. Although I have seen little evidence of the current importance of DIM in the annuals I have studied, the model just presented does assume that, as environments deteriorate beyond the tolerance of most individuals, DIM is the most potent selective force in the evolution of the annual habit. Once annuals have evolved, however, they might experience a greater proportion of deaths from DD (competitive) thinning during the growing season than is true of perennials.

Current Patterns of Postestablishment Mortality

Roughgarden (1977) has stressed that the *lore* of r- and K-selection has arisen in large part independently of theory and must stand or fall independently of theory. Stearns (1976) includes a large table (pp. 22–23) of many presumed correlates of r- and K-selection. One of them, first noted by Pianka (1970), is high frequency of Type III survivorship curves (Deevey, 1947) among "r-selected" species. That is, after germination, there is abundant juvenile mortality, with few plants surviving to reproductive maturity. In annual plants, however, when postestablishment juvenile mortality does occur, it is either from massive DIM due to disturbance or exceptionally harsh conditions or from DDM, as is assumed in the 3/2 power model of thinning. True 3/2 thinning will produce a Type II (log-linear) survivorship curve.

As analysis of plant cohorts becomes more common, data will accumulate on the frequencies of the different survivorship curves in "r-selected" plants, as well as on the correctness of the presumed correlates of age-specific mortality patterns. Figure 10.2 presents Type I postgermination survivorship curves for cohorts of three species of *Polygonum*. Data for each species were gathered weekly from 8 to 16 replicates of 5 × 5-cm plots. Individuals of *P. cascadense* and *P. douglasii* were followed, but stands of *P. kelloggii* were too dense to identify all individuals consistently. Observation was begun when seedlings had only cotyledons for photosynthetic production, that is, immediately after emergence.

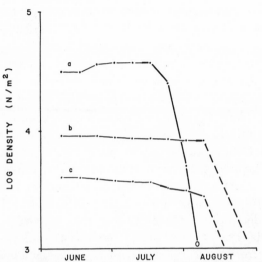

Figure 10.2. Survivorship curves for seedlings of (a) *Polygonum kelloggii* Wats., (b) *P. douglasii* Greene, and (c) *P. cascadense* Baker on a summer-dry mountain slope. Heavier line indicates mean time of first anthesis.

Polygonum kelloggii had 5 percent late recruitment after observations were begun. Fewer than 1 percent of the plants failed to set seed (died as juveniles), and those were all late recruits. Plants were all dead within 3 weeks after the initiation of flowering. *Polygonum douglasii* had 8 percent mortality after emergence. Here, too, only smaller individuals died, apparently from DD thinning. In this species, seed set is delayed in dense populations and began shortly before termination of the observations. *Polygonum cascadense* had 32 percent mortality during the course of the experiment, but only about half that proportion died as juveniles. Of the total deaths, 62 percent were from human trampling, 14 percent were from digging by rodents that disturbed root systems or buried plants, and 6 percent were from larval lepidopteran herbivores. The causes of 18 percent of the mortalities were undeterminable but were probably a combination of rodent disturbance and water stress. These preliminary data add further support to the importance of adult (as opposed to juvenile or continuous) mortality for germinated annuals but do not address the question of whether juvenile or adult mortality was important in the evolution of annuals. They also ignore the impact on survivorship statistics of seed-stage mortalities, a topic that is considered in a later section. It is intriguing to find that even very dense annuals (30,000·m⁻² for *P. kelloggii*) may experience essentially no juvenile mortality after emergence.

The distinction between DIM and DDM, although essential to theory, is quite difficult to support with field evidence. Most experimentalists have argued for a particular population sequence in the relative importance of DIM, but this either has been done with no data on mortality (Solbrig, 1971; Gadgil and Solbrig, 1972; Abrahamson and Gadgil, 1973; Gaines et al., 1974) or has involved arbitrarily categorizing some kinds of death as DI and others as DD. Unless *all* coexisting plants die in a catastrophe, which clearly implicates DIM, any given death could be either DI or DD, and it is difficult to determine, in nature and after the fact, which has occurred. This is particularly true of deaths that result from resource limitation or from the activities of other organisms whose behavior may or may not be modified by plant density. Some deaths that it may be tempting to consider DI may in fact be inversely DD. For example, the rodents that destroyed some *P. cascadense* plants regularly chose open sites of low plant cover for their burrows. Consider also death from drought stress. Water is one of the more mobile resources of the soil. Even if two root systems are not in close proximity, zones of water depletion may overlap. If a particular plant death from drought could have been hastened by overlap of depletion zones, a DD cause cannot be rejected. Similarly, a DI cause cannot be assumed unless it can be shown that the lethal level of drought stress for the phenotype would have been achieved regardless of the existence of neighbors. Both sides of this coin are susceptible to experimental confirmation in some cases, but not on the scale necessary for numerical studies of mortality.

Two other difficulties are associated with the DIM/DDM distinction. Plant plasticity forces us to decide whether we mean *density* independence or something like *cover* independence. Small plants at high genet density may have fewer "density-dependent" interactions, with less effect on the population, than do larger plants at lower genet density.

In animal population biology, density dependence is an intraspecific measure. For plants, which live together in spatially static configurations and compete for a limited set of resources, it is more appropriate to make the concept interspecific in its application. Unlike animals, with their sensory-stimulated behavior patterns, plants do not have species-specific, induced responses to potential competitors. A seedling will die as readily from being shaded by a larger plant of another species as from equal shading by a sibling. When plant population biologists use the term "density dependence," they often mean something more like "interspecific cover dependence." Care must be taken to specify the sense in which such concepts are intended. The system of hierarchical components of N discussed earlier will accommodate all possible senses of "density" dependence, including those relating to population genetics.

Given the operational difficulties of the DIM/DDM distinction, it seems unfortunate that so much theory must be tied to it. Uncertainties could be readily controlled only in the field selection experiments that are called for to test evolutionary hypotheses, including those of r- and K-selection. It seems unlikely, however, that any great proportion of work in plant population biology will move in this direction. For the most part, further work, like that of the past, will have to be focused on study of the assumed correlates of evolutionary theories rather than on the theories themselves.

Interference in Plants

Although attempting to assign a DI or DD cause to mortalities in unmanipulated populations is dangerous, study of DD interactions has become more precise as methods have proliferated. The classic approach to plant competition has been the replacement series developed by de Wit (1960). It requires establishing artificial communities, usually in greenhouse or garden, with predetermined, regular spacing and comparing production of two interactants at different frequencies. This method has been particularly valuable for studies of crop plants, where spacing may be controlled for maximizing production and the effects of competition from weeds in precisely spaced stands must be known. For natural populations and for evolutionary studies, however, the technique has obvious limitations.

Another approach is to determine plant dispersion patterns in nature, with the assumption that the more regular the pattern, the more important are interference interactions (Pielou, 1959); Woodell et al. (1969) attempted to explain pattern in *Larrea divaricata* in this way. As D. J. Anderson (1971) implies, however, this method has the danger of running afoul of the effects of different plant sizes on pattern. Simplifying the analysis by considering only one size class remains less than satisfying. Dispersion patterns are also compromises. Even when intraspecific interference is strong, plants may be clumped because of the overriding importance of positive sociability, as when aggregated flower displays best attract rare pollinators for which many species must compete.

The techniques of neighborhood analysis are based genetically on the work of Wright (1946) and for studies of interference on the methods developed by Pielou (1960). These techniques have been elaborated in different directions by a variety of authors (Levin and Kerster, 1971; Trenbath and Harper, 1973; Yeaton and Cody, 1976; Mack and Harper, 1977; A. P. Smith, unpublished data) and show great promise for detailed study of DD interactions in plant populations. For example, using as independent variables interplant distances, angular dispersions, and sizes of neighboring plants within 2 cm, Mack and Harper were able to account for about 70 percent of the total variation in size and reproductive output in several species of dune annuals.

I have attempted some preliminary experiments, with *Polygonum* species, intended to compare a simple de Wit replacement experiment with a conceptually equally simple analysis of neighborhoods. Cotyledon-stage seedlings of *P. cascadense, P. kelloggii,* and *P. douglasii* were transplanted to cleared plots previously dominated by *P. kelloggii* but supporting all three. Seedlings were planted in rectangular arrays, with individuals separated by 1 cm. Some arrays were pure and others were mixed pairwise, with pair members at equal frequencies. Plants were grown to maturity or nearly so, and interaction coefficients were determined by dividing the mean net production of plants of species A from pure culture by the mean net production of plants of species A from mixed culture with species B. The coefficient measures the impact of species B on species A relative to the intraspecific impact of individuals of species A at the same density. The coefficients are given in Figure 10.3.

Because these three species and P. *minimum* all occur in pure, uniform stands at variable neighborhood density and in mixed stands at both variable relative frequencies and variable neighborhood density, it was possible to isolate stands that show the intra- and interspecific effects of nearest neighbor distances on net production. This was done for pure stands of all four species and for all possible situations where species A was the minority species in an otherwise pure population of species B. The only pairwise combinations that did not occur naturally were those in which P. *kelloggii* should have been the minority species. For 18 to 50 plants per

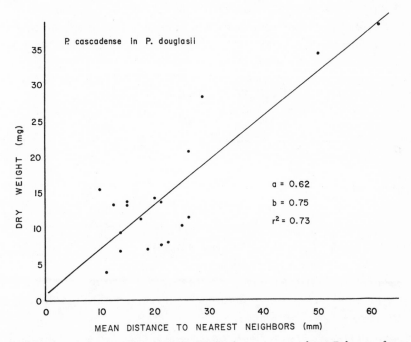

Figure 10.3. Net production as dry weight of *Polygonum cascadense* Baker as a function of mean distance to four nearest neighbors of P. *douglasii* Greene. *Polygonum cascadense* was a minor component in an otherwise pure stand of P. *douglasii*. Regression significant at $P < 0.001$.

population, distance to the four nearest neighbors was measured, and the central plant was harvested. Net production was regressed on mean distance to nearest neighbors. An example (*P. cascadense* as the minority species in a population of *P. douglasii*) appears as Figure 10.4. The coefficient of determination is 0.73, showing that proximity of *P. douglasii* plants (regardless of their size and angular dispersion) had an overriding effect on the production of *P. cascadense* individuals. No doubt a greater proportion of the production variance could be accounted for with a multiple-regression approach that incorporated size of neighbor as well.

Figure 10.5 shows the coefficients of determination for all the intra- and interspecific interactions studied. It is comforting to note that the sequence from best to worst competitor is the same for the two types of analysis: *P. douglasii* > *P. cascadense* >> *P. minimum* > *P. kelloggii*. From other correlates of r- and K-selection, however, *P. douglasii* and *P. minimum* should be considered the most K-

RELATIVE   EFFECT   OF   (b)

	cascadense	kelloggii	douglasii
cascadense	1.00	0.66	1.12
kelloggii	2.04	1.00	1.21
douglasii	0.81	0.89	1.00

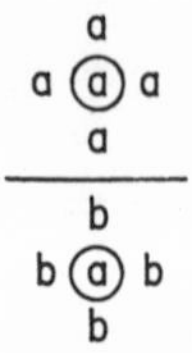

ON (a)

Figure 10.4. Experimentally determined competition coefficients from a DeWit–type array of transplanted *Polygonum* species. Plants were grown from near-cotyledon stages to maturity in cleared areas previously dominated by *P. kelloggii* Wats. but also supporting the other two species. Coefficients were derived by dividing the mean net production of a species grown in pure culture by the mean net production of the same species grown in competition with another species.

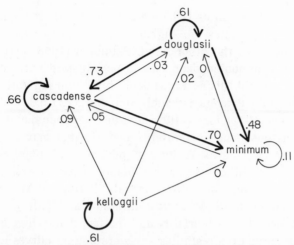

Figure 10.5. Some neighborhood effects among four coexisting species of *Polygonum*. Populations were isolated in pairwise combinations wherever one species occurred in nearly pure stands, but with a minority of neighbors of only one other species. Arrows lead from majority to minority species. Numbers at arrow points are coefficients of determination (r^2) for regressions of net dry weight production of the minority species on mean distance to the four nearest neighbors of the majority species. Heavier arrows indicate significant regressions.

selected and *P. kelloggii* and *P. cascadense* the most r-selected (cf. Hickman, 1977). At least the end points of the continuum of r–K correlates appear to be supported by this analysis.

Despite the fact that both experimental modes produced interpretable data, the effort and artificiality involved in the replacement transplants were much greater than for the neighborhood analysis. The latter, moreover, included a greater diversity of populations and environments and provided much more information. Exploration of the varied potentialities of neighborhood analysis of plant populations will provide many new insights into population structure and regulation.

Seed Biology

Even in annuals like *Polygonum kelloggii* that show little "juvenile" mortality, the vast majority of seeds fail to germinate. From comparisons in near-equilibrium populations of genet densities and seed production densities (Hickman, 1975, 1977), it appears that germination frequencies are higher in populations under greater physical stress. On the other hand, Palmblad (1968) and Linhart (1976) have documented decreases in germination with increased density and suggest this as a means of population regulation.

In spite of the existence of such compendia as Kozlowski (1972) and Heydecker (1972), far too little attention has been devoted to the seed level as the locus for mortality patterns that may help determine life-history patterns. At least 95 percent of all plant mortality occurs here, and differences among plants in patterns of mortality after germination may be entirely trivial. Sharitz and McCormick's (1973) life-table data for *Minuartia uniflora* show both the overriding importance of seed mortality and the fact that, when seeds are included in survivorship studies, postgermination curves identifiable as Type I or Type II are converted into Type III curves. Levin (1979) also provides life-table statistics for *Phlox drummondii* that include seed mortalities. His data show Type II seed survivorship and Type I postgermination survivorship. Both these studies offer some support for Cohen's assumption (1966) that pregermination survivorship is Type II. One possibly important result of considering seed survivorship, however, is the tendency to convert all plant survivorship patterns to Type III; the distinction between greater and lesser juvenile mortality that has been of such importance in the development of life-history theory is obscured by the predominance of juvenile mortality (as seeds) for all plants.

There is a choice of considering the analog of "birth" in plants to be either seed set or germination. At seed set, independent recombinant individuals are formed. At germination, seedlings gain independence of parental tissues. Life history theorists such as Schaffer and Gadgil have accepted the latter path. I think the former is more reasonable because it allows consideration of new genotypes as they are formed (and so can be adapted more readily to the needs of population genetics) and because dispersal precedes germination. Considering seed set as the equivalent of birth will allow for adapting parts of animal population theory that are suitable for plants and for accommodating genetics and evolution.

Such a choice, however, is not without cost, because it inevitably focuses attention on seeds. Longevous plants like trees generally have short-lived seeds, but short-lived plants, which are best for experimental work at the population level, frequently have longevous seeds that require attention in their own right as determinants of population dynamics. Seeds are notoriously difficult to study, but methods are improving. The work of Sarukhán (Sarukhán, 1974; Sarukhán and Harper, 1973; Sarukhán and Gadgil, 1974) is exemplary and provides a model for methods and ideas, as well as of persistence, and has yielded intriguing results. Cavers (e.g., 1974) and Baskin and Baskin (e.g., 1974a, 1975b) have also contributed important but more restricted studies of seed fates. Plant population biologists in general have remained reluctant to enter this arena, but doing so is necessary if we are to understand plant responses to spatial heterogeneity, temporally unpredictable environments, and both preemptive and interactive modes of competition.

Most of the accumulated knowledge of seed biology has been reviewed by Harper et al. (1970) and by Harper and White (1974). I will cite just a few examples of divergent approaches. Baker's (1972) synthesis of seed and environmental parameters in California shows that some insights into life histories can be achieved without detailed study of the fates of seeds. Janzen's many studies (e.g., 1969, 1970, 1971a) have highlighted the consequences of various predispersal fates of ovules and seeds, especially as determined by seed predators. Dispersibility is readily measured, at least in a relative way, and consideration of it has been incorporated into a number of studies (Werner, 1975b; Werner and Platt, 1976; Platt and Weis, 1977; Wilbur, 1977; Staniforth and Cavers, 1977). Some work (e.g., Ross and Harper, 1972) has considered the nature and effect of slight differences in germination position and timing.

There nevertheless exists a huge gap in our knowledge of what happens to seeds between dispersal and germination. Of all the failures of our knowledge at this time, this one is perhaps the most complete. The theoretical work of Cohen (1966, 1968) provides a foundation for experimental investigations that will be of great interest. Closing this knowledge gap should be one of the highest priorities for plant population biologists in the years immediately ahead.

CONCLUSION

This is intended to be a critical essay and, because I frequently have had to make critical comments about both theory and experimental design, it is important that my intent not be misunderstood. My criticisms are not aimed at theory in the abstract, which is and must remain the guidepost to useful experimentation and furthering of any discipline. Rather, I have tried to suggest areas in which some specific theories can be improved upon and have occasionally objected to instances of formally untested theories gaining acceptance because they are attractive and internally consistent.

Levins (1966) admirably laid out the dilemma facing ecological theorists. Models must be simpler than nature to be tractable, so trade-offs are necessary among generality, realism, and precision. Levins argued for a mixed strategy of model building as a way of determining sufficient parameters and constructing robust theorems. Most theory applicable to plant population biology, however, has emphasized precision first and generality second. Biological realism too often has been the attribute traded off. My own bias is similar to the view expressed by Botkin (1977), who reviewed the effectiveness of the numerous model-building attempts of the International Biological Program: He found that large-scale or general models are characteristically too vague to integrate results or guide research and that small models might be elegant and precise but are normally too unrealistic to be tested adequately under field conditions. Models that are just large enough to encompass those phenomena known to be important are those most likely to provide a testable, realistic base for an emerging discipline. Accordingly, most of my suggestions have been for improving the biological realism and hence the testability of population models. These suggestions are not intended to detract from the original efforts but rather to extend their utility.

Our most urgent need in plant population biology is to know the relative importance of the trade-offs we know or suspect to occur in natural systems of populations. This can be accomplished most effectively by broad and long-term study of populations chosen for their ability to provide the greatest possible diversity of information that will bear upon current and future thought about population dynamics. If early results inform both theoretical constructs and later experimental design in direct ways, a coherent picture is likely to emerge. To increase coherence of results, exemplary plant populations must be approached from a variety of mutually reinforcing perspectives, including those of physiology, genetics, spatial and temporal patterns, breeding systems, demography, predation and competitive patterns, and life-history parameters. Subtle but important trade-offs occur within and among these overlapping aspects of population dynamics. Broad knowledge about these compromises in diverse population systems is the surest way to isolate general patterns of behavior in plant populations.

SUMMARY

The major points addressed in this article are as follows:

1. High variance in size at reproductive maturity, which has environmental determinants in many plants, confounds comparisons of both N and K and hinders fusion of demography with population genetics. A proposed solution requires a hierarchical N, with progressively less inclusive components being number of genets, number of ramets per genet, and functional size of ramets. The concept of ramet must be enlarged to include small-scale structural units, and more attention must be paid to patterns of growth and development.

2. The most general attempt to model size plasticity, density, and mortality is the "3/2 power law" of Yoda and others. If plants are limited by substrate factors rather than light, exponents should vary around −1.0 rather than −1.5. Natural populations, in different environments and starting from different numbers of seeds, might also show the 3/2 power relationship, regardless of whether or not thinning mortality occurs. Some preliminary data are given.

3. Previous studies of life-history evolution have concentrated too heavily on fecundity patterns at the expense of mortality patterns. Recognition of age-specific mortality patterns is

basic, but there has not been general recognition of the importance of fluctuations in age-specific patterns in different physical environments, different communities, and through successional changes.

4. A graphical model of resource availability is developed that explains selection for several life-history parameters, emphasizes measuring trends in environmental quality in comparative life-history research, and incorporates a more realistic view of "juvenile" and "adult" mortality.

5. r-Selection and K-selection and other evolutionary theories have not been adequately addressed by experimentalists because only postselection traits have been studied. The required tests for evolutionary theories involve the selection process itself. A posteriori evidence, which is fallible, suggests that the trade-off between survival potential and reproductive potential assumed by all life-history theorists does not necessarily hold.

6. Analysis of plant cohorts will facilitate testing of the suggested correlations that comprise the lore, as distinct from the theory, of r- and K-selection.

7. Environmental stress diminishes mortality in many plant species by slowing growth and decreasing interference reactions. Even in very dense populations, postestablishment survivorship curves for many annuals are Type I because most mortality is end-of-season, density-independent adult mortality. When seed deaths are included, such curves may appear as Type III, which obscures fine differences in postestablishment patterns.

8. The distinction between density-dependent and density-independent mortality is essential to theory but is nearly nonoperational in nature. For plants, "density dependence" should most generally be construed as "interspecific cover dependence," unless genetic questions are involved.

9. Some aspects of interference studies in plants are reviewed, incuding de Wit replacement series, dispersion patterns, and neighborhood analyses. A comparison is made using replacement series and neighborhood analyses for four species of coexisting *Polygonum*. Neighborhood analysis holds great promise for study of natural plant populations.

10. At least 95 percent of all plant mortalities occur at the seed

stage, but knowledge of seed fates is minuscule. Much work is needed here to understand plant responses to spatial and temporal heterogeneity and to preemptive and interactive competition.

11. A most urgent need in plant population biology is knowledge of the relative importances, in various environments, of known or suspected trade-offs within and among the physiological, genetic, and demographic arenas of population response. The best strategy for the development of realistic and general theory of plant populations involves broad and long-term study of exemplary population systems.

11 THE ORIGIN OF PHENOTYPE IN A RECENTLY EVOLVED SPECIES

L. D. GOTTLIEB

TO THE ECOLOGIST, the sympatry of a newly evolved species and its progenitor is interesting because it is likely to lead to the subdivision of their habitat (Harper et al., 1961) and thereby demonstrate what minimal difference is necessary to permit cohabitation of extremely similar organisms. To the geneticist, the same juxtaposition is interesting because it reveals how the particular properties of a new species are the consequence of what was inherited from its parent, and their sympatry permits the contribution of these properties to fitness to be assessed under the same natural conditions. The coordinated study of ecological and genetic properties of such pairs of closely related species is appropriate for the synthetic discipline of plant population biology.

Very few pairs of diploid plant species have been identified which appear to have a recent progenitor–derivative relationship. As a result, most comparative studies have dealt with conspecific populations from contrasting habitats or with congeneric species from similar habitats. Conspecific populations often live in markedly different habitats and are differentiated in many morphological and physiological features. Yet they may be fundamentally alike.

Two populations of *Achillea millefolium* ssp. *borealis*, one native to stormy Kiska Island in the Aleutian chain and the other to the fruitful San Joaquin Valley of California, provide a striking example. The populations differ greatly in many features, including stem length (1 cm versus 200 cm), type of root (rhizome versus taproot), and their ability to survive when transplanted to the three Carnegie transplant stations in central California (Hiesey and Nobs, 1970). Yet their F_1 hybrid progeny was fertile and vigorous, and their F_2 progeny contained a very high proportion of individuals that did well at each transplant station. This suggests that even though the Kiska and San Joaquin populations have accumulated divergent adaptations, possibly involving large numbers of gene loci, their genomes could be recombined with little apparent disadvantage. Such evidence of genetic unity cannot be ascribed to gene flow (Ehrlich and Raven, 1969) or to similar selective regimes since the habitats of these two populations are distant and grossly dissimilar. The example leads to the suggestion that even adaptively distinct conspecific populations share certain fundamental biological properties.

If we wish to define these properties and determine how they originated, however, comparison of conspecific populations will reveal very little. Nor is comparative study of most congeneric species likely to provide this information; in general, phylogenetic separations occurred sufficiently long ago that the pathways leading to present character states have been obscured and cannot be reconstructed.

This clearly means that to understand the basic properties and organization of the genome of a species, one must compare it to its progenitor, and this must be done at a time shortly after its origin, when modifications of its genome remain limited in number and consequence. The emphasis in this type of study must not be directed merely towards the specification of the mode of origin of reproductive isolation or the identification of barriers that prevent gene exchange. It is also necessary to identify those components of the epigenetic systems of newly arisen species that are responsible for their distinctiveness (Mayr, 1963, pp. 543–48). This will ultimately involve analysis of how patterns of gene expression and regulation result in differences in development, as well as vertical studies of biochemical and physiological responses to environmental cues. Perhaps it will then be possible to understand how the phenotype of a species is assembled.

In this report, I describe certain ecological and genetic properties of a recently evolved diploid annual plant species and its sympatric progenitor. Identification of biologically significant properties of the new species leads to a hypothesis that accounts for their origin and for how these properties adapt the new species to an infrequent feature of the habitat that permits it to persist alongside its progenitor. The juxtaposition of progenitor and derivative species in the same habitat, the recent origin of the derivative, and the high probability that it went through a genetic bottleneck meet the minimal requirements for an exemplary case which may answer questions about the origin of a species phenotype, the functional constraints imposed by the genetic legacy inherited from the progenitor, and the appropriateness of the concept of genetic revolution (Mayr, 1963, p. 544) to describe the outcome of speciation.

THE STUDY SITE

Twenty-five miles south of Burns, Oregon, beside the main road to Steens Mountain is a sagebrush-covered hilltop composed of volcanic tuff, overlooking the glistening white remains of Lake Harney. The hilltop, about 150 acres in extent, is surrounded by soils derived from cracked and pocked basalt; it is like an island and is isolated by many miles from the next nearest tuff. It provides the habitat of the most northern population in the geographical distribution of *Stephanomeria exigua* ssp. *coronaria*, a member of the sunflower family frequently encountered in California, but in very different habitats including secondary sand dunes behind the ocean beach, grasslands in the arid inner Coast Ranges, openings in incense cedar forest in the Sierra Nevada, and as high as 9000 feet in volcanic soils in the southeastern Sierras (Gottlieb, 1971). *Stephanomeria malheurensis* also grows on the hilltop, but this species has been discovered nowhere else (Gottlieb, 1973a, 1978a).

Both Stephanomerias are obligate annual plants. After germination, their seedlings grow as rosettes and, after several months, send up a single stem which develops side branches that bear the flowering heads. Flowering usually begins by early July and seeds are produced until the plants die in late August.

That two Stephanomerias grow at the site has been known only since 1966, although S.e. ssp. *coronaria* was collected there in 1925 by Henderson (collection number 8604). *Stephanomeria malheurensis* would easily have been missed, even by taxonomists eager to discover new taxa; it closely resembles S.e. ssp. *coronaria* in most morphological features and, perhaps more important, it is rare. Observations made since 1968 have shown that the number of reproductive individuals of S. *malheurensis* has not exceeded 2–3 percent of that of S.e. ssp. *coronaria*. The total number of S. *malheurensis* plants has not been more than 750 in any one year (a liberal estimate, since less than half this number has actually been counted during diligent searches; Gottlieb, 1973a, 1974a, unpublished data). Although S.e. ssp. *coronaria* may have a population of about 40,000 individuals when the precipitation is sufficient in the spring months, other annual species on the hilltop have substantially larger populations.

The relatively small size of the S.e. ssp. *coronaria* population is not surprising since the locality represents the "species border," which is characterized by more extreme and less predictable environmental conditions than those found at the Californian localities. Precipitation averages only 9 inches, much of which falls as snow, and in most years the temperature drops below freezing at least once every 24 hours between mid-October and mid-April (based on records taken at the nearby U.S. Weather Bureau Station at Malheur Wildlife Refuge Headquarters). The harsh winter conditions have selected S.e. ssp. *coronaria* seeds which require freezing to break their dormancy (Gottlieb, 1973a). Field observations have shown that the seeds germinate only when the temperature rises in the spring. In contrast, the seeds of the Californian populations of S.e. ssp. *coronaria* do not require freezing and germinate with the first rains in the fall, with seedlings overwintering as rosettes. The seeds of S. *malheurensis* also appear to lack freezing requirements for their germination (Gottlieb, 1973a). Although S. *malheurensis* seeds germinate in the spring, recent observations indicate that some also germinate in the fall (G. Wing, Bureau of Land Management; personal communication); these are killed during the severe winters.

In 1966, plants of S.e. ssp. *coronaria* were found scattered over most of the hilltop in openings in the sagebrush. Since the fall of

1972, when much of the site was inadvertently burned, the total number of S.e. ssp. *coronaria* plants appears to have increased, with high densities in several specific areas (Figure 11.1). This increase in number probably has resulted from the increased soil fertility often associated with rangeland fires in this region as well as from an increase in moisture supply brought about by the removal of the sagebrush (W. K. Sandau, Bureau of Land Management, personal communication). At present, more than 95 percent of the S. *malheurensis* plants are found in a burnt area at the northern portion of the hilltop (Figure 11.1). In addition, a few dozen plants of both species

Figure 11.1. Aerial view of the hilltop, south of Burns, Oregon, which is the location of S.e. ssp. *coronaria* and its recent derivative S. *malheurensis*. The view is toward the northeast, showing a portion of State Highway 205. The dark areas are predominantly sagebrush. The fall 1972 fire burned off the sagebrush on much of the hilltop, and these burned areas, now devoid of sagebrush, appear more light colored on the photograph. *A* is the region of the site which contains about 95 percent of the *malheurensis* plants; in this region, S.e. ssp. *coronaria* outnumbers S. *malheurensis* by more than 2:1. *B* designates the regions in which S.e. ssp. *coronaria* plants have the highest densities.

have been found on a tuff outcropping on the same hill about a mile to the west of the main site. Even where S. *malheurensis* plants are most densely concentrated, they are outnumbered by S.e. ssp. *coronaria* by about two or three to one. The two species grow side by side and their branches are occasionally found overlapped.

THE PROGENITOR–DERIVATIVE RELATIONSHIP

At the hilltop location, S. *malheurensis* and S.e. ssp. *coronaria* are morphologically so similar that they cannot be distinguished with certainty until flowering and seed set have begun. The most reliable character for delimiting them is the size of the achenes. Those of S. *malheurensis* average about one-third longer (Figure 11.2), and are twice as heavy as those of S.e. ssp. *coronaria* and bear longer and more numerous pappus bristles (Gottlieb, 1973a, 1977a). The species resemble each other closely in most other morphological features, but when they are grown in uniform conditions, important quantitative differences have been discovered. Both species are diploid, with $2n = 16$.

The species differ in their breeding systems. Individuals of S.e. ssp. *coronaria* possess a sporophytic self-incompatibility system

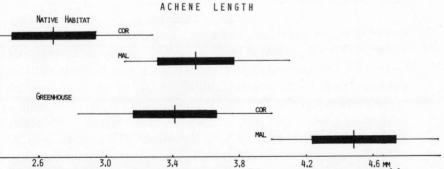

Figure 11.2. Achene length in S. *malheurensis* and S.e. ssp. *coronaria* measured for achenes collected on plants growing in the natural habitat and in the greenhouse. The bar diagrams show the mean, one standard deviation on each side of the mean, and the range. Sample sizes: native habitat—S. *malheurensis*, 75, and S.e. ssp. *coronaria*, 30; greenhouse—S. *malheurensis*, 69, and S.e. ssp. *coronaria*, 39.

that prevents pollen from germinating on stigmas of the parent plant, making the species an obligate outcrosser. *Stephanomeria malheurensis*, in contrast, is self-compatible and highly self-polli-nating. In one analysis, done in the greenhouse during the long warm days of June, approximately 58 percent of its florets set seed by self-pollination (unpublished data). Its ligules are narrower than those of *S.e.* ssp. *coronaria*, presumably a consequence of reduced selective coefficients to maintain attractiveness to pollinators.

Reproductive isolation is maintained by three factors: the restriction of pollen exchange resulting from the differences in breeding system; a crossability factor(s) which reduces seed set from interspecific cross-pollinations compared to conspecific ones by about one-half; and several differences in chromosomal structural arrangement, including a reciprocal translocation. The latter reduces fertility of F_1 hybrids to 25 percent, should they be produced (Gottlieb, 1973a). Although a few interspecific hybrids have been identified in nature, the reproductive isolation of *S. malheurensis* appears to be highly effective even though it grows interspersed with the outcrossing, highly polymorphic and much more abundant population of *S.e.* ssp. *coronaria*. *Stephanomeria exigua* ssp. *coronaria* is also unlikely to receive genes from *S. malheurensis* since its outcrossed stigmas remain closed and unreceptive to pollination until about noon on most days, by which time the heads of *S. malheurensis* have closed.

An electrophoretic analysis of variation in 12 enzyme systems, specified by 25 gene loci, has been carried out. The genes of *S. malheurensis* are present in moderate to high frequency in its parent (except for a single rare allele), whereas the parent has many alleles not found in *S. malheurensis* (Gottlieb, 1973a, 1976a). Sixty percent of the genes assayed in *S.e.* ssp. *coronaria* were polymorphic, with a total of 45 alleles, and only 12 percent of the genes, with seven alleles, were polymorphic in *S. malheurensis*. The average individ-ual of *S.e.* ssp. *coronaria* is about 15 percent heterozygous, but only two heterozygous genes were detected in the more than 600 individ-uals from about 100 progenies of *S. malheurensis* which were exam-ined. Thus, in terms of structural genes specifying enzymes, the *S. malheurensis* genome appears to have been extracted from relatively common genetic elements of the rich gene pool of S.e. ssp. *coronaria*.

The electrophoretic analysis was done with starch gels made up at a single concentration and pH. Recent studies with *Drosophila* (Singh et al., 1976) and *Colias* (G. Johnson, 1976c) have demonstrated that the use of multiple electrophoretic conditions can reveal additional enzyme heterogeneity within what appear to be single "bands" on starch gels. With this possibility in mind, 6 enzyme systems in 13 lines of S. *malheurensis* were assayed again (alongside extracts from individuals of S.e. ssp. *coronaria*), using acrylamide gels made up at two concentrations and with different pH conditions for each concentration. No additional variants were observed in S. *malheurensis*, which suggests that the high degree of electrophoretic homogeneity initially observed on the starch gels is not likely to be reversed as more individuals and enzymes are analyzed by more rigorous techniques (Roose and Gottlieb, unpublished data).

The restriction of S. *malheurensis* to a single locality, the species' very high morphological and electrophoretic similarity to the sympatric population of S.e. ssp. *coronaria*, and the type of reproductive isolating barriers between them suggest that the ssp. *coronaria* population is its progenitor. The possibility that S. *malheurensis* originated at a different locality and then migrated to its present site is unlikely since the soils on which it grows are azonal and found only in widely separated outcrops. Indeed, the nearest known locality for S.e. ssp. *coronaria* is 30 miles south.

Stephanomeria malheurensis could have evolved after a rapid and abrupt series of events initiated by the occurrence of a mutation at the self-incompatibility locus, which eventually led to the formation of a self-pollinating individual of S.e. ssp. *coronaria*. Presumably, an inbreeding lineage developed, in which one or two chromosomal rearrangements took place, giving S. *malheurensis* reproductive isolation in the midst of its progenitor population. That only a small number of genetic changes were involved in the origin of its reproductive isolation is indicated by the substantial increase in mean pollen viability from an average of 25 percent in interspecific F_1 progenies to 62 percent in an F_2 from one of them, which also included a significant proportion of individuals that were fully fertile (Gottlieb, 1973a). In addition, recombination of the important parental character differences of self-pollination and seed length and the recovery of individual F_2 segregants with character expressions of both parents suggest that only a small number of genes control many of the phenotypic differences between S. *malheurensis* and S.e. ssp. *coronaria*.

"PERFORMANCE" OF PROGENITOR AND DERIVATIVE SPECIES

The particular events presumed to account for the origin of S. *malheurensis* are so profound that their genetic consequences should be relatively easy to identify and evaluate. Thus, a priori, the rapid and abrupt mode of origin has genetic effects very similar to those described by the founder principle (Mayr, 1963). The original S. *malheurensis* lineage would have contained only a small fraction of the allelic variation of S.e. ssp. *coronaria*. In addition, because of its predominant self-pollination, the reduction in variability was necessarily associated with the drastic change from the high heterozygosity characteristic of the S.e. ssp. *coronaria* genome to a nearly complete homozygosity affecting all gene loci. Some alleles were probably lost by sampling errors, and those that were fixed would have been exposed to different selection pressures. Mayr suggests that the new genetic background, reflecting new combinations of alleles at a large number of loci, would have in itself constituted a feedback effect, changing the selective values of particular alleles. Mayr refers to such changes as a "genetic revolution" and considers that it is likely to facilitate the evolution of a "new system of gene interaction or different epigenetic system" (1963, p. 544). It should be possible to determine whether the theoretical concept of genetic revolution provides an appropriate description of the properties of the S. *malheurensis* genome relative to the genome of S.e. ssp. *coronaria*.

My approach to this problem has been to determine the extent of divergence between the species in a wide range of phenotypic characters. In addition to evidence from morphology and electrophoretic comparisons that served to establish the overall genetic similarity of the species, the characters investigated have included duration of vegetative growth (number of days to initiation of stem elongation), dry matter production and allocation, growth rates, competitive responses when grown in pure and mixed species stands, gas exchange, seed dormancy, and phenotypic variability. Most of the studies have been done in the laboratory under controlled experimental conditions. It is anticipated that the significance of divergence between the species will eventually have to be evaluated in the natural habitat. The criterion of interest will be the extent that a particular difference contributes to separating their niches.

In an early study, both species were grown (one plant per pot) in 11 different experimental conditions obtained by manipulating soil composition, temperature, and photoperiod regimes to simulate usual, stressful, and atypical environments (Gottlieb, 1977a). Species differences in patterns of growth became apparent during the first weeks following germination. The seedling rosettes of S. malheurensis initially increased in total leaf length more rapidly than those of S.e. ssp. coronaria, probably helped by reserves stored in their larger cotyledons. However, by the end of the fourth week, S.e. ssp. coronaria achieved a faster rate of increase and within the next two weeks initiated stems (bolted), an important developmental change which marks the onset of events leading to flowering and reproduction. Bolting occurred in S.e. ssp. coronaria about 7 to 10 days earlier than in S. malheurensis. Thus, since leaf formation continues until bolting, the S.e. ssp. coronaria rosettes were smaller when they bolted than those of S. malheurensis. Nevertheless, the S.e. ssp. coronaria stems reached the same height as the S. malheurensis stems, and bore more and longer branches with twice as many florets per individual (Table 11.1).

The more rapid vegetative growth of S.e. ssp. coronaria, permitting it to flower first, and its increased floret number combine to provide the potential to produce significantly more seeds than S. malheurensis (both species live about the same length of time). The reduction in number of florets and the presumed consequent reduc-

TABLE 11.1.

COMPARISON OF S.e. ssp. coronaria AND S. malheurensis FOR CHARACTERS OF PLANT SIZE AND FECUNDITY UNDER VARIOUS TREATMENTS[a]

Character	Treatment								Overall mean
	1	2	3	4	5	6	7	8	
Rosette diameter	0.90	1.05	0.69	1.10	0.90	0.79	0.89	0.92	0.90
Stem height	0.90	1.14	1.09	0.99	0.84	0.92	1.03	0.86	0.97
No. of branches	1.75	1.78	1.96	1.48	1.36	1.13	1.24	1.14	1.48
Total branch length	0.69	1.12	1.54	1.51	0.96	1.06	2.94	1.12	1.37
Total no. of florets per plant	1.76	1.76	2.40	1.72	2.05	2.09	2.41	1.75	1.99

[a]Modified from Gottlieb, 1977a. Values represent means for S.e. ssp. coronaria divided by means for S. malheurensis.

tion in seed number in S. *malheurensis* may be necessary to compensate for the increased weight of its seeds (two times heavier), on the hypothesis that the proportion of metabolic resources that can be allocated to seed production in these annual species is more or less the same (Gottlieb, 1977a). The ecological equivalence of such a tradeoff is uncertain, however, since the proportion of florets that actually set seed in the species under natural conditions and the relative viability and germinability of the seeds are not known.

In these species, stem height, floret number, and other measures of adult size are correlated with the size and vigor of the seedling rosettes. Consequently, the growth of the seedlings was examined in more detail. Particular emphasis was paid to dry matter accumulation and allocation, growth rates, and gas exchange. Approximately 50 plants of each species were tested after growth under each of three thermoperiods (30/20, 20/12, and 12/5° C). Values for relative growth rate (RGR), leaf area ratio (LAR), and net assimilation rate (NAR) were obtained from two measurements of each individual during its rosette growth, and then averaged to give population means and variances. Net CO_2 uptake was measured on whole rosettes with an infrared gas analyzer, and dark respiration of leaf tissue was determined manometrically on leaf segments.

The results (Gottlieb, 1978b) provided a more explicit description of the differences in rosette growth observed in the previous study. In all three tested conditions, S. *malheurensis* individuals produced leaves with a smaller surface area and greater dry weight per unit area than those of S.e. ssp. *coronaria*. In addition, the S. *malheurensis* roots were always heavier and their root/shoot ratios substantially higher. Both species had similar RGR values, but in S. *malheurensis* this resulted from the combination of a lower LAR with a higher NAR, whereas in S.e. ssp. *coronaria* it resulted from the opposite combination (Table 11.2). Thus, the pattern of vegetative growth of the species does not appear to differ in growth rates, but rather in the extent of leaf expansion, specific leaf weight, relative amount of assimilates allocated to roots, and duration.

Both species showed similar mean rates of apparent net photosynthesis and dark respiration (Table 11.2). Strong associations were not found between the gas exchange and growth parameter values.

TABLE 11.2.
MEANS, AND COEFFICIENTS OF VARIABILITY (CV) FOR DERIVED GROWTH PARAMETERS, NET PHOTOSYNTHESIS, AND DARK RESPIRATION IN S.e. ssp. coronaria [COR] AND S. malheurensis [MAL] GROWN IN CONTROLLED ENVIRONMENTS UNDER THREE TEMPERATURE REGIMES[a]

Parameter	Set 1 (High)			Set 2 (Medium)			Set 3 (Low)		
	n	X̄	CV	n	X̄	CV	n	X̄	CV
LAR[b]									
COR	52	0.42	8.7	49	0.43	8.9	54	0.35	15.9
MAL	44	0.35***	7.4	55	0.33***	5.4	54	0.29***	7.9
NAR[c]									
COR		0.33	20.8		0.29	14.3		0.20	30.6
MAL		0.34	15.9		0.32**	10.0		0.25***	20.0
RGR[d]									
COR		0.14	15.8		0.12	13.2		0.07	35.3
MAL		0.12	11.6		0.11	8.4		0.07	24.3
Net CO$_2$ uptake[e]									
COR		5.9	27.2		6.3	21.7		4.0	31.9
MAL		5.8	41.8		7.4*	15.8		4.2	22.4
Net O$_2$ uptake[f]									
COR		56.8	30.6		32.4	27.8		28.4	31.2
MAL		60.2	19.1		32.3	30.0		28.0	32.9

[a]Temperature regimes were as follows: high, 30/20°C; medium, 20/12°C; low, 12/5°C. Photoperiod was 10 hours; gas-exchange measurements were taken at the temperature of the photoperiod.

[b]Values for LAR (leaf area ratio) are in square centimeters per milligram.

[c]Values for NAR (net assimilation rate) are in milligrams per square centimeter per day.

[d]Values for RGR (relative growth rate) are in milligrams per milligram per day.

[e]Values for CO$_2$ uptake are in milligrams of CO$_2$ per square decimeter of leaf area per hour at 75 watts per square meter.

[f]Values for dark respiration (net O$_2$ uptake) are in microliters of O$_2$ per milligram dry weight of leaf segment per minute × 1000.

*P < 0.05; **P < 0.01; ***P < 0.001.

Similar absence of high correlations between the two character sets is the case with many crop plants (Wallace et al., 1972). Wallace et al. (1972) point out that a high correlation between a physiological component such as net CO_2 uptake and yield is most unlikely since it would indicate that all genetic variability affecting yield occurred in the former component. Consequently, the growth phenotype of individual plants results from different combinations of the expressions of many different characters, some being more and others less favorable. Since the species are annuals, the phenotypic balance of these characters can be selected, depending on heritabilities, to track short-term environmental fluctuations. (However, this would be counteracted by the genetic dampening affect of seed storage in the ground; Gottlieb, 1974a). For the present purposes, this means that fitness differences between the species will have to be formulated in terms of whole-plant phenotypes at the population level, and not on the basis of the mean expression of single characters.

Thus, an important finding was that individuals of S. *malheurensis* do not attain the large physical size in nature that many plants of *S.e.* ssp. *coronaria* do. This was discovered by measuring shoot dry weights of 104 adult plants of *S.e.* ssp. *coronaria* and 83 of *S. malheurensis*, collected the same day within a 1 by 400 m belt transect. The distributions of the shoot dry weights of single individuals of both species were strongly skewed, with a very large number of small plants and relatively few large ones (Gottlieb, 1977a,b). The average weight of the *S.e.* ssp. *coronaria* plants was twice that of *S. malheurensis*, though their median weights were similar. However, 13 percent of the *S.e.* ssp. *coronaria* individuals were heavier than the heaviest one of *S. malheurensis*. Shoot dry weight of adult plants is primarily comprised of stem, branches, and flowers, making the entire shoot an inflorescence. Thus, a substantial fraction of the parental species is both more vigorous and more fecund in the same local habitat.

In combination with its rarity, the smaller shoot dry weight of S. *malheurensis* in nature appears to be a serious disadvantage which could limit its persistence. Consequently, another study was carried out to examine the effects of competition within and between the species. Specifically, this study tested the hypothesis that the apparently lower fitness of S. *malheurensis* results, at least in part, from inability to compete against its progenitor. The two species were grown by themselves (pure stand) and with each other (mixed stand) at two densities (4 and 16 plants per pot), and harvested at weekly or biweekly intervals during a 12-week period between germination and full-flower. Several important results are summarized here, and more detailed analysis will be presented elsewhere (Gottlieb, Bennett, and Taillon, unpublished data).

The pattern and relative duration of rosette growth of the species were similar to those observed when they were grown singly in pots. *Stephanomeria exigua* ssp. *coronaria* plants bolted earlier and reached 15 to 20 cm stem height when six weeks old whereas S.

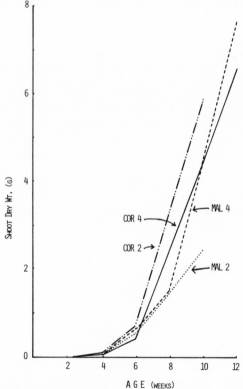

Figure 11.3. Mean shoot dry weight, on a per plant basis, of *S. malheurensis* and *S.e.* ssp. *coronaria* grown in pure stand (C4 and M4) and mixed stand (C2 and M2) at a density of four plants per pot. Harvests 1, 2, and 3 were made at weeks 2, 3, and 4, and harvests 4, 5, 6, and 7 at weeks 6, 8, 10, and 12, respectively.

malheurensis individuals at this age had just begun to bolt. In pure stands, at both planting densities, the mean shoot dry weights of the species were similar (Figures 11.3 and 11.4). However, when both species were grown in the same pots, the S.e. ssp. *coronaria* shoots showed increased weight relative to that in pure stand and S. *malheurensis* showed a decrease in weight in the late harvests at both densities. Since shoot dry weight indicates vigor and floret number (reproductive potential), the increased weight of S.e. ssp. *coronaria* when grown alongside S. *malheurensis* suggests that it enjoys a competitive advantage against its derivative. The data on partitioning reveal that a higher proportion of the shoot weight in S.e. ssp. *coronaria* is comprised of branch weight (the secondary branch category includes flowers and unopened flower buds), a difference which is particularly evident in mixed stands (Figure 11.5). At each harvest, S. *malheurensis* roots were heavier, resulting in higher root/shoot ratios (Figure 11.6), and also similar to the results when the plants were grown singly.

The competition study documented that the distinctive phenotypic properties of S. *malheurensis* do not provide competitive advantages in terms of plant size or fecundity. This result further suggests that the persistence of the species probably depends upon

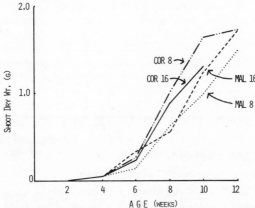

Figure 11.4. Mean shoot dry weight on a per plant basis, of S. *malheurensis* and S.e. ssp. *coronaria* grown in pure stand (C16 and M16) and mixed stand (C8 and M8) at a density of 16 plants per pot. See caption to Figure 11.3 for details of harvests.

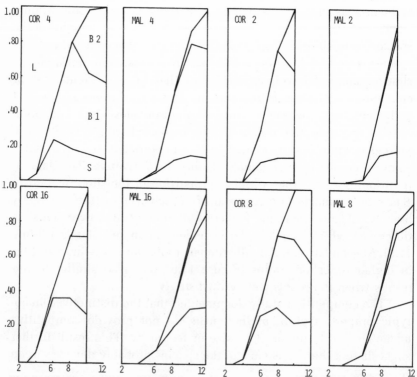

Figure 11.5. Partitioning of mean shoot dry weight in S. *malheurensis* and S.e. ssp. *coronaria* grown in pure stands (C4, M4, C16, M16) and mixed stands (C2, M2, C8, M8) at 4 and 16 plants per pot. The ordinate is percentage of total shoot dry weight, the abscissa is age of plants in weeks. See caption to Figure 11.3 for details of harvests. For each diagram, the designation of plant is as shown in C4: L, leaves; S, stems; B1, primary branches; B2, secondary branches, flowers, and unopened buds.

the utilization of a niche which is not available to S.e. ssp. *coronaria*. If this were not the case, the relative performance of the two species indicates that S. *malheurensis* would soon be eliminated.

The delay in onset of bolting and seed set is likely to be ecologically disadvantageous in the rigorous environment of eastern Oregon, with its summer drought and frequent late-summer frost. Such climatic conditions increase the probability of death as summer progresses, and plants which start to set seeds at a relatively later date may thereby set fewer of them.

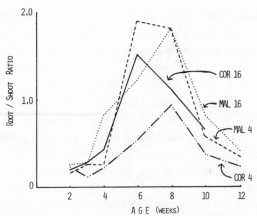

Figure 11.6. Root/shoot ratio, on a per plant basis, of *S. malheurensis* and *S.e.* ssp. *coronaria* grown in pure stand at 4 and 16 plants per pot. See caption to Figure 11.3 for details of harvests.

PHENOTYPIC VARIANCE OF THE PROGENITOR
AND DERIVATIVE SPECIES

Comparison of the ranges of variation of the measured characters can reveal which aspects of the phenotype of *S. malheurensis* have been assembled from common genetic building blocks present in *S.e.* ssp. *coronaria*, which represent rare or infrequent features in the progenitor, and which are unique to *S. malheurensis*. This comparison could lead to an estimate of the proportion of the phenotype of *S. malheurensis* that was extant within its progenitor at the time of its origin. If the proportion were high, this would suggest that relatively little epigenetic change accompanied its origin. If the proportion were low, this would be taken to mean that its genome had come through a genetic revolution.

A total of 33 quantitative characters were measured on each of more than 100 individuals of each species, grown in the 11 different experimental treatments described in the previous section. These traits included 14 seedling characters, 11 stem and branch characters, and 8 floret and seed characters. The data were subjected to an analysis of variance using a completely randomized, single-classification design. The between-environment, or treatment, variance measured the similarity of the responses of the species to the treatment conditions, and the between-plant, or error, variance measured all the genetic and microenvironmental sources of variation contributing to the differences among replicate plants.

The between-plant variances were larger in S.e. ssp. *coronaria* than in S. *malheurensis* in 29 of the 33 characters, with 11 differences significant at 1 percent and five others at 5 percent (Gottlieb, 1977a). Nearly all the size and growth rate characters of the seedlings of the progenitor were more variable, whereas the floret and seed traits, as a group, had the fewest significant differences between species. The larger coefficients of variation of S.e. ssp. *coronaria* for the growth parameters, dry weights, and other characters measured under the three temperature regimes (Gottlieb, 1978b), indicate that the differences in variability between the species probably extend throughout their phenotypes. For many characters in S. *malheurensis*, the ranges of variation were nearly entirely included within those of S.e. ssp. *coronaria* and, for other characters, the mean expression of S. *malheurensis* was well within the S.e. ssp. *coronaria* range. Overall, the analysis of quantitative characters upheld the validity of the previous electrophoretic comparison which had shown that the assayed alleles of S. *malheurensis* represent a highly limited extraction from the richly polymorphic gene pool of its progenitor.

Another aspect of the similarity of the species was shown by ranking their between-environment variance components, from the one with the smallest value to the one with the largest value. The rankings were nearly identical (Spearman's coefficient of rank correlation, $r_s = 0.90$, $P < .001$; Gottlieb, 1977a). The least plastic characters in both species were those having to do with the florets and seeds, and the most plastic characters were seedling growth rates. The result suggests that both species maintained similar patterns of developmental integration across the diverse experimental environments, and it is particularly interesting in view of the demonstrated difference in dry matter partitioning.

Complete absence of overlap in the ranges of variation was observed in very few characters. One was the length/width ratio of the ligules, a change often correlated with change in breeding system. The other was seed length, which was measured both in the native habitat and under the more uniform conditions of the greenhouse (Figure 11.2). The remarkable separation of seed lengths in the two species, juxtaposed with the opposite condition for nearly every other character measured, including both morphological and growth characters, underscores the critical importance of seed differences in the overall ecology and adaptation of the species.

SUMMARY AND CONCLUSIONS

The comparative studies of S. *malheurensis* and S.e. ssp. *coronaria* represent an attempt to understand how the phenotype of a newly evolved species was assembled from the genetic legacy of its progenitor. The species were chosen for two reasons: their phylogenetic relationship appears unambiguous, with S. *malheurensis* originating recently; and the probably rapid and abrupt origin of S. *malheurensis* (which involved a change from high genetic heterozygosity to nearly complete homozygosity as a result of its predominant self-pollination increased the likelihood that it went through a genetic revolution and acquired a distinct epigenetic system (*sensu* Mayr). Novel epigenetic patterns, presumably, are the basis of the distinctive phenotypes of species. But why a species has one set of properties and not some other set has not been specifically studied. The problem is appropriate to population biology because species originate in an ecological context, so that the characteristics of their environment, at the time of their origin, probably play a major role in the initial selection of their epigenotype. Once evolved, the epigenetic pattern may constitute a highly conserved feature that is not readily disturbed by divergent local adaptation (e.g., *Achillea*).

The morphological features of S. *malheurensis*, the particular alleles it possesses coding enzymes detected by electrophoresis, the mean values of its growth rates, and the increased duration of its vegetative growth prior to stem elongation represent character expressions which have been found within the wide ranges of variability of its highly polymorphic progenitor. For example, a small proportion of *S.e.* ssp. *coronaria* plants has been observed with reduced branch number and more widely spaced flowering heads. Other individuals display slow seedling growth rates, high number of days to bolting, and high root/shoot ratios. The most significant divergence between the species, in addition to the change in breeding system and loss of the freezing requirement for seed germination, is the marked difference in the length (weight) of their seeds.

This difference may be the basis for many of the morphological and developmental differences between the species, on the simple hypothesis that the larger seeds of *S. malheurensis* require an increased supply of assimilates, which is achieved by reducing seed number. Number is reduced by reducing branch number and spacing adjacent flowering heads farther apart. The distinctive rosette growth of *S. malheurensis* seedlings, with their larger root/shoot ratios, may be important for the bolting of its stems, which are stouter and bear stouter branches than stems of equivalent height in *S.e.* ssp. *coronaria* (Figure 11.7).

An alternative hypothesis based on an ecological perspective is that the heavier roots of *S. malheurensis* extend deeper into the soil and tap water levels not ordinarily utilized by *S.e.* ssp. *coronaria*. This might be important during seedling establishment. However, little difference in length was found between the roots of adult plants of the species which were excavated in nature. Even if the heavier roots of *S. malheurensis* facilitate seedling establishment, such a capability furnishes only a limited advantage since its population size is very small relative to that of *S.e.* ssp. *coronaria*.

The difference in seed weight might be the basis for niche separation if it permits the large *S. malheurensis* seeds to utilize safe-sites in the soil which are not suitable for the smaller seeds of its parent. Differences in requirements for germination and establishment could represent a meaningful subdivision of their common hilltop habitat and may permit *S. malheurensis* to persist. However, it appears that only a few sites are suitable for large seeds since the species have an overwhelmingly disproportionate abundance.

This model suggests that large seeds were selected in the new species because they adapted it to a different niche. The safe-site hypothesis is a critical one because the studies to date have ruled out

Figure 11.7. Representative plants of S. *malheurensis* (right) and S.e. ssp. *coronaria* (left) of the same age grown under uniform conditions. Note that both species are the same height and that S. *malheurensis* has a thicker stem and fewer, thicker branches. A number of flower heads have opened in S.e. ssp. *coronaria*, but not in S. *malheurensis*.

the possibility that the persistence of S. *malheurensis* is the consequence of advantage of size, growth rates, gas exchange, competitive ability, or several components of fertility. The changes in morphology, allocation, and rosette duration presumably reflect developmental responses to this selection pressure. Since most of the basic genetic elements were already present within its progenitor, it is not necessary to view the present phenotype of S. *malheurensis* as the consequence of a genetic revolution. Relatively minor readjustments may have sufficed.

Alternatively, the S. *malheurensis* phenotype may not entirely represent the consequences of selection for increased seed size. For example, it may simply reflect the genetic characteristics of the particular individuals of *S.e.* ssp. *coronaria* from which *S. malheurensis* descended. The *S. malheurensis* phenotype may represent a founder effect. If this were true, then the large seed phenotype has since been selected against in *S.e.* ssp. *coronaria*.

The differences between progenitor and derivative probably have to do with changes in the regulation of genes rather than in the products they specify; in this sense, the present results are consistent with recent thinking about the nature of the genetic differences between species (for example, King and Wilson, 1975). The present example demonstrates, however, that many of the regulatory patterns characteristic of a newly evolved annual plant species are not unique but are already present within the progenitor population.

Acknowledgments

Different aspects of this research were supported by grants from the National Science Foundation and by a fellowship from the John Simon Guggenheim Memorial Foundation, for which I am very grateful. The gas exchange studies were conducted in the laboratories of Drs. C. F. Eagles and D. Wilson, Welsh Plant Breeding Station, Aberystwyth. I thank them for their interest and graciousness while I was at the station.

12 COMPETITION AND COEXISTENCE OF SIMILAR SPECIES

PATRICIA A. WERNER

A BASIC QUESTION in ecology is Hutchinson's (1959) classic, Why are there so many animals (or plants!)? The way in which this question is answered depends to some extent on the scientific background of the researcher. The systematist or geneticist might attack the problem from the viewpoints of sympatric speciation, variations in breeding systems, ecotypic variation, phenotypic plasticity, patterns of gene flow, phenology, etc. Almost all of these approaches attempt to explain how species or ecotypes do or do not remain isolated from each other (often with hypotheses relating to the evolutionary development of the phenomena), but not how they manage to coexist in the same habitat. To a geneticist, the problem of species range, abundances, and cooccurrence may be mainly a genetics problem; i.e., each species is viewed as limited by its genetic "equipment" (see Antonovics, 1976b). What quality of the environment selects for various genetic schemes may be of interest only secondarily. An ecologist may approach the problem by finding some phenotypic character, morphological or physiological, that differs systematically among a group of similar plants and then ascribe to those differences a mechanism(s) that allows coexistence. This approach seems more promising in relating environmental factors to organism constraints, but, without additional experimental manipulation or genetics study, it cannot be demonstrated that the proposed mechanism is in fact operating to produce the species distributions observed in the field. Indeed, there are few studies (for plants or animals) which demonstrate the relationship between potential distributions of a population across some resource gradient, given the absence of competitors, and actual distributions, given the presence of competitors. The area where the potential distributions of closely related forms overlap is where competition can occur; whether the competition is among conspecifics, congeners, or guild members, it is a powerful molder of life-history characteristics and is of interest to the systematist, geneticist, and ecologist.

The answer, or answers, to Hutchinson's question also depends to some extent on what scale is being considered. Ecologists' attempts to answer the question of coexistence on a local scale have resulted in extensive theoretical and empirical studies relating to interspecific competition and niche. The most recent arguments have centered on niche shifts, niche compression, predation–competition interactions, the effects of regular disturbances, niche overlap, and the degree of limiting similarity of species (e.g., Mac Arthur and Levins, 1967; Whittaker, 1969, 1972; Roughgarden, 1972; Schoener, 1974a; Levin and Paine, 1974; Werner and Hall, 1976; Whittaker and Levin, 1976; E. E. Werner, 1977). In general, the theory in this field is based upon assumptions of life histories that best fit mobile animals; only now are researchers examining how and which aspects of current theory apply to plants (e.g., Schaffer and Leigh, 1976). Schoener (1974b) reviewed the problems of coexistence in animals and summarized many of the important questions regarding the evolution and mechanisms of competitive relationships among cooccurring heterotrophs. Although he did not attempt to deal with autotrophs, many of his conclusions apply to plants on an abstract level and differ only (but significantly) in the details of the competitive mechanisms.

In this paper I briefly explore the nature of plants relative to competitive mechanisms, discuss current research in life-history adaptation and cooccurrence of several *Solidago* species, present some patterns of trade-offs in such characteristics of weedy plants, and suggest paths for future work. I do not attempt to distinguish or discuss such terms as coexistence (implying a stable equilibrium) versus cohabitation (implying a sharing of habitat or resources) versus cooccurrence (implying only a mutual presence of species). Differences in meaning between terms stem mainly, if not entirely, from differences in temporal and/or spatial scale.

THE NATURE OF PLANTS, COMPETITION AND COOCCURRENCE OF CONGENERS

The terrestrial higher plant system is basically a plane with portions of sessile organisms physically emergent above and below the soil surface. In general, all individuals are "born" at or near the surface of the plane and require the same major resources—light, carbon dioxide, water, and nutrients. On the basis of these "food" resources alone, it is not easy to imagine a sufficient number of niches to explain a species-rich plant community (see extended discussions along this vein in Harper et al., 1961; Whittaker, 1969; Grubb, 1977). Since plants are sessile and require basically the same resources, they are competitors for space. Any plant that can hold a space succeeds in preempting the resources required by other plants. And, if competition for space is important, then it becomes crucial to our understanding of long-term coexistence that we refine our ideas on how space opens up, how it is colonized, and how it is held by a plant against potential invaders.

Empirical studies have revealed various mechanisms adapted by plants that reduce competition and explain patterns of coexistence. These mechanisms have been reviewed by Grubb (1977) and include differences in life forms (Sarukhán, 1974; Turkington, 1975; Thomas and Dale, 1976; Grubb, 1976; P. A. Werner, 1976, 1977; Parrish and Bazzaz, 1976a; Hickman, 1977), phenological separation (Hurlbert, 1970; Bratton, 1976), fluctuations in the environment (Cantlon, 1969; Rabotnov 1974; Thomas and Dale, 1976), balanced mixtures (Marshall and Jain, 1969b), and variation in competitive ability with age (Watt, 1955).

Grubb (1977) pointed out that most of the mechanisms studied to date involve interactions among adult plants. Although studies of adult–adult interactions are enlightening and will continue to be important, there is increasing awareness that they may be insufficient in explaining the diversity of plants found in a local area. This awareness is not only a logical deduction that derives from considering competition for space among plants but is also based to a great extent on the findings of plant demographic studies of the past decade (see Hickman, article 10; Cook, article 9; Harper and White, 1974.)

In most cases, the probability of death is highest in the seed, seedling, and juvenile periods of the life cycle, and drops to a very low, often constant, level during the adult period (up to senescence) (Harper, 1967; Hett and Loucks, 1971; Hett, 1971; Antonovics, 1972; Thomas, 1972; Sharitz and McCormick, 1973; Sarukhán and Harper, 1973; Hawthorn and Cavers, 1976; King, 1977a; Werner and Caswell, 1977). Once an individual plant survives early life, it can usually hold the site as an adult. Often the species arrays seen in the field are the result of events that occurred during a time when seedlings were present but that are no longer apparent to even the most careful observer. Harper (1977) cites examples.

Another pattern that has emerged from recent ecological studies concerns the places where seedlings survive. In general, seedlings of most species survive in various types of openings. The type of opening required depends on the species; in some cases, openings are completely bare ground, in others, disturbances of the litter layer, discontinuities in a low herbaceous ground cover, and so on (e.g., Cavers and Harper, 1967; Livingston and Allessio, 1968; Summerfield, 1972; Beimborn, 1973; Watt, 1974; Marks, 1974; P. A. Werner, 1975c; Platt, 1975; Jalloq, 1975; Raynal and Bazzaz, 1975; Forcier, 1975; Grubb, 1977; Fox, 1977; T. J. King, 1977a). Within such openings there are a myriad of even more subtle, species-specific, sometimes biotype-specific, "safe-site" (Harper, 1967) requirements for seedling emergence and survival, which are under varying intrinsic and environmental controls (see Harper, 1977). The result is that, for almost any terrestrial plant species (or other taxonomic group, even an individual), the environment can be conceived of as a matrix where seeds do not germinate (or seedlings do not survive), with patches throughout wherein seeds do germinate and seedlings do survive to adulthood. Since the spatial and temporal pattern of the openings regulates the presence and placement of the plants, the interaction of such heterogeneity with the biology of the species is of great importance for understanding coexistence of plant species and should give insight into one aspect of natural selection in plants.

The above discussion of past work on mechanisms of competition among adult plants and the recognition of the importance of looking at the recruitment phase of the life cycle suggest that, for convenience, competition among terrestrial plants may be divided into two broad categories: preemptive competition, or competition to reach (or to be at) an open site first and preempt space; and interactive competition, operationally defined as the reduction in performance of an established plant by the activities of another. In the first type of competition, often associated with fugitive or colonizing species (Baker and Stebbins, 1965), dispersal ability and dispersal pattern of seeds (or other diaspores) carry a very high premium for survival. The particular strategy evolved depends upon the distribution and predictability of safe-sites in space and time, which select for dispersal distance, timing of release of seeds, dormancy (dispersal in time), longevity of seeds, etc. Undoubtedly, some researchers would prefer to think of this first type of competition not as competition at all but as escape from competition. Given that fugitive plants may initially escape competition with slow-growing, larger-seeded, vegetatively reproducing plants (which often replace them eventually), they certainly do compete with other, similar, fugitive plants for openings. The successful individual is the one that arrives first. Examples of fugitive guilds are found in the work of Platt (1975), Grubb (1976), and Parrish and Bazzaz (1976a).

The second type of competition assumes that plants are in the physical presence (sphere of influence) of each other and are mutually interfering with one another's ability to gather resources. Here such characteristics as growth rate or size and position of leaves and underground structures convey competitive superiority. As with most sessile organisms, a plant does not "read" or perceive interactive competition unless its space is violated. The competitive effect on other plants which enter the physical border of a plant can be very strong. For example, Mack and Harper (1977) found that 69 percent of the variation in size and reproductive output of individual plants of dune annuals could be attributed to neighboring plants within a radius of 2 cm of the plant in question.

In order to study, in an empirical manner, plant adaptations which reduce either type of competition and which permit coexistence, a researcher must have a group of closely related plants living in the same area. Ideally, the plants should be congeners, or genotypic races, that remain genetically isolated when growing in the same habitat. In actuality, such situations are not often found: The more taxonomically related the plants, the less likely they are to be found in the same habitat. Because of this, many of the ideas we have concerning the interaction of particular selective factors and particular plant characteristics have been derived from comparative studies of ecologically similar (but taxonomically very different) plants growing in the same habitat (Grubb, 1976; Jain, 1976a; P. A. Werner, 1977; Parrish and Bazzaz, 1976a; Harper, 1977). Alternately, we have learned much from the comparative studies of congeners living in different habitats, e.g., species of *Rumex* (Harper and Chancellor, 1959), *Papaver* (Harper and McNaughton, 1962), *Trifolium* (Harper and Clatworthy, 1963), *Solidago* (Abrahamson and Gadgil, 1973), *Agropyron* (Tripathi and Harper, 1973), *Ranunculus* (Sarukhán and Harper, 1973; Sarukhán and Gadgil, 1974; Sarukhán, 1974), *Helianthus* (Gaines et al., 1974), *Hieracium* (Thomas and Dale, 1976), *Plantago* (Hawthorn and Cavers, 1976), *Polygonum* (Hickman, 1977), and *Asclepias* (Wilbur, 1976). Some of the most promising insights have come from studies of genetically distinct groups of one species living in the same area, e.g., *Taraxacum officinale* (Solbrig, 1971; Gadgil and Solbrig, 1972; Solbrig and Simpson, 1974, 1977) and *Trifolium repens* (Turkington, 1975), from studies of congeners living in the same habitat, e.g., *Avena* (Marshall and Jain, 1967), and from the few studies that demonstrate the relationship between potential distributions of a population, given the absence of competitors, and actual distributions, given the presence of similar forms (e.g., for annuals, Sharitz and McCormick, 1973, and Pickett and Bazzaz, 1976; for biotypes of *Taraxacum officinale*, Solbrig and Simpson, 1974; for *Hieracium* species, Thomas and Dale, 1976; for algae species, Titman, 1976).

LIFE-HISTORY ADAPTATIONS AMONG COOCCURRING GOLDENRODS

Currently in progress is a long-term study of cooccurring goldenrods (*Solidago* spp., Compositae). Data from past and ongoing studies are presented to demonstrate the sort of patterns in life-history adaptations and interactions that can be found among cooccurring congeners and what sorts of implications can be drawn from them. Several species of *Solidago* (up to 6 or 7, and rarely 10) may cooccur in a single field; in most localities, the species are quite distinct from each other, although on a regional scale, morphological characteristics intergrade.

The study reported here mainly involves five species of perennial goldenrods, *Solidago nemoralis* Aiton, *S. missouriensis* Nuttall, *S. speciosa* Nuttall, *S. canadensis* Linnaeus, and *S. graminifolia* (Linnaeus) Salisbury, which are all native to North America and are widely distributed east of the Rocky Mountains. Their habitats are abandoned fields, roadsides, waste places, and grasslands.

Morphologically, the goldenrods differ in height at maturity and in extent of vegetative propagation from "rhizome-runners" (so called because they are horizontal stems that are usually below the soil but may also "run" on top of the ground if there is a moist litter layer) (Figure 12.1, Table 12.1). Both *S. nemoralis* and *S. speciosa* form clumps of stems at a single node, but the stems are shorter than those of the other species. *Solidago nemoralis* forms almost no new rhizomes; reproduction is almost entirely by seeds; *S. speciosa*, on the other hand, produces some new buds of rhizomes on the parent rhizome, which results in small clones less than 0.5 m in diameter. *Solidago missouriensis, S. canadensis,* and *S. graminifolia* produce

TABLE 12.1.

CHARACTERISTICS OF THE ADULT PHASE OF FIVE SPECIES OF GOLDENROD (Solidago)
IN A 25-YEAR-OLD FIELD[a]

	S. nemoralis	S. missouriensis	S. speciosa	S. canadensis	S. graminifolia
Flowering stems					
Height (cm)	42	64	59	102	87
No. per node	3.1	1.0	2.0	1.1	1.0
Vegetative cloning					
New buds per ramet[b]	0.1	7.4	3.6	11.3	3.2
Resource allocation (percentage biomass)	0	3.9	2.7	10.5	14.1

[a]Values are means. After Werner, 1976.
[b]Mean number produced annually per ramet, i.e., per plant unit comprised of a rhizome, roots, and erect stem(s) with leaves, arising from a single node of the rhizome. Each new rhizome bud has the potential for producing a new ramet.

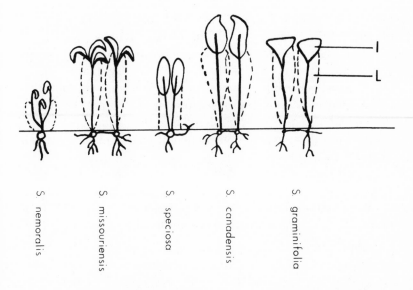

Figure 12.1. Schematic drawing showing the morphological differences in adult plants of five goldenrod species in Michigan. *I*, outline of the area of inflorescence; *L*, outline of the area of leaves.

only one erect stem at a rhizome node, and their stems are taller than those of the other two species. (*Solidago canadensis* is the tallest of all.) *Solidago missouriensis* and *S. canadensis* produce a very high number of new buds per parent rhizome, whereas *S. graminifolia* produces fewer new buds. In abandoned fields in Michigan, clones up to 2.5 m in diameter for *S. missouriensis* and *S. graminifolia* and up to 3.5 m in diameter for *S. canadensis* are not uncommon.

The five species cooccur in a 15-ha field in Kalamazoo County, Michigan, abandoned from pasture and some prior cultivation about 25 years ago. The vegetation is composed mainly of a mixture of forbs and grasses with scattered patches of *Rubus* and seedlings of oaks and cherry that began growth immediately after abandonment. The soil is rather poor in nutrients. The field has a gentle slope of 5 to 10°, which creates a soil-moisture gradient, with a maximum frequency for S. *nemoralis* at the drier end, followed successively by S. *missouriensis*, S. *speciosa*, S. *canadensis*, and S. *graminifolia* with increasing moisture. Zones of species overlap along the gradient are extensive. In general, the first two are dry-habitat species, the last two are wet-habitat species, and S. *speciosa*, with the widest range in distribution, is intermediate (top of Figure 12.2). It is recognized, of course, that a "gradient" of soil moisture is a complex dimension and that several environmental factors change over the gradient simultaneously.

The goldenrods in the 25-year-old Michigan field exist in a semiequilibrium environment where changes in vegetational composition are rather slow (Werner, unpublished data). The abundances and distributions of the goldenrods along the soil-moisture

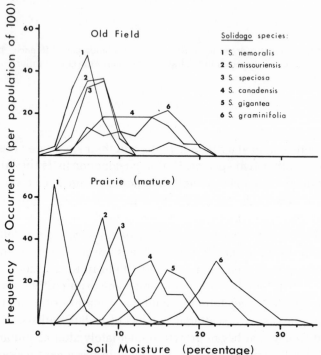

Figure 12.2. Distribution (frequency per sample of 100 per population) of individual plants of goldenrod populations along a soil-moisture gradient in an old field and a virgin prairie (Werner and Platt, 1976).

gradient depend upon the physiological capabilities of the species, the independent growth rates of the populations, and the interaction of the goldenrods with each other as well as with other plants and animals. Further, the outcome of interaction among goldenrod species will depend upon the probability of arriving at safe-sites and the relative competitive ability during interactions of adult–adult, seedling–adult, and seedling–seedling.

Adult–adult interaction takes place whenever one individual or vegetatively reproducing clone grows into the space of another. Such interaction is possible whenever both plants are physiologically capable of surviving the abiotic environment, but the ranges in physiological capabilties in relation to soil moisture have not been determined to date for the separate species. Nevertheless, for adult–adult interaction among goldenrods, it is instructive to compare the two dry-habitat species and the two wet-habitat species. From stem height and potential numbers of new rhizomes, we expect that S. missouriensis will be competitively superior to S. nemoralis in the range of soil moisture that both can tolerate. Similarly, we expect that S. canadensis will be competitively superior to S. graminifolia. Hence, if competitive interaction in the adult stage occurs, we expect that, with time, S. nemoralis will be restricted to the extreme driest end and S. graminifolia to the extreme wettest end of the continuum.

In fact, this is the case for goldenrods in a virgin tall grass prairie, where populations have coexisted for hundreds of years (Cayler Prairie, Dickinson County, Iowa). The species in the prairie are distributed uniformly along the same moisture gradient (bottom of Figure 12.2). In the old field, however, the species are broadly distributed over the soil-moisture gradient, with much overlap among species, which represents initial colonization events by seeds introduced from outside the field (site preemption). Field experiments are under way to determine whether the prairie distributions represent active competitive displacement or genetic shifts (reduced variation and shifts in the distributional means of the extreme wet-end and dry-end species); early results indicate that both factors play a role in determining the distributional breadth of the adult plants.

The observed distributions of adult plants support the current

theory of species packing. However, the changes in species distributions and overlap under different selective pressures can be interpreted in light of current theory only if the species distributions do
in fact represent competitive interaction. Since, in plants, competition is a neighborhood phenomenon (see Schaffer and Leigh, 1976);
it must be questioned whether these plants indeed enter each other's
neighborhoods. In actuality, few individual clones of adult plants in
either the old field or prairie (other than for S. *canadensis* and S.
graminifolia at the wet end of the field) abut another clone. The
explanation of species cooccurrence that derives from an examination of only adult clones may be insufficient, and an examination of
the seed and seedling stages is called for.

   In any seedling–adult interaction among goldenrods, there is no
contest. The seedling will die. In a seedling–seedling encounter, the
successful individual will usually be more physiologically suited to
the particular abiotic properties of the site, and, if the seedlings are
physiologically similar, the one with the larger embryo will survive
(see related work of Marshall and Jain, 1967; Stebbins, 1976a).

   The probability that a seedling will come in contact with
another goldenrod seedling in the first place will be determined by
seed number, dispersal ability, and dormancy properties. In terms of
number of seeds only, the probability that a goldenrod species will
have at least one of its offspring reach any particular site in the field
increases with increasing soil moisture. The same trend, but less
obvious, is found in dispersal ability (Table 12.2). Differential dormancy can be dismissed as a factor in goldenrods because seeds are
rarely viable more than 1 year under field conditions. Although S.
graminifolia adults have the highest chance of leaving seeds in an
open site, the seedlings of the larger-seeded species (e.g., S. *nemoralis*) have a greater probability of surviving either seedling–seedling
interaction or a dry period. There is a trade-off between the probability of a seed arriving at a safe-site (dispersability and numbers) and
the probability of its seedling surviving (seed weight), which is
balanced differently among the goldenrod species in each habitat
(see Table 12.2).

   Within a species, most of the prairie populations had seeds that
were heavier (averaging three times heavier) and fewer in number

TABLE 12.2.

CHARACTERISTICS OF THE SEED PHASE OF FIVE SPECIES OF GOLDENROD (Solidago) IN A 25-YEAR-OLD FIELD AND IN A VIRGIN PRAIRIE[a]

	S. nemoralis	S. missouriensis	S. speciosa	S. canadensis	S. graminifolia
Old field					
No. of seeds per ramet[b]	2,300	4,200	9,100	13,000	17,700
Weight of one seed (μg)	26.7	17.6	19.5	27.3	24.5
Relative dispersal ability[c]	1.0	1.7	0.7	1.7	2.3
Prairie					
No. of seeds per ramet	200	1,100	500	1,100	7,800
Weight of one seed (μg)	104.0	39.3	146.3	58.3	10.6
Relative dispersal ability	1.0	0.8	1.0	1.3	1.8

[a]Values are means. After Werner (1976) and Werner and Platt (1976).
[b]The term "seed" is used in the broad sense; a diaspore of Solidago is actually an achene with attached pappus. Values are mean numbers produced annually per ramet, i.e., per plant unit comprised of a rhizome, roots, and erect stem(s) with leaves, arising from a single node of the rhizome.
[c]Relative dispersal ability in either habitat is the ratio of the amount of time a seed remains airborne relative to S. nemoralis. The larger the value, the longer the seed remains airborne and the greater the potential dispersal distance.

(averaging about one-third reduction) (Table 12.2; Werner and Platt, 1976). In fact, when the same species are examined in habitats other than the 25-year-old field and the virgin prairie, the same trend holds—smaller and more numerous seeds with increasing environmental disturbance. This relationship is especially striking in S. *nemoralis* (Figure 12.3), where there is no overlap in the seed weight/number characteristics of the 25-year-old field and the virgin prairie populations. Another population, found on a dirt roadway adjacent to another prairie, had seed numbers and weights intermediate between the other two populations. In S. *canadensis* (Figure 12.4), the seed weight/number characteristics changed in the direction one would predict over four different habitats, but there was a great deal of overlap. We are only now doing the work that would indicate to what extent such differences in seed weight and number are phenotypically plastic in these and other *Solidago* species.

We do know that there is some advantage to larger seeds of

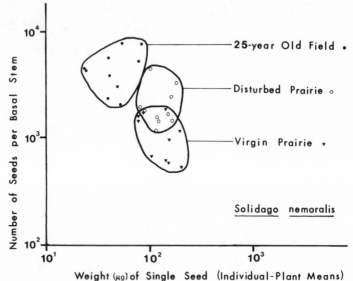

Figure 12.3. Relationships of the number of seeds (achenes) per basal stem and the mean weight (in micrograms) of a seed for individual plants (clones) of *Solidago nemoralis*. Plants are from a 25-year-old field in Michigan, a heavily used dirt roadway at the edge of a prairie in Wisconsin, and a virgin upland prairie in Iowa.

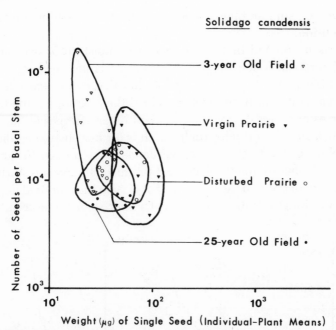

Figure 12.4. Relationship of the number of seeds (achenes) per basal stem and the mean weight (in micrograms) of a seed for individual plants (clones) of *Solidago canadensis*. Plants are from a 3-year-old field in Michigan, a 25-year-old field in Michigan, a heavily used dirt roadway at the edge of a prairie in Wisconsin, and a virgin upland prairie in Iowa.

goldenrod. Within a population, larger seeds have a higher probability of germinating (as determined by laboratory germination tests) (Figure 12.5), although a population with larger seeds does not have higher average germination than another population of the same species with smaller seeds (compare populations 12, 13, and 14 in Figure 12.5, all *S. canadensis*). In all comparisons (across populations and within populations), the heavier the seed, the larger the emerging embryo. Only a few seeds were unfilled; none was dormant, as determined by tetrazolium tests. If the results of work on other congeners can be extrapolated to *Solidago* (Stebbins, 1976a), then the larger embryos, i.e., those from heavier seeds, should be competitively superior, at least in the initial month of life, to smaller embryos of similar plants whenever the two meet in interactive competition.

It is postulated that, with decreased moisture or over successional time, selection operates to produce larger seeds (Stebbins, 1971b; Baker, 1974; Salisbury, 1974) and that coselected with this trait is a reduction in number, provided there is not a reduction in resources allocated to some other plant organ (Harper et al., 1970; Stebbins, 1971b; Gottlieb, article 11). It is further postulated that the counter-selective force against ever-larger seeds is tied to the seed weight/number trade-off. Plants that produce larger seeds must produce fewer of them. If the environment is such that survival depends upon reaching a suitable open spot rather than being a superior

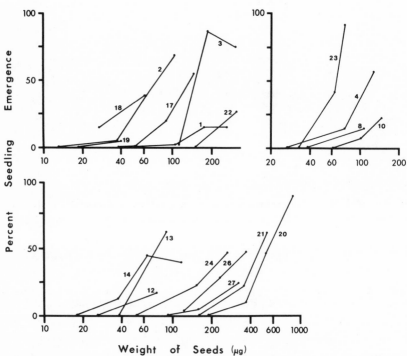

Figure 12.5. Percentage seedling emergence as a function of seed (achene) weight (in micrograms) for various populations of goldenrods (*Solidago* spp.). See legend to Figure 12.6 for species and habitat.

seedling competitor, then the plant with seeds just the minimum
size necessary to survive the abiotic rigors will reach more open
spots (i.e., have more genotypes represented in the next generation)
than will the plant that produced seeds larger than necessary and
(because of the seed number/weight trade-off) fewer in number.

Interestingly, when the study of seed characteristics is extended
to include other species of *Solidago*, we learn that there is a trade-off
in individual seed weight (W) and number per stem (N) such that
$\log N = k + b \log W$, or $N = kW^b$ (Figure 12.6). Since b is not
significantly different from -1, the equation reduces to $NW = k$,
where k is a constant. This means, roughly, that the mean biomass of
seeds (NW) per stem (i.e., per ramet) among populations of goldenrods is a constant value. This finding is in contrast to NW for
milkweeds (Wilbur, 1976, 1977). Individual perennial plants that
reproduce by both seed and vegetative cloning (that is, most goldenrods, but only one of the milkweed species) may set an upper limit
on the absolute amount of biomass per stem devoted to the production of seeds and allocate the "remainder" to clonal expansion. Such
a pattern of resource allocation is indicated in *S. canadensis* (Table
12.3), and is likely to be important in initial colonization of early-
successional environments in which competition will increase if
there is no abiotic disturbance. Similar patterns of great clonal
expansion in low-density situations have been found in *Tussilago
farfara* (Ogden, 1974), *Rubus hispidus* (Abrahamson, 1975), *Hiera-*

TABLE 12.3.
PERCENTAGE OF BIOMASS ALLOCATED TO TWO TYPES OF
REPRODUCTION IN *Solidago canadensis* L.[a]

Habitat	Percentage to sexual reproduction	Percentage to new rhizomes
Weeded garden	3.7%	28.8%
3-year-old field	9.2	32.5
Dirt roadway, edge of prairie	9.6	18.3
25-year-old field	9.0	10.5
Virgin prairie	11.5	13.2

[a]Values represent means of 15 plants.

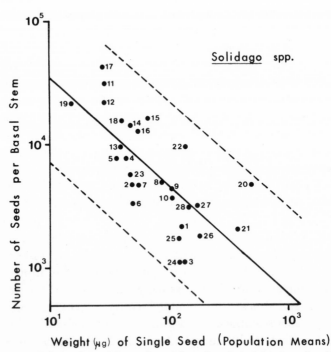

Figure 12.6. Relationship of the mean number of seeds (achenes) per basal stem and the mean weight (in micrograms) of an individual seed for 28 populations of goldenrods (*Solidago* spp.) in various habitats. 1–3: *Solidago nemoralis* (1, RP; 2, OF2; 3, VP). 4–7: *Solidago missouriensis* (4, OF2; 5, GG; 6, DP; 7, VP). 8–10: *Solidago speciosa* (8, RP; 9, OF2; 10, VP). 11–15: *Solidago canadensis* (11, OF1; 12, RP; 13, OF2; 14, GG; 15, VP). 16: *Solidago gigantea* (RP). 17–19: *Solidago graminifolia* (17, RP; 18, OF3; 19, VP). 20, 21: *Solidago rigida* (20, GG, 21, VP). 22: *Solidago sempervirens* (UP). 23: *Solidago remota* (DP). 24: *Solidago caesia* (EW). 25: *Solidago rugosa* (EW). 26: *Solidago houghtonii* (WP). 27: *Solidago riddellii* (WP). 28: *Solidago uliginosa* (WP). Letters in parentheses above represent the habitats for each sample number: (UP) urban parking lot, Chicago; (OF1) 3-year-old field, Michigan; (OF2), 25-year-old field, Michigan; (RP) heavily used dirt roadway at the edge of a prairie, Wisconsin; (GP) permanent grassland, formerly heavily grazed, Iowa; (DP) disturbed prairie, Illinois; (EW) edge of oak woods, Michigan; (WP) virgin wet prairie, Chilwaukee Prairie, Wisconsin; (VP) virgin upland prairie, Cayler Prairie, Iowa.

cium spp. (Thomas and Dale, 1976), *Fragaria virginiana* (Holler and Abrahamson, 1977), and *Agropyron repens* (Werner and Rioux, 1977).

Selection pressures are different in the prairie and old-field habitats; in today's world the safe-sites for the prairie populations would have to be within the 160 acres of undisturbed native vegetation, not in any of the plowed surrounding land. The safe-sites for the old-field population are outside the old field and are probably some other field or roadside in the township (there are almost no new seedlings within the 25-year-old field) (Figure 12.7). In both cases, the adult is a perennial and lives a long time. (Plants as old as 35 and 100 years are found in the old field and prairie, respectively). Over the lifetime of the adult plant, many seeds are produced, but most perish since suitable openings in space are rare. There is no way of knowing how closely the selective pressures of today's world match those under which the plants evolved. This is unfortunate; nevertheless, it is interesting and encouraging that seed characteristics of species along soil-moisture gradients and of populations of single species in different habitats follow expected patterns. The degree to which knowledge of seed characteristics, seedling interactions, and adult interactions will explain species cooccurrences remains to be seen.

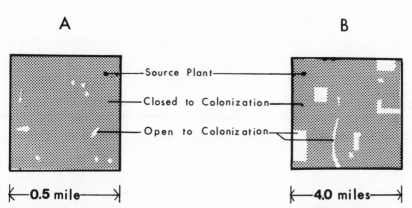

Figure 12.7. Schematic drawing illustrating the relative scale of "openings" for a goldenrod source plant in virgin prairie (*A*) and a 25-year-old field (*B*).

PATTERNS IN SEED CHARACTERISTICS
OF WEEDY PLANTS

Have other cloning perennial plants evolved the same seed characteristics as have the goldenrods, i.e., lack of dormancy and high dispersal from the parent? I compiled seed characteristics of over 200 weedy plants listed in weed manuals of the northeastern United States and those herbaceous plants found in secondary succession seres in Michigan. I found clusters of characteristics that occurred repeatedly. In general, plants were (1) annuals whose seeds lacked morphological adaptations for long-range dispersal (and which dropped near the parent plant), but with high capacity for longevity in the soil, (2) herbaceous perennials without vegetative cloning, whose seeds landed near the parent but could live dormant for long periods, (3) herbaceous perennials with some form of vegetative cloning and whose seeds had morphological adaptations for more long-range dispersal (e.g., a pappus or an attractive fleshy fruit) but almost no ability to live beyond 1 or 2 years (Table 12.4).

Whether a seed was dispersed far away from the parent was correlated with whether or not the adult had vegetative cloning. For many of the type 3 plants, cloning is highly plastic, a means of covering much space quickly under conditions of low competition

TABLE 12.4.
TYPES OF FUGITIVE PLANTS AS CHARACTERIZED BY ADULT HABIT, SEED DISPERSAL, AND LONGEVITY OF DORMANT SEEDS

Adult habit	Morphological adaptations for dispersal distance	Longevity of dormant seed
Annual, no vegetative cloning	nil (only 9 percent are wind dispersed)	50 or more years not uncommon
Herbaceous perennial, no vegetative cloning	nil (usually drop near the parent)	variable, usually somewhere between 5 and 50 years
Herbaceous perennial, vegetative cloning	various types (pappus, fleshy berries, sticky burs, mucilaginous coats, etc.)	most often 1 year, sometimes 1 to 5 years

(see Table 12.3, and earlier references). Given a high resource level (e.g., low density or low competition), such colonizing species seem to put "extra" resources into cloning. It makes sense that these space-grabbers distribute seeds far from the parent plant.

The first two types of plants, (1) annuals and (2) herbaceous perennials without cloning ability but with longer-living seeds, probably represent differences in selective pressure on the adult length of life (as well as on the seedlings). The environments wherein type 1 would be selected for would be constantly disturbed or otherwise inhospitable for adult survival, whereas those environments which select for type 2 plants would allow more long-term adult survival but less probability of seed survival.

There are some indications that, among early successional woody plants (e.g., *Lonicera*, *Sassafras*, *Rhus*), the clonal habit is also correlated with long-distance dispersing seeds and the non-clonal habit is correlated with dropping seeds. Further, some non-cloning annual weeds which are associated with crops (especially grains) have lost long-term dormancy, and have no overt morphological adaptations for dispersal, yet are transported great distances by man annually to new open sites. Although they are not part of the pattern found in Table 12.4, the seed characteristics are nevertheless related to where and when open spots occur (and their predictability).

INTERSPECIFIC COMPETITION IN PLANTS AND CURRENT THEORETICAL APPROACHES

Given the sessile nature of plants, the necessary competition for space, and the importance of seed and seedling characteristics in preemptive competition, do the simple assumptions underlying current models ever apply to terrestrial plants, and if so, when? Perhaps they apply only in the following two situations:

1. *Where a limiting resource is mobile and comes to the plants.* An example is competition among plants for pollinators. Thus,

annual plants in an old field may separate common general pollinators over a diurnal cycle (Parrish and Bazzaz, 1976b) and bog ericads
may separate pollinators over a seasonal cycle (Reader, 1975). In bee-
pollinated *Delphinium*, two congeners that flower simultaneously
are separated in length of corolla tube by a ratio of 1.23 (Inouye,
1976), whch supports current theory of species packing.

2. *Where the individual plants are so densely distributed over
the environment that they may be considered ubiquitous (at some
fine scale) relative to the size of the individual plants.* Here, interactive competition will be assured. The plants may be ubiquitous in
the form of seeds, such as in a yearly-plowed field where seed pools
are built up to "saturation levels." In such habitats, annual species
separate in space along a moisture gradient (Pickett and Bazzaz,
1976); such distributions may represent competitive displacement.
At the opposite end to the yearly-plowed field in the successional
continuum, seed pools of some species build up to the point where
potential competitors may be considered ubiquitous (Werner,
unpublished data). (This is not likely to be the case in the prairie
goldenrods, however, since properties of seed dormancy and longevity are not well developed in *Solidago*.)

In intermediate-successional systems, which include a great
part of the current landscape, plants of a single species are certainly
not ubiquitous. In most plant systems, therefore, dispersal must be
considered in any hypothesis attempting to explain plant species
distributions.

Even when resources are mobile or when plants are ubiquitous,
competition can still only be inferred from the distributional patterns of species seen in the field. To determine whether patterns
along some environmental gradient or resource axis are in fact the
result of ongoing interactive competition and to predict what the
distributions and relative abundances of each will be when the
species cooccur, it is necessary to know the following for each
species or population:

1. *The interactive competitive abilities of individuals at each
point along the resource gradient,* i.e., the plant's performance when
grown alone and when grown in the presence of other plants. The

competitive abilities may be related to morphological structure (as in goldenrods) or physiological capabilities (as in groups of annuals; Sharitz and McCormick, 1973; Raynal and Bazzaz, 1975; Friedman and Elberse, 1976). For a first approximation of competitive ability, it may be sufficient to examine adult–adult interaction (e.g., Mooney, 1976). However, the next step is the quantification (even in relative terms) of competition ability over all stages in the life cycle (for an example of this approach in fish, see E. E. Werner, 1977; I found none for plants).

2. *The distribution of resources.* Only a few studies have identified the distribution of resources in competitive studies (e.g., E. E. Werner, 1977; Platt and Weis, 1977). To date, in almost all plant studies, resource axes have been identified only as an environmental gradient (e.g., Werner and Platt, 1976; Pickett and Bazzaz, 1976), with no indication of the frequency (abundance) of each part of the resource gradient in the environment. Hence it is impossible to make predictive statements about the expected abundances of each competitor in the environment at any point in time or with any change in level of resources. [An exception: Platt and Weis (1977) present frequency distribution across a resource gradient in a study of a guild of prairie fugitives and were able to make such predictions.]

An alternate approach is to deal directly with patches in which a plant can survive, distributed in some fashion in a matrix of nonlivable space. Such theory has been developed to deal with sessile organisms (Horn and Mac Arthur, 1972; Levin and Paine, 1974; Slatkin, 1974). Intutitively, it would seem that this approach would be more useful in approaching problems of coexistence, competition, and the evolution of those plant life-history characteristics associated with colonization and competition. One problem with applying the patch approach, however, is that the current theory only allows frequency data, not absolute abundance data, and is unable to handle spatial patterning per se (S. A. Levin, 1976a). Furthermore, except for a few notable cases, such as animal-produced disturbances (Watt, 1974; Platt, 1975; Jalloq, 1975; Grubb, 1976; King, 1977a,b,c) and tip-up mounds in a forest (Cook, article 9), we currently know very little empirically about what constitutes a patch, the distribution of patches in the environment, or the dynamics within a patch.

Acknowledgments

Several persons made helpful comments on an earlier draft of the manuscript, especially O. T. Solbrig, E. E. Werner, and N. Fowler. Many hours of field and laboratory work relating to seeds and seedlings of goldenrods were provided by K. A. Erdman, L. Goostrey, and S. Henwood. Several of the seed samples were collected by or with W. J. Platt. This chapter is contribution number 362, Kellogg Biological Station, Michigan State University. I am grateful to the National Science Foundation for supporting the research on goldenrods (Grants No. BMS 75-10602 and DEB 77-14811).

3 ENERGY HARVEST AND NUTRIENT CAPTURE

INTRODUCTION

THE PROBLEM of adaptation has to be addressed in a rigorous way if the ultimate goal of population biology—development of models that predict the fate of a population or species in different environments—is to be realized. As Gates (1975) has pointed out very eloquently, the flow of energy, the flow and cycling of minerals, and the gains and losses of biomass are the fundamental mechanisms that control and regulate the growth of populations. Consequently, population biologists must become conversant with the approaches that measure these fluxes—the area popularly known as physiological ecology—and incorporate them into their models.

Energy and mass exchange are ultimately related through the process of photosynthesis (Figure 1). Light, water, CO_2, and minerals are the primary independent variables (resources) that affect plant growth. The shape of the plant, including leaves, roots, and leaf canopies, and the biochemistry of the photosynthetic apparatus determine the way that energy and mass exchange of the plant takes place. These fundamental physiological and biophysical mechanisms determine the chemical energy that the plant will dispose of for further leaf and root growth and for reproduction.

We first present a discussion of the environmental and evolutionary constraints on the photosynthetic characteristics of higher plants (Mooney and Gulmon, article 13). Carbon-gaining capacity is of basic importance to evolutionary success. Plants that are able to maximize carbon gain within the constraints of the environment are presumably able to gain advantages in competition with other plants, provide better defenses against predators, and have more energy to devote to reproduction.

The availability of water has a fundamental influence on the metabolism and growth of the plant. Water economy and carbon

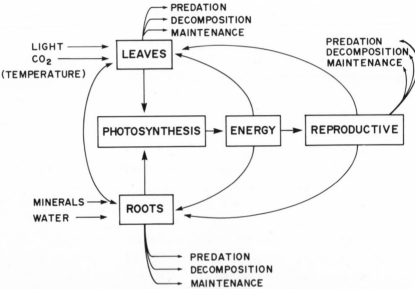

Figure 1. Generalized physiological model of a plant. The information of concern is chemical energy its generation from the four principal independent variables (light, CO_2, minerals, water); its allocation to vegetative growth (roots, leaves) and reproductive structures and its loss through predation, death of tissues, and use in maintenance.

balance are intimately linked, as Chabot and Bunce point out (article 14). Species differ in their ability to maintain a positive carbon balance under different degrees of environmental drought. Because of the inevitable link of water loss and CO_2 gain through stomata, generally, the more drought resistant a species is, the lower its relative carbon-gaining capacity under high water levels. This sets up limits (both upper and lower) to the range of humidity conditions under which a given drought adaptation confers competitive superiority. In article 15, J. A. Teeri explores this problem in relation to C_4 plants. Specifically, he investigates the correlation of a number of macroclimatic factors with the distribution and abundance of species with the C_4 photosynthetic pathway in the United States.

Leaves and roots are the structures that confer drought resistance and determine the shape of the carbon-acquisition curves. T. A. Givnish (article 16) presents a mathematical optimality model of leaf form, and M. Caldwell (article 17) discusses the costs of root construction and root function. The inevitable conclusion is that, although there are heuristic advantages to studying them separately, leaves and roots constitute an integrated system in the ultimate sense.

Part 3 is closed by a discussion of canopy function and environmental interaction in which P. Miller and W. Stoner present a model that accurately predicts phenological behavior and microhabitat distribution of species of the California chaparral and incorporates the concepts discussed in the preceding articles.

Twenty years ago, the subjects dealt with in this section were being explored cautiously by a small number of physiologists and biophysicists. Today, physiological ecology has grown into a very exciting and conceptually rich field. As these articles show, enough knowledge has been gained to begin integrating this outlook with the more traditional view of population biology.

13 ENVIRONMENTAL AND EVOLUTIONARY CONSTRAINTS ON THE PHOTOSYNTHETIC CHARACTERISTICS OF HIGHER PLANTS

H. A. MOONEY AND S. L. GULMON

PLANTS DIFFER in their photosynthetic capacity by over two orders of magnitude (Figure 13.1). Relative carbon-gaining capacity is a critical parameter in plant competitive relationships (Mooney, 1977). It is understood that factors other than carbon gain per se, such as reproductive output or success, may determine long-term fitness, but these in turn cannot be completely divorced from carbon-gaining capacity. In view of this dependence, we explore here a theoretical framework for the physiological and evolutionary basis of this variation in photosynthetic capacity among plant species. We then consider the implications of these constraints on photosynthesis for carbon allocation within the plant.

To structure this analysis, we employ the electrical analogy in which the photosynthetic rate (P) of a leaf is determined by the difference in the CO_2 concentration between the bulk air (C_a) and the leaf interior (C_i) divided by the various transport resistances to CO_2, which include a boundary-layer resistance (r_a), a stomatal resistance (r_s), and a "biochemical" or mesophyll resistance (r_m) (Gaastra, 1959):

$$P = \frac{C_a - C_i}{r_a + r_s + r_m} \tag{1}$$

The limiting resistances are determined in a complex manner by environment, leaf morphology, and biochemistry. We consider the effect of the relationships between environment and physiology on each of these resistances in turn. As the boundary-layer and stomatal resistances have already been examined extensively in terms of water-use efficiency, we review these ideas briefly and focus on the mesophyll or "biochemical" resistance.

Throughout the remainder of the discussion, we use the term "conductance," the reciprocal of resistance, in place of resistance. The boundary-layer conductance decreases with increased leaf size and increases with wind speed, as indicated in Figure 13.2. At wind speeds above 50 cm·sec^{-1}, the boundary layer becomes quite small for most leaves. Generally, under natural conditions, the boundary-layer conductance is not a significant limitation to CO_2 diffusion into leaves.

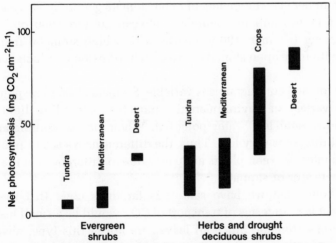

Figure 13.1. Rates of photosynthesis of diverse plant types measured under natural conditions. Tundra values from Tieszen, 1975 (herbs, 26 species; evergreens, 3 species); Mediterranean from Mooney, 1977 (evergreens, 5 species; drought deciduous, 7 species); crops from Elnore et al., 1967 (7 species). Desert values for Death Valley plants from Mooney et al. (1977), and unpublished data.

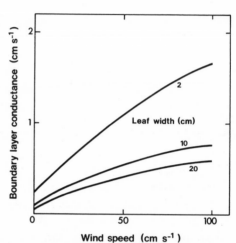

Figure 13.2. Influence of leaf size and wind speed on boundary-layer conductance of a flat leaf to CO_2. The boundary-layer conductance in still air equals $(8 \, Dd_L^{-0.6})/\pi$, where d_L is the effective leaf diameter and D is the CO_2 diffusion coefficient (0.147 $cm^2 \cdot s^{-1}$). In moving air, the boundary-layer conductance in flat leaves has been found empirically to equal $(U^{0.56}/0.89 \, d_L^{0.44}) \, D$, where U is wind speed (Bannister, 1976).

The stomatal conductance is dependent on stomatal depth, area, and number. The diameters of open stomata generally vary between 3 and 12 μm and may number between 20 and 1000 per square millimeter (Kramer, 1969). Leaves with a high stomatal frequency and with fully open stomata have a high diffusion conductance (Fig. 13.3).

Stomatal conductance is variable. Stomata close in response to a wide variety of environmental parameters, including light, CO_2, humidity, and leaf water potential. When closed, conductance to CO_2 diffusion is very low. Thus, the differences we see in photosynthetic rates between plants are partly due to differences in the kinds and behavior of stomata.

From what we have said thus far, it is clear that the most efficient leaf, in terms of transport of CO_2, would be a small leaf with numerous stomata. Since all leaves are not of this type, obviously there are other considerations in the evolution of leaf form, as pointed out by Parkhurst and Loucks (1972). We explore what these considerations might be before returning to a consideration of the other major component causing variability in photosynthetic capacity, the mesophyll conductance.

If leaves were nothing but chlorophyll-packed, moist cells exposed to the atmosphere, there would be little resistance to CO_2 diffusion from the atmosphere to the sites of carboxylation in the chloroplasts, and high photosynthetic rates would prevail, given a full complement of the photosynthetic machinery. Conversely, there would be little resistance to the diffusion of water from the mesophyll to the atmosphere. Water is lost from leaves via the same pathway as CO_2 enters, except that the pathway is shorter since there is no mesophyll component. Transpiration can be expressed as

$$T = \frac{C_a - C_i}{r_a + r_s} \quad \text{or} \quad \frac{(C_a - C_i)C_aC_s}{C_a + C_s} \tag{2}$$

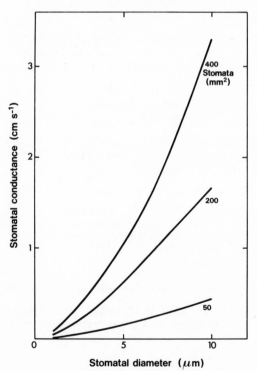

Figure 13.3. Influence of stomatal diameter and number on stomatal conductance to the diffusion of CO_2. Stomatal conductance equals the reciprocal of $(1/nD)$ $[a_s d_s/(L_s d_s + a_s)$, where L_s is stomatal depth (taken as 5 μm) is, a_s is stomatal pore area, d_s is stomatal diameter, D is diffusion coefficient of CO_2, and n is stomata per square millimeter (from Bannister, 1976).

using conductance, where C in this case is the water-vapor concentration of the bulk air (C_a) and the leaf interior (C_i).

Water loss from plants is several orders of magnitude greater than the simultaneous carbon gain, even though the same general pathway is used. This is true for two reasons: The diffusivity of CO_2 is 1.5 times greater than H_2O; and, of considerably more importance, the normal concentration gradient between bulk air and the site of CO_2 carboxylation can be no more than about 0.02 percent, whereas the concentration gradient of water vapor from the saturated mesophyll cell walls to the atmosphere on a sunny day is several percent. It is this commonality of the diffusion pathways of both CO_2 and H_2O that leads to constraints on leaf form and function which have implications to the diffusion-limited aspects of photosynthesis.

In most terrestrial habitats, atmospheric water loss precludes the viability of water-filled, permeable cells. Putting an impermeable layer containing variable conductor stomata over the leaf mesophyll has been the general evolutionary solution to this problem. This gives the plant control over water loss and the possibility of utilizing differing strategies to optimize the ratio of carbon gained to water lost. This particular optimization process has recently been discussed by Parkhurst and Loucks (1972), Mooney (1975), Givnish (1976), Miller (1978), and Orians and Solbrig (1977).

Here we review the variety of strategies that plants have used to achieve high water-use efficiencies, that is, how plants have maximized carbon gain per unit of water loss. Since water-use efficiency is of greatest evolutionary significance in arid environments, we consider, in particular, plants from such environments.

TEMPORAL MECHANISMS TO ENHANCE WATER-USE EFFICIENCY

Crassulacean Acid Metabolism

It has been shown that those succulents that have crassulacean acid metabolism (CAM) have a high water-use efficiency (WUE, ratio of carbon gained to water lost), since they open their stomata at night, when vapor pressures are low and water loss is minimal. Exogenous carbon dioxide fixed into organic acids at night is refixed into carbohydrate during the day through utilization of light energy while the stomata are closed. This process is quite inefficient as a photosynthetic process, and rates of carbon gain are low. CAM plants are, of course, most prevalent in desert environment where there is limited light competition but where long-term survival to drought is important. Not all succulents use CAM, nor do all of those that use it do so exclusively. It has been found that those species having flexible photosynthesis—that is, employing Calvin cycle (C_3) photosynthesis part of the time and CAM during other periods—are generally leafy succulents (Mooney et al., 1974, 1977a). These plants generally do not have extensive water-storage capacity and can become desiccated during extended droughts. While water is available, they may have unusually high photosynthetic rates for a succulent (Bloom, unpublished data).

Stomatal Control

As shown above for succulents, closure of stomata during daylight periods of high vapor-pressure deficit (VPD) can result in high WUE, but with a significant concomitant loss in photosynthetic efficiency. Can plants manipulate stomata during the day to enhance WUE? We examine this possibility using a plant native to Death Valley, *Prosopis glandulosa,* which is known to have stomata sensitive to a number of environmental parameters, including vapor-pressure deficit.

We determined the effect of stomatal conductance on WUE using an energy balance model. Light and temperature response curves for *Prosopis* were used in conjunction with micrometeorological data for a mid-June day in Death Valley to determine net photosynthesis, transpiration, and water-use efficiency throughout the day at different stomatal conductances.

In Figure 13.4, the daily courses of transpiration and net photosynthesis are shown for the conductances giving maximum water-use efficiency. Stomatal conductance increased during midday but was relatively high throughout the day. Water-use efficiency was highest during the morning and declined steadily thereafter. Two

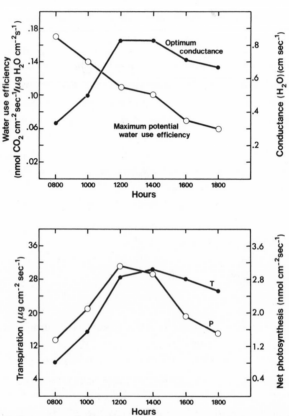

Figure 13.4. Daily courses of stomatal conductance to water vapor, transpiration, and photosynthesis for *Prosopis glandulosa*, where stomatal conductance has been optimized to obtain maximum water-use efficiency. Daily courses of temperature, solar irradiation, infrared irradiation, and water-vapor density of air were measured in Death Valley in mid-June. Leaf temperatures as a function of stomatal conductance were then obtained from the energy balance equation, assuming a boundary-layer resistance of 0.05 sec · cm^{-1}. Using photosynthetic light and temperature response curves for *Prosopis*, the photosynthetic rates were determined for the incident Phar over the possible range of stomatal conductances.

factors account for maximum water-use efficiencies occurring at such high stomatal conductances. First, the light response curve of the species does not saturate, even with full noon sun, and the calculated mesophyll conductance is quite high (>1). Thus, during midday, stomatal conductance has an approximately multiplicative effect on photosynthesis, and this offsets the increased transpiration. In the late afternoon, mesophyll conductances are lower owing to the decreased light level. However, temperature increases until after 1600 hours, and leaf temperatures are about 10° above the optimum for photosynthesis. Under these conditions, even small increases in leaf temperature owing to stomatal closure reduce photosynthesis to a relatively greater extent than transpiration. The responses just described appear to be typical for many desert plants and suggest that, in high light environments, increasing photosynthesis rather than decreasing water loss will result in the greatest water-use efficiency. We must emphasize, however, that this mode is effective only if water is available to the plant to sustain relatively high rates of transpiration.

C_4 Plants

As noted above, the way to achieve optimum water-use efficiency is to have a higher photosynthetic rate. In general, plants using the C_4 photosynthetic pathway have high photosynthetic rates owing, in part, to their lack of photorespiration. Furthermore, since photosynthesis is saturated at a relatively low internal CO_2 concentration, partial closure of stomata will not result in decreased photosynthesis, although transpiration is reduced. These features partially explain the prevalence of C_4 plants in hot regions of the world.

The question has often been posed, why C_4 plants are not more prevalent than they are, because of their generally superior carbon-gaining properties. Ehleringer and Björkman (1977) have found that, in contrast to C_3 plants, the quantum yield (moles of CO_2 fixed per mole of photosynthetically active radiation) of C_4 plants is temperature-insensitive; C_3 plants have a higher quantum yield than C_4 plants at lower temperatures, and a lower yield at higher temperatures. Utilizing these facts, Ehleringer (1977) has noted that C_3 plants would have a carbon-gaining advantage in cool and shady habitats. He successfully predicted distribution patterns and seasonal activity displacements between C_3 and C_4 grass species (Figure 13.5).

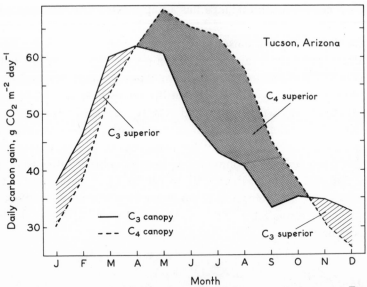

Figure 13.5. Simulations of daily carbon gain by C_3 versus C_4 grasses at Tucson, Arizona, during different months (from Ehleringer, 1977).

LEAF-TEMPERATURE CONTROL AND PHOTOSYNTHETIC CAPACITY

As noted above, in hot habitats, closure of stomata may not result in an increase in water-use efficiency, since leaf temperature may rise above the optimum for photosynthesis. A number of mechanisms have been found that reduce potential leaf temperature during periods when stomata close owing to drought stress in the desert. One example is the mechanism utilized by *Atriplex hymenelytra*, another native of Death Valley. This species conserves water through effective leaf-temperature reduction by decreasing its radiant energy absorption during the drought period and by having steeply angled leaves (Figure 13.6) (Mooney et al., 1977b). *Atriplex* modifies its leaf absorptance characteristics by utilizing salt glands. Another native of Death Valley, *Encelia farinosa*, accomplishes the same thing by changing leaf hair length and density (Ehleringer et al., 1976). In both cases, however, the amount of photosynthetically active radiation absorbed by the leaf is also reduced.

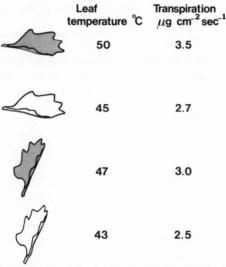

Air temperature, 45 $^\circ$C
Radiation, 1.5 cal cm^{-2} sec^{-1}
Conductance, 0.05 cm sec^{-1}
Dew point, 10 $^\circ$C

	Leaf temperature $^\circ$C	Transpiration μg cm^{-2} sec^{-1}
	50	3.5
	45	2.7
	47	3.0
	43	2.5

Figure 13.6. Leaf temperatures and transpiration rates of leaves of *Atriplex hymene-lytra* having different angles and reflectivities. Values determined from the energy balance equation for physical and physiologicial parameters prevailing in Death Valley during the summer (from Mooney et al., 1978).

THE LIMITING ROLE OF THE CARBOXYLATING ENZYME IN PHOTOSYNTHESIS

It can be shown that plants that have comparable stomatal and boundary-layer conductances may still have quite different photosynthetic rates owing to dissimilarities in their mesophyll conductances. Tieszen and Wieland (1975) note that mesophyll conductances are generally only one-third to one-tenth as high as total leaf conductances $1/(r_a + r_s)$. The mesophyll conductance is principally biochemical rather than diffusional and most likely relates directly to the concentration of the photosynthetic carboxylating enzyme. Treharne (1972) and Tieszen and Wieland (1975) have shown a strong relationship between optimal photosynthetic rates and carboxylase activity. Since the principal carboxylating enzyme of plants, ribulose bisphosphate carboxylase, makes up a substantial fraction of the total leaf protein (Björkman, 1968), it is not surprising to find a strong relationship between leaf photosynthetic rate and leaf nitrogen content under equal carbon dioxide conductivities (Fig. 13.7) (see also Natr, 1975).

In view of the relationship between carbon-gaining capacity and fitness noted initially, we now consider why all plants do not have high leaf protein contents, and hence high photosynthetic rates. We propose a theoretical framework based on the carbon-gain return on

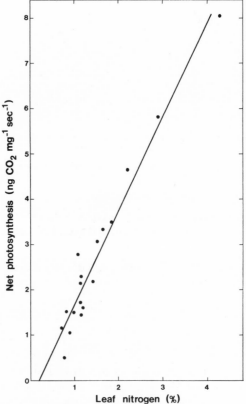

Figure 13.7. Relationship between photosynthetic rate and leaf nitrogen content in *Eucalyptus* (Mooney et al., unpublished data).

protein invested, the time course of resource availability,[1] and, possibly, the degree of predation pressure to answer this question.

Shade plants have lower light-saturated photosynthetic rates than sun plants (Björkman, 1968). This is most likely due directly to the lower contents of enzymes, especially the carboxylating enzyme, and electron carrier of the shade plants (Björkman, 1968, 1973; Gauhl, 1968), although alternative hypotheses have been suggested (Nobel et al., 1975). The relationship between leaf enzyme content, light intensity, and photosynthetic rate is illustrated in Figure 13.8. Plants grown under conditions of either low light or low nitrogen have lower light-saturating photosynthetic rates than plants grown under high light and high nitrogen. This correlates with the content of carboxylating enzyme maintained in the leaf under these conditions (Medina, 1971).

We propose that a low investment in leaf protein is adaptive in shade plants because there is insufficient light energy to efficiently utilize higher levels for photosynthesis. This relationship is expressed graphically in Figure 13.9, showing the net photosynthetic rate P as a function of leaf enzyme content E in shade (low P) and sun (high P) habitats. These curves are derived from light response curves of plants grown in sun and shade conditions. The net gain dP/dE of carbon return for an incremental unit of protein invested approaches zero as the photosynthesis curves level off. When dP/dE equals zero, additional protein will not result in any additional carbon gain and would thus represent "wasted" energy. In the low-light habitat, dP/dE approaches zero more quickly,

[1]To clarify our use of such terms as "resource," "gain," or "loss" in regard to plants, certain intrinsic differences in the meaning of these terms when applied to plants and animals must be explained. In the case of animals, the resource sought, such as food or nesting space, is directly identifiable with the resource gained. With plants, however, we can consider the primary resource to be carbon dioxide, which is reduced with light energy to form organic compounds. Light thus becomes a secondary or indirect resource, because it is not possible to measure a gain in light, only an increase in fixed carbon. Similarly, limitations of water, temperature, and, to a greater extent, mineral nutrients are most meaningfully measured in terms of carbon gained, or net productivity. Further, we also consider tertiary relationships. For example, water limitation may induce a plant to restrict the gas diffusion pathways to the extent that increases in the level of carboxylating enzyme do not result in additional carbon gain equal to the cost of construction. In this case, productivity is actually limited by carboxylating enzyme levels, but ultimately by lack of water.

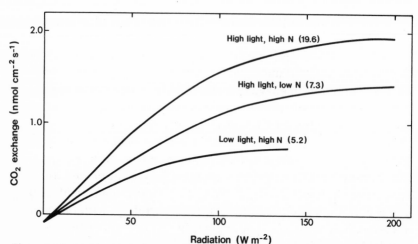

Figure 13.8. Light-related photosynthetic rates of *Atriplex patula* ssp. *hastata* grown under different light levels and nitrogen concentrations. The numbers in parentheses represent the activity of ribulose bisphosphate carboxylase in micromoles of CO_2 per square centimeter of leaf area per second (from Medina, 1971). Radiation is photosynthetically active radiation (400–700 nm; 200 W m^{-2} = 92 nE cm$^{-2} \cdot$ s^{-1}).

and thus the optimum enzyme level is lower than in the sun habitat.

We now suggest a similar relationship between photosynthetic rate, protein content, and the water resource in the habitat (Mooney and Gigon, 1973). In Figure 13.10, the photosynthetic rates P as a function of enzyme content are compared for moist and dry conditions. In the dry condition, photosynthesis is primarily diffusion-limited because stomata are closed to conserve water. Here we introduce the net marginal gain G, which includes the added costs in connection with any incremental increase in enzyme content. We have explicitly added a cost term dC/dE, such that net marginal gain G equals $dP/dE - dC/dE$. The cost C is considered to be a rate of expenditure for the entire leaf, including enzyme synthesis. Fixed costs would be averaged over the life of the leaf. Consider that, when G equals zero, additional enzyme would represent a nonadaptive allocation of plant energy. In the moist region, the optimal enzyme level ($G = 0$) is higher than in the dry habitat.

The leaf-cost term includes the direct carbon cost of construction, dark respiration, and indirect costs of concentrating the necessary minerals, additional root growth, and possible increased preda-

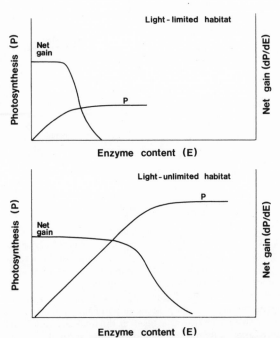

Figure 13.9. Hypothetical rates of photosynthesis of plants versus carboxylating enxyme content in light-limited and light-unlimited habitats. Curve shapes are derived from light response curves of plants grown in sun and shade habitats. The net gain in photosynthesis per unit enzyme investment (dP/dE) fall to zero at a lower enzyme concentration in the light-limited habitat. Further enzyme investment under these conditions would be nonadaptive; in fact, shade-adapted plants are found to have lower enzyme concentrations than sun-adapted plants.

tion losses. The shape of the relationship between total cost and enzyme content is unknown but would be at least a linear, and most likely an upward-bending, function. Figure 13.11 illustrates the effect of two hypothetical cost functions on net marginal gain, and thus optimal enzyme content of the leaf. (Figure 13.8 contained an example of increased leaf cost.) In the low-nitrogen condition, the cost of taking up nitrogen for protein was increased, and the resulting enzyme content was lower than in the high-nitrogen condition.

Leaf cost will be affected by environmental parameters such as temperature, soil moisture, and available nutrients, and by predation intensity. By decreasing the effective leaf size, predation loss scales up all the other leaf costs in proportion to its severity. There is some

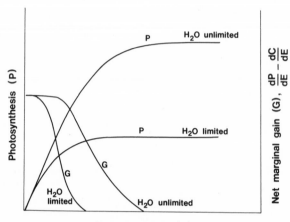

Enzyme content (E)

Figure 13.10. Hypothetical light-saturated rates of photosynthesis of plants versus carboxylating enzyme content in water-limited and water-unlimited habitats. The net marginal gain (G), or the difference between increased photosynthesis and increased cost per incremental increase in enzyme concentration ($dP/dE - dC/dE$), varies in these two habitats. Where water is limited, the net marginal gain reaches zero at a lower leaf enzyme content than in plants living in habitats where water is unlimited. In computing the cost of the leaf, C is taken as the total energetic rate of expenditure of leaf synthesis and maintenance, including mineral uptake, etc. Fixed costs are averaged over leaf life span.

evidence to indicate that predation intensity increases with increased leaf protein contents. Soo Hoo and Fraenkel (1966a,b) have shown that, for a typical polyphagous insect (*Prodenia eridania*), plants most readily accepted by larvae had higher protein contents than plants least accepted. Such a pattern, if general, would result in an upward-bending cost function for leaf protein, such as shown in Figure 13.11a. Furthermore, it has been suggested (Mattson and Addy, 1975) that plant foliage provides only marginally adequate nutrition to its usual insect consumers. To the extent that this is true, small variations in leaf protein content would produce relatively greater feeding and growth responses by the insect grazers. An evolutionary alternative to predation loss is production of predator protection mechanisms such as secondary compounds lethal or distasteful to predators or physical structures which protect the leaves mechanically or make the resources unavailable to predators. These mechanisms would have an associated cost of construc-

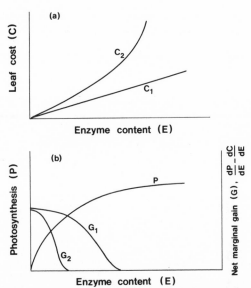

Figure 13.11. (a) Hypothetical relationships between leaf cost and enzyme content in two contrasting environments. (b) For C_2, incremental leaf costs increase at higher leaf enzyme contents as a result of, for example, increased predation on protein-rich leaves or decreased nitrogen availability. The differences in the cost function result in different zero intercepts of the net marginal gain in the two environments.

tion, and maintenance and possibly indirect costs of increased nutrient uptake and root growth. Whether or not they result in an optimum productive system depends in part on the actual photosynthetic rate and predation intensity as functions of enzyme content.

This is illustrated for an hypothetical example in Figure 13.12. We consider, as in previous sections, two photosynthetic functions of enzyme content, high photosynthesis (P_1) and low photosynthesis (P_2), and two cost functions, in the absence of predation (C_1) and with predation (C_2). In this case, the difference between C_1 and C_2 is the proportion of leaf lost to predation. As shown in Figure 13.12, the optimum leaf enzyme content for photosynthetic function P_1 and predation-cost function C_2 is substantially lower than that for P_1 and C_1 (no predation). Further, the slope of the cost function C_2 at the optimum enzyme content (shown as a dashed line) defines an upward bound for expenditure on predator protection. If the cost of protection plus the residual predation lies between this bound and

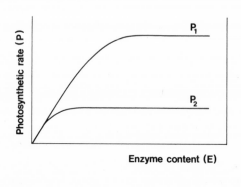

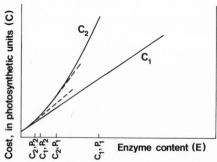

Figure 13.12. (a) Two photosynthetic functions of enzyme content, high (P_1) and low (P_2). The low function represents a plant that is light- or water-limited. (b) Solid lines represent two hypothetical cost functions, C_1 (no predation) and C_2 (predation). The dashed lines are the slopes of the cost function C_2 at the optimum enzyme contents for P_1 and P_2 ($dP/dE = dC_2/dE$). The difference between C_1 and the dashed line for each photosynthetic function P defines a potential expenditure for predator protection which would result in higher net productivity than no protection in the presence of predation (C_2).

C_1, in the vicinity of C_2P_1, then protection will result in greater net productivity than no protection. If the lower photosynthetic curve P_2 is analyzed in a similar manner, we also get an upward bound for predator-protection cost (in the vicinity of C_2P_2), but it is much closer to the minimum leaf cost C_1. That is, the additional amount that the P_2 (low P) plant could spend on predator protection with a net increase in carbon gain is lower than the expenditure available to the P_1 (high P) plant. Thus, the net marginal gain for investment in predator protection mechanisms depends upon the photosynthetic rate, the predation intensity, and the relation between cost and effectiveness of the actual mechanism used. Note that the net gain would also be affected by the basic cost C_1. Thus, longer-lived leaves, with a lower mean cost, could "afford" more predator protection. Similarly, leaves that are short-lived but have very low fixed costs would also have greater margins for predator protection costs.

Resources and predation intensity vary temporally (e.g., available soil moisture) in most habitats, and thus enzyme contents and intrinsic photosynthetic rates should also vary, since investment in enzyme can be reclaimed and reinvested in functions with greater return for a specified season (Mooney, 1972). For example, *Larrea divaricata*, a desert shrub common in Death Valley, changes its temperature-related photosynthetic capacity with the changing seasonal thermal conditions (Fig. 13.13). The mechanism for this thermal shift in photosynthetic capacity is complex and involves an increase in thermal stability of the photosynthetic apparatus in high-temperature-grown plants and apparently an increase in a rate-limiting enzymatic step in cold-temperature-grown plants (Mooney et al., 1977c). Since cost functions may also vary in time, the model predicts that enzyme levels should fluctuate in response to enzyme costs as well as to environmental limitations on photosynthesis.

Alternatively, plants may construct leaves specifically adapted to a band of the total environmental regime and shed them at other

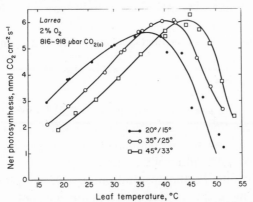

Figure 13.13. Temperature-related photosynthetic response of the desert shrub *Larrea divaricata* from three different thermal regions. Analyses were performed under low oxygen and high CO_2 to eliminate diffusion limitations and photorespiration effects (from Mooney et al., 1977).

periods. In habitats where resource availability varies greatly during the season, plants may construct leaves that have a very high photosynthetic rate or low cost (high G) for the brief time of nonlimiting resources and then shed them during resource-limited periods, rather than investing more carbon in protecting the leaves (from limited water, for example) during periods of resource limitation (Orians and Solbrig, 1977). Alternatively, plants may invest the additional carbon in protection that will be repaid by the greater time period of carbon gain by the leaf (Mooney and Dunn, 1970a,b). In many habitats, particularly those that are light-unlimited, these plant types coexist. Thus, physical conditions of the habitat can greatly influence the intrinsic rate of photosynthesis of a leaf through influences on enzyme production both directly and indirectly through an adaptive response by the plant.

PHOTOSYNTHESIS AT THE PLANT LEVEL

In addition to being dependent on the photosynthetic capacity of each leaf, the productivity of a plant is also controlled by the proportion of the photosynthate that is reinvested into expansion of the carbon-gaining system, the leaves. The effect of carbon allocation on subsequent carbon gain has been discussed by Monsi (1968), Ledig (1969), Mooney (1972), and others.

Monsi starts with a simplest case model of continuous exponential growth,

$$W_t = W_0 e^{k(mP-nr)t} \tag{3}$$

The weight at time $t(w_t)$ is given as the initial weight W_0 times an exponential function of construction efficiency k, proportion of production allocated to the photosynthetic system m, proportion allocated to support and root systems $n(m + n = 1)$, the net photosynthetic rate P, the respiration rate of nonphotosynthetic tissue r, and time t. The factors n and m are constant and equal the distribution of biomass at $W = W_0$. Thus, productivity increases as m approaches 1. If $m = 1$, the plant has no roots or stems. Epiphytes and vines are examples of adaptations maximizing m, but both these groups require other plants or physical structures for support. Root and stem parasites have also maximized m, and these groups depend on other plants for nutrients.

Even among free-standing plants, however, the ratio of photo-synthetic tissue to nonphotosynthetic tissue varies greatly and in a systematic fashion among plants of different environments. For example, plants from environments characterized by low soil-nutrient or water availability will invest less into leaves than plants growing where light or nutrients are abundant (Mooney, 1972). This variation hinges upon the relationship between net photosynthetic rate P and leaf allocation fraction m. The growth rate is proportional to the product mP, so that, if an increase in m causes an equal or greater decrease in photosynthetic rate, no net gain in productivity will result. Photosynthesis may be limited by the availability of water or minerals, and these resources are supplied through roots and stems. Thus, where one or more resources are strongly limiting, decreasing the proportion of nonphotosynthetic tissue may result in a reduction of net photosynthetic rate which overbalances the poten-tial gain of an increase in leaf area.

A hypothetical relationship between leaf net photosynthesis and the structural allocation to roots (n) in a normally wet and normally dry environment is shown in Figure 13.14. The photosyn-thetic rates $P(n)$ are considered functions of the form

$$P(n) = b(1 - e^{-an}) \qquad (4)$$

There is no experimental evidence for the precise shape of such functions. Our argument derives solely from the assumption that, in a water-limited (or nutrient-limited) condition, the leaf photosyn-thetic rate increases more slowly as a function of root allocation n than in a less limited condition. This is equivalent to saying that the specific root activity (rate of uptake of an essential nutrient per unit root weight) is lower under nutrient- or water-limited conditions, and this has been experimentally demonstrated (Troughton, 1968; Davidson, 1969; Hunt, 1975). We can substitute the $P(n)$ from Figure 13.14a into the growth rate term $R = k(mP - nr)$ from Equation (4),

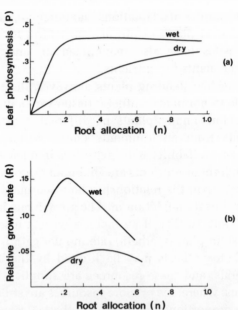

Figure 13.14. (a) Hypothetical relationship between leaf net photosynthetic rate P and structural allocation to roots n in normally dry and wet habitats. Functions are of the form $P(n) = b(1 - e^{-an})$. (b) Calculated relative growth rates $[R = k(mP - nr)$, assuming k of 0.5 and r of 0.05, and P as above] in the same habitats as a function of the root allocation factor. In the dry habitat, fractional allocation to roots of 0.4 is optimal for growth, and in the wet habitat 0.2 is optimal.

so that

$$R(n) = k\,[(1 - n)b(1 - e^{-an}) - nr] \tag{5}$$

Maximum growth occurs when $dR(n)/dn = 0$. The growth rates $R(n)$, derived from Figure 13.14a, are plotted in Figure 13.14b. Where water or nutrients are limited, there will be a greater return in carbon by investment in the means (i.e., roots) for gathering more of the limiting resource than in making more resource-limited leaves.

Since Equation (4) is based on continuous growth, it does not take into account the frequency of leaf replacement or leaf life-length. The longer the life of the photosynthetic tissue, the greater the return in the invested carbon of construction or the lower the mean cost. This relationship has been discussed in general by Monsi (1968) and for specific systems by Mooney and Dunn (1970a,b) and

Miller and Mooney (1974). Leaves vary in their lifespan from days or weeks, in the case of such plants as desert annuals, to decades, in the case of certain conifers. For a given habitat type, there is a strong inverse relationship between the length of life of a leaf and its inherent photosynthetic capacity (Johnson and Tieszen, 1976; Harrison et al., 1971; Mooney et al., 1976a).

SUMMARY

The wide variation in photosynthetic rates among plant species can be partitioned primarily into variation in stomatal conductance and mesophyll conductance. Stomatal conductance has been considered mainly in terms of water-use efficiency. We have focused on mesophyll conductance, which appears to be closely correlated with carboxylating enzyme activity and hence total protein in the leaf. Through a theoretical framework based on maximization of net carbon gain (gain–cost) and existing knowledge of the photosynthetic process, we have demonstrated unique optima for leaf enzyme contents in different environments or in the same environment at different times. Insofar as net carbon gain can be considered a primary component of fitness, deviation from these optima would be maladaptive. The theory provides testable relationships between environmental parameters and plant responses.

Acknowledgment
Material presented in this article resulted from work supported by National Science Foundation grant DEB78-02067.

14 DROUGHT-STRESS EFFECTS ON LEAF CARBON BALANCE

BRIAN F. CHABOT AND JAMES A. BUNCE

VARIATION in water supply has been shown to affect structure and productivity of land plant communities, the distribution patterns of individual species, and evolution of a number of adaptive traits. Water deficits have a fundamental influence on metabolism and growth of the individual. To a significant degree, moisture-related phenomena at the population and community levels can be explained in terms of the behavior of individual plants. In this regard, maintenance of a positive carbon balance is critical to the survival of individuals. Species differ in their ability to continue net uptake of carbon under drought conditions. After a brief review of water stress effects on leaf CO_2 exchange, we present evidence that, in order to understand the value of many specific adaptations to water stress, we need to focus greater attention on plant carbon balance within real environments and over ecologically relevant time scales.

STRATEGIES FOR SURVIVAL

It is possible to describe quantitatively degrees of drought resistance among species by measuring the ability of individuals to recover from some specified water stress conditions. This procedure defines a spectrum of drought resistance. At one end are those species which lose water easily during a drought and whose physiological processes are sensitive to very slight depressions in internal water content. At the other end are species which can either physiologically withstand or prevent extensive drying.

Plant species can maintain physiological activity during periods of drought through a variety of mechanisms. These mechanisms have been grouped into generalized strategies of either avoidance or tolerance of dehydration (Levitt, 1972). Avoidance refers to the ability of the plant to maintain high tissue water content in the face of high transpiration demand. This is achieved through utilizing underground water at a rapid rate, the "water spenders" of Maximov (1929), or by controlling water loss through stomates and cuticle, a "water-saving" strategy. Drought tolerance is the ability to maintain physiological activity or to survive tissue drying. Relatively few species possess extensive drought tolerance in the vegetative phase. Thus, it is generally found that avoidance strategies, including conversion to a dormant phase, become more important as environmental moisture stress of the habitat is increasingly severe. However, any particular species will usually possess both avoidance and tolerance mechanisms, so that classification of species according to drought resistance strategies becomes difficult, if not misleading.

Drought Tolerance

As leaf water potential decreases under drought conditions, net photosynthesis rates also decrease. Not all species show the same degree of sensitivity to low leaf water potential (Figure 14.1). In general, dry-habitat species are able to maintain photosynthesis at more negative water potentials than species from mesic to wet habitats. Factors contributing to the decline in photosynthesis are clearly diverse and have been described in numerous reviews (Levitt, 1972; Hsiao, 1973; Slavik, 1975; Boyer, 1976). Both stomatal closure and direct disruption of biochemical processes and cell structure occur simultaneously during a drying cycle. Most data indicate, however, that stomatal closure inhibits photosynthesis before nonstomatal changes become important. The relationship between stomatal and nonstomatal control of photosynthesis is of

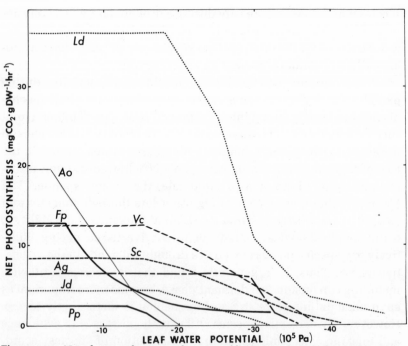

Figure 14.1. Net photosynthesis as a function of leaf water potential in eight perennial species from the Santa Catalina Mountains, Arizona: Ao, *Alnus oblongifolia*; Fp, *Fraxinus pennsylvanica* spp. *velutina*; Pp, *Pinus ponderosa*; Jd, *Juniperus deppeana*; Vc, *Vauquelinia californica*; Sc, *Simmondsia chinensis*; Ag, *Acacia greggii*; Ld, *Larrea divaricata*.

obvious ecological importance but has not been fully resolved as yet (H. G. Jones, 1973a; Boyer, 1976).

Evidence for nonstomatal inhibition of photosynthesis comes from two sources: measurement of residual, or mesophyll, resistance in whole-leaf gas-exchange studies and the activity of isolated organelle and enzyme systems. Slatyer (1973) found that mesophyll resistances increased in cotton and corn, but not in three other, more drought-tolerant species, when all were subjected to water stress. Drought-sensitive varieties of dry beans showed greater changes in mesophyll resistance with water stress than did drought-tolerant varieties (O'Toole, 1975). In a comparison of 12 species from a range of habitats, Bunce (1977) observed an increase in mesophyll resistance with increasingly negative leaf water potentials in all cases. Changes in mesophyll resistance paralleled the decreases in both net photosynthesis and stomatal conductance. In general, though, the dry-habitat species were able to maintain photosynthesis rates at lower leaf water potentials than the wet-habitat species. These and similar studies of whole-leaf gas exchange provide the best evidence for substantial differences between species in their sensitivity to cellular inhibition of photosynthesis.

Studies of water stress effects on organelle systems deal with a small number of relatively drought-sensitive species. Activity of isolated chloroplasts paralleled decreases in net photosynthesis and stomatal conductance in pea, sunflower, and spinach (Boyer and Bowen, 1970; Plaut and Bravdo, 1973). Boyer (1971) demonstrated that photochemical activity of chloroplasts, rather than carboxylase activity or stomatal conductance, limited photosynthesis in sunflower. This appears to be the only reported instance in higher plants where change of stomatal conductance during dehydration does not have an overriding influence on photosynthesis. Other evidence suggests that disruption of photochemical activity occurs at relatively low leaf water potentials (Nir and Poljakoff-Mayber, 1967; Bourque and Naylor, 1971; Keck and Boyer, 1974). Damage to chloroplast and cellular membranes occurs in water-stressed tissue, but primarily at severe water stress (Giles et al., 1974, 1976; Fellows and Boyer, 1976; Vieira da Silva, 1976). Ribulose-1,5-bisphosphate carboxylase activity does decrease with water stress, but does not appear to be limiting at the point where net photosynthesis is zero (Huffaker et al., 1970; H. G. Jones, 1973b; R. R. Johnson et al., 1974).

Changes in respiration rates have a substantial impact on leaf carbon balance during water stress. Both dark respiration and photorespiration are known to be affected by leaf water status, though the relationship is less consistent than is the case for photosynthesis. When subjected to slight water stress, dark respiration either increases, decreases, or remains constant (Brix, 1962; Mooney, 1969; Boyer, 1970; Dina and Klikoff, 1973; Koeppe et al., 1973; Bunce and Miller, 1976). There appears to be no consistent relationship between pattern of initial change in dark respiration and drought sensitivity. At more severe water stress, all species eventually show a decline in dark respiration. Water stress effects on photorespiration have received less attention. Photorespiration has been shown to decrease with water stress in spinach, sunflower, and wheat (Boyer, 1971; Plaut and Bravdo, 1973; Lawlor and Fock, 1975; Lawlor, 1976); however, Bunce and Miller (1976) found that photorespiration decreased in four wet-habitat species and increased in four dry-habitat species.

The ability of the plant to resume photosynthetic activity rapidly when rewatered may be an important component of drought tolerance in nature. Maintaining photosynthetic capacity even though CO_2 exchange is inhibited by stomatal closure appears to be one way of avoiding permanent metabolic damage. Recent work by Bunce and Miller (1976) has shown that increased rates of photorespiration with increasing tissue water stress are coupled with rapid recovery of photosynthesis. When photorespiration was inhibited while plants were in a stressed condition, recovery of photosynthesis was retarded. These workers suggest that this beneficial effect of photorespiration might result from removal of NADPH, which builds up when carboxylation reactions are starved for CO_2. Zabadal (1974) has provided evidence that drought-tolerant species have a greater differential between the water potential of stomatal closure and lethal water potential than is the case for less drought-tolerant species. This could be interpreted as a buffering mechanism to protect metabolic capacity under conditions where continued water loss through the cuticle would occur in spite of stomatal closure. This behavior, along with low cuticular conductance, could significantly lengthen the "specific survival time" (defined by Pisek in Larcher, 1975) of plants under drought.

Modifications that improve cellular resistance to water stress have not been clearly identified. Levitt (1972) reviews evidence that drought injury results primarily from mechanical stresses induced by dehydration of cell contents. Reduced cell size, increased wall thickness, and increased osmotic concentrations of cell contents would all tend to retard cell collapse. These anatomical and osmotic changes are known to occur as adaptive responses to water stress (Walter and Kreeb, 1970; Quarrie and Jones, 1977). Levitt also suggests that changes in bonding structure of the proteins might also inhibit protein denaturation.

Drought Avoidance

Although it is true that many species possess some degree of drought tolerance, this alone is not sufficient. Drought avoidance appears to be essential for the survival of most plants. Avoidance involves control over the relationship between plant and environmental water status. Since atmospheric water concentrations are usually well below the tolerance limits of nondormant plants, it is advantageous for the plant to maintain internal water potentials as close as possible to soil water potentials. Numerous biological factors can affect the relationship between leaf and environmental water potentials. Leaf water potential at a particular moment represents a balance between the rate of water supply from soil and roots and the rate of loss through stomata and cuticle (Kaufman and Hall, 1974). Each species achieves this balance through a unique and complex combination of characteristics. Factors that affect supply include structure of the root system (density, volume of soil exploited), root/shoot ratios, and conductances of root, stem, and leaf xylem tissue. Loss rates are determined by total leaf area, individual leaf size, and stomatal, cuticular, and mesophyll conductances. The situation is complicated by the fact that none of these biological factors is constant; they may vary adaptively in response to changes in the environment, often over fairly short time scales ranging from minutes to weeks. With few exceptions, we have only enough information concerning these variables to make tentative generalizations about ecological patterns. All this is important because interpreting the significance of a particular relationship between photosynthesis and leaf water potential requires that we know the range of leaf water potentials that the species experiences in nature.

The stomatal complex is a sophisticated mechanism for the short-term control of water loss. As envisioned by Raschke (1975), a system of feedback relationships permits stomatal aperture to respond to both the internal and external plant environment. Some of the variables known to affect somate functioning include light intensity and quality, CO_2 concentration within and outside the leaf, external relative humidity, wind, leaf water potential, and root water stress (Raschke, 1975; Hall et al., 1976). Stomatal closure is seen in many species at periods of high water demand. This permits the lag in absorption to catch up to transpiration and prevents leaf water potentials from dropping to a point where irreversible physiological damage occurs. Thus, species from habitats as different as deserts (Schulze et al., 1972) and swamps (Schlesinger and Chabot, 1977) show midday stomatal closure. Of ecological significance is the fact that the stomata of different species can show different sensitivities to their environment. Both Camacho-B et al. (1974) and Bunce et al. (1978) found that leaf water potentials in dry-habitat species were influenced very little by vapor pressure deficit, in contrast to species of moist environments, at least partly as a result of strong stomatal control in the dry-habitat species. Hall and Kaufman (1975) argue that stomatal sensitivity to vapor pressure gradients in dry-habitat species is advantageous since it allows these species to avoid desiccation and might increase water-use efficiency during stressful parts of the day. Wuenscher and Kozlowski (1971) showed that stomatal resistance and sensitivity to temperature were greater in dry-site tree species. Several studies have shown that species from xeric habitats close their stomates at lower leaf water potentials than more mesic species (Sanchez-Diaz and Kramer, 1971; Al-Ani and Strain, 1972; Detling and Klikoff, 1973; Dina and Klikoff, 1973; Zabadal, 1974; Stoner and Miller, 1975; Johnson and Caldwell, 1975; Bunce et al., 1978). This, coupled with drought tolerance, permits xeric-adapted species to maintain photosynthesis for a longer period during extended drought conditions.

    Evidence of differential use of abscisic acid to regulate stomatal conductance among species was found by Zabadal (1974). Abscisic acid production is triggered at a species-specific water potential during a drying cycle. Its presence in the leaf results in prolonged stomatal closure since it is inactivated slowly and in proportion to its initial concentration. Wet-site species, which experience primarily short-term drought, produce high levels of abscisic acid at relatively high leaf water potentials. Since photosynthesis is inhibited by stomatal closure, this strategy works only when there is a high probability of the drought being relieved quickly, so that a positive carbon balance can be maintained. Dry-site species, by contrast, produce much lower levels of abscisic acid and at more negative water potentials, thus maintaining photosynthesis for longer periods during a drought.

Although stomatal regulation of water loss serves to maintain leaf water status in the short-term, additional modifications conserve water on a more continuous basis. Cuticular transpiration, which in some species is a substantial fraction of total leaf water loss, can be reduced through changes in the thickness, structure, or chemical composition of the lipid layers (Clark and Levitt, 1956; Martin and Juniper, 1970; Schonherr, 1976). Reduction in stomatal frequency and size can reduce water loss, particularly if leaf area and longevity do not simultaneously increase (H. G. Jones, 1977; Quarrie and Jones, 1977). Kozlowski (1976) has reviewed the importance of leaf shedding to reduce transpiration surface during drought. Dunn (1975) has suggested that mesophyll resistances to water flow can be varied, perhaps through changes in cell wall structure. Leaf size and pubescence can affect leaf–air temperature differences, thereby altering the water-vapor concentration gradient (Taylor, 1975; Ehler-inger et al., 1976).

Increasing water uptake is an alternative way of avoiding water deficits in leaf tissue. The distance between veins and photosynthesizing cells may be important in increasing water flow to mesophyll tissue (Dengler and MacKay, 1975) and is known to be subject to phenotypic modification, as well as variation among species (Maximov, 1929). Rate of water flow through stem tissue increases in proportion to the cross-sectional area of xylem vessels (Oppenheimer, 1949). The amount of root surface area in proportion to leaf area directly affects transpiration rates (Kausch and Ehrig, 1959), and well-developed root systems can tap larger soil water reserves (Caldwell, article 17).

Both tolerance and avoidance mechanisms contribute to a plant's ability to survive drought; their relative importance depends upon the frequency and severity of drought periods. Where a species faces a favorable growing period free from frequent, prolonged drought, stomatal control may be sufficient to avoid daily or short-term water stress. Extensive physiological tolerance would not appear to be necessary. As the growing season is shortened by drought, there is an advantage to prolonging the effective period of carbon gain by maintaining open stomata at more negative water potentials. Drought tolerance is of obvious importance under these circumstances. Additionally, a greater differential between the point of stomatal closure and lethal water potential would help to avoid cellular damage, particularly where the daily amplitude of water stress is potentially great. H. G. Jones (1973b) also has stressed the importance of coordination between avoidance and tolerance characteristics. His analysis of water stress effects in cotton showed that, although stomatal closure had a dominant impact on reducing photosynthesis, nonstomatal inhibition also occurred, including quantitative reductions in enzymatic activity, soluble protein, and chlorophyll. He suggested that coordination among all processes affecting photosynthesis has adaptive value—that inhibition of photosynthesis by stomatal closure means that energy investment in biochemical capacity can be reduced. A similar economic argument has been used to explain the highly coordinated responses of plants to light intensity (Björkman, 1975; Boardman, 1977). More recently, Orians and Solbrig (1977) have related physiological capacity under drought to the anatomical and structural costs of maintaining that capacity. They postulated a direct trade-off between high photosynthetic rates and the ability to sustain photosynthesis at low soil water potentials. This cost–income model integrates a number of the specific adaptations to drought outlined previously by using a carbon-balance approach fundamentally similar to that described in the next section.

PLANT CARBON BALANCE

The relevance of specific drought-resistance mechanisms becomes most clear when considered in an ecological context. Of particular value is the comparative approach that evaluates relative performance of a group of species possessing different degrees or types of drought resistance. As mentioned earlier, a number of such studies have pointed to stomatal control as a key character in determining the ecological tolerance to water stress. In most cases, wet-habitat species close their stomata at relatively high water potentials. However, given the existence of numerous drought-resistance mechanisms, it is unlikely that a single mechanism will be consistently emphasized, particularly if a large group of species is compared. Indeed, in some instances a pattern of stomatal control opposite to the general case cited above has been found (Jarvis and Jarvis, 1963; Lopushinsky and Klock, 1974) or the trend is inconsistent among a group of species (Bunce et al., 1977).

Recent studies utilizing a group of species distributed along an elevational-moisture gradient in the Santa Catalina Mountains of Arizona sought to establish a relationship between several drought-avoidance and drought-tolerance mechanisms. We analyzed photosynthesis and respiration as a function of leaf water status in eight species that reached their distribution limits at different points along a gradient of increasing aridity. Results such as those in Figure 14.1 allowed us to calculate a number of common measures of drought resistance (Table 14.1). To evaluate the physiological data, we had, as measures for comparison, direct climatic evidence for a gradient of moisture availability along the transect, the actual distribution limits for all of the species, and an independent measure of relative drought adaptedness for each species (Ecological Index, EI) developed by Whittaker and Niering (1964) from the phytosociological associations of the species. The EI is an estimate of the ability of species to occupy increasingly dry habitats. Taken as a whole, the experimental data support existing conclusions that, in general, dry-habitat species are able to maintain positive photosynthesis rates at low water potentials through a combination of physiological tolerance and an ability to maintain open stomata in the face of water stress. When examined closely, however, the data present several ambiguous situations. For example, *Pinus ponderosa* reached zero net photosynthesis at a higher leaf water potential than either *Alnus oblongifolia* or *Fraxinus pennsylvanica*, even though it occupies

TABLE 14.1.

MEASURES OF PLANT RESPONSE TO ENVIRONMENTAL MOISTURE IN EIGHT SPECIES OF WOODY PLANTS FROM THE SANTA CATALINA MOUNTAINS[1]

Species	E.I.[2]	Slope of P_n decline with decreasing ψ_l[3]	ψ_l of initial P_n decline[4]	ψ_l of P_n compensation point	ψ_l of initial increase in r_r[5]	r_r at -15×10^5 Pa	Percentage P_g at 20 percent T_{max}[6]	ψ_l of incomplete recovery of P_n[7] After 24 hours	After 1 week	Lethal ψ_l	Dependence of ψ_l on ψ_{soil}[8]	ψ_l at 80 percent RWC[9]
Alnus oblongifolia	0	5.6	-3	-20	-5	24	20	-13	-16	-20	1.7	-18
Fraxinus pennsylvanica ssp. velutina	2	<3.7	-5	<-32	-5	18	31	-21	-28	-40	1.5	-20
Pinus ponderosa	2.5	16.7	-13	-18	-10	65	—	-14	-18	-22	0.6	-26
Juniperus deppeana	4	6.7	-17	-32	-15	18	71	-40	-54	-60	1.0	-32
Vauquelinia californica	5.5	4.2	-15	-39	-10	12	89	-45	-55	-75	1.2	-46
Simmondsia chinensis	6	3.9	-15	-42	-10	14	70	-50	-65	-85	1.2	-34
Acacia greggii	6.5	9.1	-25	-36	-15	8	75	—	—	—	0.9	-26.5
Larrea divaricata	8	<3.1	-18	<-50	-15	2	47	-42	-63	-90	0.8	-17.5

[1] See Bunce (1975) and Bunce et al. (1978) for details of methods.
[2] Ecological Index, from Whittaker and Niering (1964).
[3] P_n, Net photosynthesis; ψ_b, leaf water potential. Units are 10^5 Pa.
[4] Units are 10^5 Pa.
[5] r_r, Residual resistance of CO_2 uptake; units are seconds per centimeter.
[6] Percentage of maximum gross photosynthesis (P_g) at the leaf water potential of 20 percent maximum transpiration rate (T_{max}).
[7] Incomplete recovery, P_n remained at 10 to 20 percent of P_n^{max} when rewatered after exposure to ψ_l lower than stated value. Complete recovery $=P_n^{max}$ achieved in 1 day or 1 week after rewatering.
[8] Coefficient from multiple regression of ψ_l vs ψ_{soil} and vapor pressure deficit.
[9] RWC, Relative water content.

much drier sites. Also, the ranking of species according to relative drought adaptedness changes when different traits are used (Table 14.1).

The results became less confusing when we considered both the range of water potentials which might actually be experienced along the mountain slope and the impact of seasonal variation in water stress on annual leaf carbon balance. The relationship between leaf and soil water potential for each species was determined, along with seasonal changes in environmental moisture at several elevations along the transect. When these factors were combined in a simple model to predict annual leaf carbon balance for each species at regular increments of elevation, we were able to determine the elevation where each species failed to maintain a positive leaf carbon balance. These predicted elevations corresponded to the ranking of actual lower elevation limits and to the relative drought-resistance rankings for each species (Table 14.2). Such findings resulted from differences in drought tolerance between species and differences in the ability of each species to maintain leaf water

TABLE 14.2.
ACTUAL LOWER ELEVATION LIMITS OF SPECIES DISTRIBUTION ON OPEN SLOPES IN THE SANTA CATALINA MOUNTAINS, ARIZONA, AND PREDICTED ELEVATION OF ZERO ANNUAL LEAF CARBON BALANCE[a]

Species	Actual lower elevation limit (m)	Predicted elevation of 0 carbon balance
Larrea divaricata	<710[b]	525
Acacia greggii	760	675
Simmondsia chinensis	915	725
Vauquelinia californica	1,070	875
Jatropha deppeana	1,370	1,175
Pinus ponderosa	1,680	1,400

[a]The two strictly riparian species are omitted. See Bunce et al. (1978) for details of leaf carbon balance model. Predicted elevations are lower than actual in each case. Including whole-plant carbon requirements or the decreasing length of the frost-free period with elevation adjusts the predicted limits to higher elevations.

[b]This is the lowest elevation in the Tucson area.

potentials when faced with declining soil moisture, and to change in the relationship between tolerance and avoidance. Thus, the ecological behavior of these species is most completely explained by a combination of attributes rather than by a single limiting mechanism. For drought-tolerant species, these attributes include: photosynthesis rates sustained longer into a dought period by stomata remaining open at low leaf water potentials and reduced sensitivity of leaf water potential to soil water potential; stomatal closure well before lethal water potentials are reached; low lethal water potentials, i.e., well-developed drought tolerance; and maintenance of photosynthetic capacity while stomates are closed, allowing rapid recovery when drought is relieved. These traits appear to be directed toward maintaining a favorable leaf carbon balance such that determining leaf carbon balance over an ecologically meaningful time period provides the most effective integration of drought-resistance capability.

The conclusions from our Arizona study were reinforced when the drought-stress behavior of a group of deciduous forest tree species was compared (Bunce et al., 1977). Different elevations and slope aspects around Ithaca, New York, produce habitats with a range of moisture availability. Community patterns associated with these moisture gradients had been studied by Lewin (1974), who also ranked the major tree species according to their drought adaptedness. We later subjected several of these tree species to the same sort of analysis of drought tolerance and avoidance mechanisms that was performed on the Arizona species. Results on net photosynthesis as a function of leaf water potential (Figure 14.2), along with other drought resistance characters (Table 14.3), showed inconsistent trends among the species studied.

Because species distributions overlap along environmental gradients, each site along a moisture gradient will contain species which differ in their degree of drought adaptation. We selected a xeric community and followed, during the course of a summer, the behavior of five species growing at the site. Three species were native to the community and two others, *Alnus rugosa* and *Fraxinus pennsylvanica*, occur naturally in moister habitats and were transplanted to the study area. Leaf water potentials measured throughout the summer (Figure 14.3) changed uniquely in each species, as

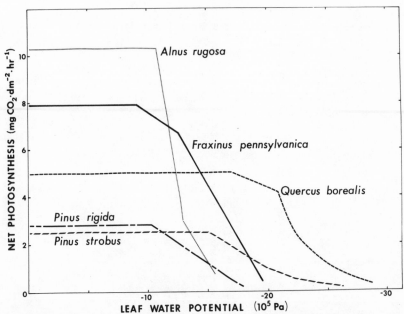

Figure 14.2. Net photosynthesis in relation to leaf water potential for five tree species from Ithaca, New York.

was expected from laboratory data on the relation between soil and leaf water potential. On any particular sampling date, species water potential differed significantly, despite insignificant spatial variation in soil water potential within the community. This suggests that leaf water potentials are in part determined by biological factors intrinsic to each species rather than being a simple function of soil and atmospheric moisture conditions, which were common to all species studied. Early in the summer, *Alnus rugosa* died as a result of water stress. The other species experienced various periods where transpiration and photosynthesis were restricted by water stress conditions.

Both leaf carbon balance and transpiration rates were calculated for the summer period. It became clear that the more drought-adapted species were able to achieve a greater fraction of their total possible carbon gain in comparison with less drought-adapted species (Table 14.4). This was due in part to higher stomatal conductances relative to maximum conductance in the drought-adapted

TABLE 14.3.
MEASURES OF RESPONSE TO MOISTURE IN TREE SPECIES FROM ITHACA, NEW YORK[a]

Species	E.I.	Slope of P_n decline with decreasing ψ_l	ψ_l of initial P_n decline	ψ_l of P_n compensation	ψ_l of initial increase in r_r	r_r at -15 bars	Percentage P_g at 20 percent T_{max}	ψ_l of incomplete recovery of P_n After 24 hr	After 1 wk	Lethal ψ_l	Dependence of ψ_l on ψ_{soil}	ψ_l at 80 percent RWC	Leaf area/soil volume exploited
Alnus rugosa	1	20.9	-11	-17	-10	35	24	-15	-17	-21	1.2	-18	0.19
Fraxinus pennsylvanica	2	9	-9	-20	-10	20	23	-21	-31	-40	0.9	-20.5	0.14
Quercus borealis	8	7.6	-17	-30							1.2	-32	0.09
Pinus strobus	10	7.6	-15	-28							0.4	-24.5	0.18
Pinus rigida	11	12.4	-10.5	-18	-10	90			-25		0.5	-19.5	0.03

[a]See Bunce (1975) and Bunce et al. (1978) for details of methods. Symbols are as defined in Table 14.1.

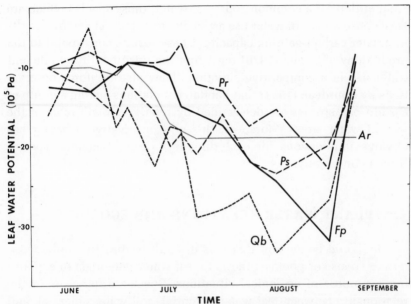

Figure 14.3. Leaf water potentials during the summer (1974) for five species growing at a common site on South Hill near Ithaca, New York: Ar, *Alnus rugosa*; Fp, *Fraxinus pennsylvanica*; Qb, *Quercus borealis*; Ps, *Pinus strobus*; Pr, *Pinus rigida*.

species when averaged for the summer period. Stomatal behavior was coupled to a syndrome of attributes similar to that described for the Arizona species. Higher carbon gain during the summer was coupled to higher water use (Table 14.3). Apparently, where there is

TABLE 14.4.
PERCENTAGE OF POTENTIAL SUMMER LEAF CARBON SUPPLY
AND POTENTIAL TRANSPIRATION RELATIVE TO
Pinus rigida IN FIVE TREE SPECIES ON SOUTH HILL NEAR
ITHACA, NEW YORK

Species	Leaf carbon supply	Transpiration
Alnus rugosa	40	15
Fraxinus pennsylvanica	56	49
Quercus borealis	68	61
Pinus strobus	72	84
Pinus rigida	84	100

competition for a common source of water, those species which are least conservative in water use are at an advantage, since this results in greater carbon-gaining capacity. These results correspond to the predictions of Cohen (1970) regarding the unsuitability of efficient water use in a competitive situation. They also reinforce observations by Bourdeau (1954) that habitat associations among a group of eastern oak species are determined by relative growth rates in the face of water stress. Moreover, this nonconservative behavior has obvious implications for excluding less drought-adapted species from a community.

PLANT WATER RELATIONS AND ECOLOGY

It should be evident that it is difficult to directly apply short-term response of photosynthesis to leaf water potentials in explaining a number of ecological situations. Species differences in the relationship between leaf water potential, soil water potential, and atmospheric vapor pressure deficit, along with differences in water uptake and transport, determine in important ways the conditions under which leaf metabolism operates. Beyond this, species differences in leaf duration (Mooney and Dunn, 1970a,b; Harrison et al., 1971; Kozlowski, 1976; Schlesinger and Chabot, 1977), acquisition of nutrients to support metabolism (Chapin, 1974; Woodwell et al., 1975; Mooney and Gulmon article 13), and relationship to other environmental variables will be expected to have some impact on the relative performance of species along an environmental moisture gradient. Additionally, the fact that most growth processes involving cell differentiation and expansion are considerably more sensitive to water stress than are the aspects of basic metabolism discussed here (Hsiao, 1973; Bunce, 1977) has considerable importance for plant population biology.

The complex interactions involved in plant water relations are being emphasized here because a dominant theme in studies of the adaptive biology of plants is identification of those traits which limit performance under a particular selective regime. The concept of limiting factors, in the sense of Blackman (1905), can be applied to such situations only with difficulty (H. G. Jones, 1973a; Boyer, 1976). Although we do not yet have a complete understanding of the adaptive puzzle, there is enough evidence to suggest that successful inhabitation of a particular environment rests upon a coordinated set of traits rather than one critical trait. It may be possible to demonstrate for conditions at any one moment that a particular plant is limited by one facet of its biology. As the environment changes through time, however, other features of the plant and the genetic structure of the population become important. This point is significant for at least two reasons. First, studies in population biology and physiological ecology must turn increasingly to understanding how functional and structural traits are integrated and coordinated at all levels in the organism. These studies must also deal with ecologically meaningful time periods and explain how a species operates in temporally varying rather than static environments. Second, if coordination among traits is an important aspect of adaptive evolution, then emphasis shifts from physiological and genetic analysis of parts of an adaptive syndrome to a more fundamental explanation of the value of the syndrome itself.

15 THE CLIMATOLOGY OF THE C₄ PHOTOSYNTHETIC PATHWAY

J. A. TEERI

THE C_4 pathway of photosynthetic carbon fixation is now known to occur in at least 15 families of flowering plants. Accurate techniques (Smith and Epstein, 1971; Tregunna et al., 1970) have been developed that permit the classification of both living and dried plant material into the C_3 or C_4 pathway category. Surveys of the taxonomic distribution of the C_4 pathway have been published (e.g., Downton, 1975; Smith and Brown, 1973), and C_4 species are known or suspected to occur in approximately 150 genera, with over half of these genera in the Gramineae. The C_4 pathway has fascinated ecologists because of the apparent high degree of sensitivity of C_4 species to both spatial and temporal variations in climate. It is now possible to compare the kinds of climates in which large numbers of C_4 species are successful with functional interpretations of the C_4 pathway based on laboratory investigations at the cellular and molecular levels. The primary functional difference between all C_4 plants and all C_3 plants is the ability of C_4 plants to have more efficient CO_2 fixation at low intercellular CO_2 concentrations (Björkman, 1976). As a result of this ability, C_4 plants are generally observed to have high photosynthetic rates at high light intensities, a high rate of CO_2 uptake per rate of transpirational water loss, and the ability to maintain high photosynthetic rates at high leaf temperatures (Björkman, 1976).

There appears to be good agreement between the climatic preferences of C_4 plants as predicted from laboratory investigations and the observed natural climatic distributions of large numbers of C_4 species. It is clear from laboratory and controlled-environment experiments that the performance of photosynthesis and growth of most, but not all, C_4 species is favored at high temperatures—greater than about 30°C. The water-use efficiency, that is, the number of grams of dry weight gained per kilogram of water transpired during the growing season, is on the average about twice as large for most C_4

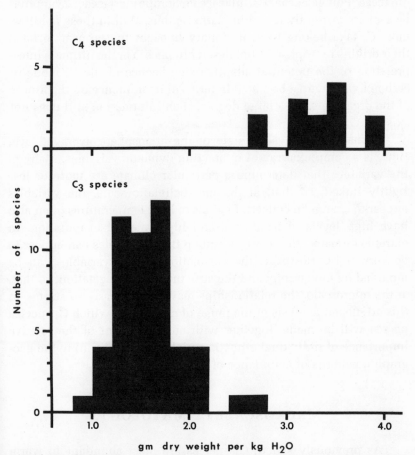

Figure 15.1. The comparative water-use efficiencies of C_4 and C_3 species grown in field experiments (data from Maximov, 1929). The test plants included both cultivated and noncultivated species.

species as that of most C_3 species from nonarid temperate climates (Figure 15.1). This suggests that C_4 species would be at an advantage in warm environments with limited water availability. Geographically, the largest numbers of C_4 species occur in warm climates. In North America, the flora with the greatest proportion of C_4 species is that of the Sonoran Desert. The eighteen C_4-containing families (Acanthaceae, Aizoaceae, Amaranthaceae, Asclepiadaceae, Boraginaceae, Capparidaceae, Caryophyllaceae, Chenopodiaceae, Compositae, Cyperaceae, Euphorbiaceae, Gramineae, Liliaceae, Nyctaginaceae, Polygalaceae, Portulacaceae, Scrophulariaceae, Zygophyllaceae) are primarily found in warm regions. Within these families, most C_4 taxa belong to evolutionary lineages that appear to have their origins in tropical or hot desert climates. Yet the ultimate interpretation of the potential adaptive significance of the C_4 mode of carbon fixation must be stated in terms of its influence on the fitness of the organisms. To a large degree, such information still does not exist for C_4 taxa.

Climate, the physical state of the atmosphere averaged over time, is a complex array of interacting dynamic variables. Many of the variables that determine a particular climate are more or less tightly linked, so that, at the macroclimatic scale, the variables appear to change in concert. Regions of high temperatures often also have high levels of irradiance and high rates of evaporation. In extreme climates, the linkages among these variables can appear to be very tight. However, the correlation among variables can be modified by topography and the structure of the vegetation, so that, at the microscale, the relationships may not be nearly so strong. In this article, an analysis of the range of climates in which C_4 species persist will be made, together with an assessment of the relative importance of individual climatic variables in determining the geographic patterns of abundance of C_4 plant species.

MACROSCALE CLIMATOLOGY

As previously stated, C_4 species are most abundant in warm climates. There are differences, however, in the precise kinds of warm climates occupied by the C_4 species in different plant families. In North America, the relative abundance of C_4 grass species is very closely correlated with measures of summer air temperature. The grass floras of the Arctic tundras have no C_4 species. At the other extreme, the grass flora of the Sonoran Desert is 82 percent C_4 species, and the grass flora of southern Florida is 80 percent C_4. There is a general trend of increasing abundance of C_4 grass species with decreasing latitude, subject to topographic modification. An analysis of the macroclimatic variability in irradiance, temperature, and moisture availability (Teeri and Stowe, 1976) showed that the single variable most highly correlated with the relative abundance of C_4 grass species in North America (Table 15.1) is normal July minimum temperature (r, 0.972). The three other variables most highly correlated ($r > 0.90$) with the abundance of C_4 grasses are all measures of the environmental thermal regime (Table 15.1). For all four of the highly correlated variables, there was a positive correlation between increasing heat and abundance of C_4 grass species. There are few comparable data for other regions of the globe, but a survey of the relative abundance of C_4 grasses among major climatic regions (Table 15.2) suggests that the macroscale relationship with temperature may be a general phenomenon. Most kinds of climates are characterized by a relatively distinct abundance of C_4 grass species. Tundra climates have few or no C_4 species. Mediterranean climates, in which the growing season is primarily during the winter moist period, have a relatively low abundance of C_4 species. In contrast, climatic regions such as hot deserts and tropical lowland forests have high abundances of C_4 grasses. In these latter kinds of climates, plant growth can occur during periods of high temperature. The growing season thermal regimes of the continental climate regions of North America range from cool to hot, and this temperature range is paralleled by a changing abundance of C_4 species (Table 15.2; see also Figure 15.5) in continental regions. In the United States, 31 percent of the grass flora of North Dakota is C_4, whereas for Oklahoma the value is 61 percent. The latitudinal sensitivity to temperature of the distribution of C_4 grass species is paralleled along elevational gradients, with few C_4 species found at high elevations. In the White Mountains of California, the rate of decrease in the abundance

TABLE 15.1.
COMPARISON OF ENVIRONMENTAL VARIABLES MOST HIGHLY CORRELATED WITH THE RELATIVE ABUNDANCE OF C₄ GRASS SPECIES AND C₄ DICOT SPECIES IN NORTH AMERICA[a]

C₄ Grasses		C₄ Dicots	
r	Variable	r	Variable
0.972	Normal July minimum temperature	0.947	Mean summer pan evaporation
0.955	Log mean annual degree-days	0.934	Mean annual pan evaporation
0.925	Mean annual degree-days	0.931	Mean annual dryness ratio
0.919	Normal July average temperature	0.911	Mean annual lake evaporation

[a]Relative abundance in each category is calculated as the percentage of C₄ species in the total grass flora and the percentage dicot species in total spermatophyte flora, respectively.

TABLE 15.2.
PERCENTAGE OF C₄ GRASS SPECIES IN DIFFERENT CLIMATES [a]

Arctic-alpine climates		Continental climates		Mediterranean climates		Hot desert climates		Tropical forest climates	
N. Slope Alaska	0	N. America	12–68	California	13	Sonoran desert	82	Jamaica	ca. 82
Greenland	0			N. Africa[b]	14	N. Saharan desert[b]	88	Costa Rica	75
N. Sweden	0								
Venezuelan páramos	ca. 19								

[a]Data from Adams (1972), Böcher et al. (1968), Hedberg et al. (1952), Hitchcock and Chase (1971), Standley (1937), Teeri and Stowe (1976), Vareschi (1970), and Winter et al. (1976).
[b]Based on partial sampling of grass flora

of C_4 species with increasing elevation is much greater than the rate of decrease in total $(C_3 + C_4)$ grass species (Figure 15.2). This decline in C_4 species generally follows the prediction based on the temperature lapse rate.

In contrast to the Gramineae, the macroscale distribution of C_4 species in the Dicotyledonae in North America appears to be controlled primarily by moisture availability. The North American Arctic tundras have no C_4 dicot species. Unlike the more latitudinal gradient for the grasses, however, there is a general trend of increasing abundance of C_4 dicot species as the arid regions of southwestern United States are approached. The greatest relative abundance of C_4 dicot species is again in the Sonoran Desert, where 4.38 percent of the spermatophyte flora is C_4 (Stowe and Teeri, 1978). The spermatophyte flora of southern Florida, however, has only 2.54 percent C_4 species, in contrast to the relatively high percentage of C_4 grass

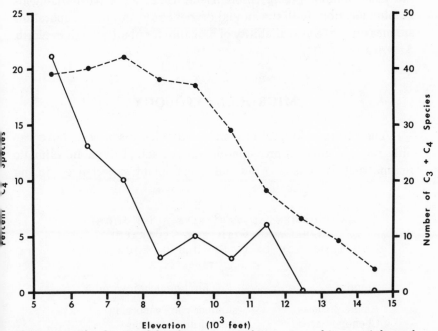

Figure 15.2. The elevational distribution (○——○) of C_4 species and (●– – –●) the total of C_3 plus C_4 species in the Gramineae in the White Mountains, California and Nevada (data of Lloyd and Mitchell, 1973).

species in that region. An analysis (Stowe and Teeri, 1978) of the macroclimatic variables potentially correlated with the abundance of C_4 dicot species resulted in the finding that mean summer (May–October) pan evaporation is the single variable (Table 15.1) with the highest correlation (r, 0.947). The three other environmental variables most highly correlated (r > 0.90; Table 15.1) with the abundance of C_4 dicot species were also measures of moisture availability. The four most highly correlated variables all had a positive correlation between increasing aridity and increasing abundance of C_4 dicot species. Among the C_4 dicot families, there are differences as to which particular measure of moisture availability is most highly correlated with the geographic distribution of C_4 species (Table 15.3), with the species in the Chenopodiaceae being most highly correlated with the annual dryness ratio. The annual dryness ratio is the ratio of net radiation to the latent heat of vaporization of the mean annual precipitation (Hare, 1972) for a particular geographic location. Both the annual dryness ratio and pan evaporation are measures of the availability of moisture for plant growth in North America.

MICROCLIMATOLOGY

The macroscale environmental variables discussed above can only provide gross approximations of the actual dynamic climatic regime that the tissue of an individual plant experiences during

TABLE 15.3.
ENVIRONMENTAL VARIABLES MOST HIGHLY
CORRELATED WITH THE GEOGRAPHIC
DISTRIBUTION OF C_4 SPECIES IN FOUR PLANT
FAMILIES

Family	Environmental variable
Amaranthaceae	summer pan evaporaion
Chenopodiaceae	annual dryness ratio
Euphorbiaceae	summer pan evaporation
Gramineae	normal July minimum temperature

growth. At the microclimatic scale, within a single plant community or at a single topographic microsite, it is likely that additional variables will be important in determining the relative performance of C_3 and C_4 species. In the Gramineae, such microclimatic differences are evident from the comparison of the local habitat preferences of the species. With regard to levels of irradiance, both C_3 and C_4 species are broadly distributed among habitats ranging from nonshaded to shaded sites. The relative proportion of the C_4 grass species occurring in open sites, however, is greater than the corresponding proportion of C_3 series (Table 15.4). This difference is significant at the 95 percent level by a Chi-square test. Along gradients of microsite substrate moisture regime, the relative proportion of C_4 grass species living in drier sites is greater than the relative proportion of C_3 species in drier sites (Table 15.4). Again, the difference is significant at the 95 percent level by a Chi-square test. The distribution of C_3, C_4, and Crassulacean Acid Metabolism (CAM) species was analyzed by Syvertsen et al. (1976) in communities of differing aridity in the norther Chihuahuan Desert (Figure 15.3). According to those authors, the driest microenvironment (bajada) contained the smallest number of C_4 species. Additionally, the biomass of the C_4 species (Figure 15.3) made up the smallest portion of the total standing crop in the driest site. In the absence of further information on the temporal variability in moisture, however, further interpretation of these findings is difficult. These data also show that there can be a very poor correspondence between the proportion of C_4 species in a particular community and the amount of the total biomass contributed by those species.

TABLE 15.4.
THE DISTRIBUTION OF C_3 AND C_4 GRASS SPECIES IN
RELATION TO MICROCLIMATIC VARIABLES[a]

	Level of irradiance		Moisture regime	
	Open sites	Shaded sites	Drier sites	Moister sites
C_3 Species	126	165	143	198
C_4 Species	117	89	151	156

[a]Data from Hitchcock and Chase (1971).

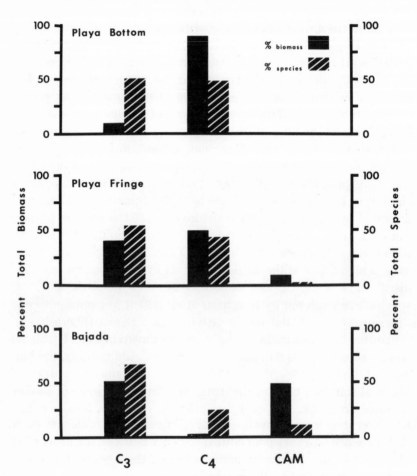

Figure 15.3. The percentage of total species and biomass (total standing crop) contributed by species with different photosynthetic pathways in three plant communities of the northern Chihuahuan desert (data of Syvertsen et al., 1976).

Another measure of microclimatic preferences is provided by the seasonality of growth patterns of C_4 species. The summer-active plant species of Death Valley, California, are nearly all C_4 (Pearcy et al., 1971). In the northern Chihuahuan Desert, over 50 percent of the total number of summer-flowering plant species in three studied communities is C_4 (Syvertsen et al., 1976). Phenologic data for the flora of the northeastern United States and adjacent Canada show that anthesis of both C_4 dicots and C_4 grasses peaks in the late summer (August and September, Figure 15.4, data from Fernald, 1950). In the southern Great Plains, the periods of vegetative growth of C_4 grasses is distinctly limited to the warm season, whereas the C_3 grasses grow during the cool part of the year (Table 15.5; data from USDA, 1948). In this regard, the agronomic classification of grasses into warm-season and cool-season grasses many years ago was a perfect classification of pathway type.

Climatic variables undergo more or less continuous fluctuation

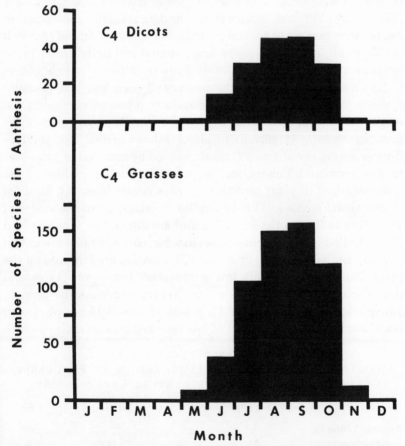

Figure 15.4. The periods of anthesis of the C_4 species in the Gramineae and the Dicotyledonae in the northeastern United States and adjacent Canada.

TABLE 15.5.
GROWTH PERIODS OF GRASSES OF THE SOUTHERN
GREAT PLAINS

	April to October	September to June
C_4 Species	34	0
C_3 Species	0	15

over a very wide range of frequencies and amplitudes of oscillation. With the exception of the annual cycles, the effect on plant growth of periods of fluctuation less than or greater than 24 hours is poorly understood. Yet leaf temperatures undergo significant recurrent fluctuations present at several periods. For example, in the Midwest (Table 15.6), in addition to the large annual and daily cycles in air temperature, there are recurrent oscillations at periods of about 2 to 6 days duration caused by the passage of high and low pressure systems. These variations are particularly pronounced during the winter season but also occur regularly during the early part of the growing season. At periods of about 1 hour or less, temperature fluctuations of about 1 or 2°C occur, caused by convection processes in the microenvironment and at organism surfaces. Most of the understanding of plant growth responses comes from two kinds of experimental studies: (1) field studies in which all of the scales of variability influence the organism and are difficult to separate, and (2) controlled-environment studies in which usually only one or two (usually annual and daily) periods of fluctuation are imposed on the plant. Periods of relatively low temperature lasting one to several days during the growing season are known to reduce the growth, and photosynthetic rates, of C_4 plants in the field. A phytotron simulation studying the effect of low temperatures on maize growth

TABLE 15.6.
AMPLITUDE OF TEMPERATURE FLUCTUATION AT REGULARLY
OCCURRING PERIODS IN THE CENTRAL GREAT PLAINS

	1 Year	2–6 Days	24 Hours	<1 Hour
Average range in air temperature fluctuation (°C)	25	14	12	2

shows that there are large genotypic differences in the sensitivity of maize lines to such events (Table 15.7). The most consistent finding in the above studies seems to be that the precision of adaptation of C_4 species to temporal (seasonal) variability may be as great or greater, at the microscale, than adaptation to spatial heterogeneity in the environment.

Regardless of the level of other climatic variables, nearly all C_4 species are found in climatic regions with relatively high air temperatures during the C_4 growing season. There are many potential mechanisms by which temperature could influence growth and thereby interact with the ability of a C_4 species to persist in a particular environment. For most, but not all, C_4 species, there appear to be a number of positive correlations between temperature and photosynthetic performance and growth. Maximum photosynthetic rates are greater at higher thermal optima for C_4 species. In this regard, Ehleringer and Björkman (1977) have found that the quantum yield of all C_4 species studied is insensitive to leaf temperature over a broad range of daytime temperatures. The C_3 species studied, in contrast, all exhibited decreasing quantum yield with increasing leaf temperature. The C_3 plants were generally more efficient at leaf temperatures below 25 to 30°C, and the C_4 species were more efficient above this range. However, in none of the C_4 families studied has daytime temperature been found to be the climatic variable most highly correlated with the distributional patterns of C_4 species. The

TABLE 15.7.
THE INFLUENCE OF SIMULATED OUTBREAKS
OF ARCTIC AIRMASSES ON THE EARLY
GROWTH OF TWO MAIZE HYBRIDS

Maize hybrid	Aboveground production after exposure to 0, 1, or 2 simulated airmass outbreaks (g dry wt of plant)		
	0	1[a]	2[b]
DK22	15.4	11.6	10.2
XL43	10.1	10.0	8.3

[a]Exposure at Days 18 through 21.
[b]Exposure at Days 18 through 21 and Days 37 through 40.

macroclimatic analyses showed that in only one case, the Gramineae, was a temperature variable the most strongly correlated with distribution. In that case, it was not maximum daily temperature but minimum daily temperature that was most highly correlated with distribution. This is an apparent discrepancy with the quantum yield data, which suggest a daytime temperature correlation. This difference may be due to several causes.

First, plant tissue temperature at any instant is a product of the interacting variables in the leaf energy balance equation. As the relative magnitudes of these variables change over the day, leaf temperature also changes. Thus, the thermal environment of leaf metabolism is dynamic. The problem then becomes, What is the best estimate of the thermal regime of the sites of metabolism, both above- and belowground, of a plant during its growth and reproduction? Ideally what is needed is a quantitative measure of the dynamic temperature regime of the leaf tissue over the period of growth. Such data do not exist for most species or climatic regions. It is possible that minimum daily temperature is more highly correlated with actual tissue temperature during growth than is maximum daily temperature.

Second, it is known that the growth of most C_4 species is inhibited by exposure to low temperature. This low temperature sensitivity appears to be correlated with the geographic origins of the species, rather than with the C_4 pathway (Björkman, 1976). Minimum daily temperature may best integrate the effects of daylight temperature on quantum yield and nighttime low temperature on growth.

Finally, it is obvious that no single climatic variable exerts sufficient influence on the growth of C_4 plants to completely negate the influence of other kinds of climatic variables. Thus, it appears that the C_4 dicots as a group are combining photosynthetic efficiency at high temperature with the efficiency of growth in the presence of restricted water supply and that C_4 grasses are responding primarily to temperature.

There does not always appear to be a close relationship between the abundance of C_4 species and total primary production in different climates. The C_4 species are favored in environments where the capacity for high rates of productivity during periods of high temperature is selectively advantageous. Because of potential environmental limitations at certain periods of the year (e.g., extreme drought or low temperature), such environments may or may not have high rates of productivity on an annual basis. The annual net primary production of those grazing ecosystems in which the dominant species are C_4 ranges from about 50 $g \cdot m^{-2} \cdot yr^{-1}$ to about 2200 $g \cdot m^{-2} \cdot yr^{-1}$ (Caldwell, 1975). A similar range of production values occurs in C_3-dominated ecosystems. Gifford has pointed out that maximum short-term growth rates are similar for both C_3 and C_4 crops when the species are grown in their respective optimal environments. On a longer time scale, however, the maximum annual yield of C_4 crop species is about two- to threefold that of C_3 crop species, which Gifford (1974) attributes to the longer period of time available for carbon fixation in the more tropical climates occupied by C_4 species. It is apparent that in environments that select for high rates of productivity, the C_4 pathway is not the only possible evolutionary response. The highest natural rate of net photosynthetic carbon fixation reported (93.5 mg of $CO_2 \cdot dm^{-2} \cdot hr^{-1}$) is for a C_3 species, *Camissonia claviformis*, in Death Valley (Mooney et al., 1976), which grows during the cooler part of the year.

WEEDINESS AND CLIMATE

The C_4 pathway provides potential competitive advantages to species that are invaders of disturbed or open sites, such as agricultural weed species (Black et al., 1969; H. G. Baker, 1974). In terms of economic impact, some of the most important weeds (Crafts, 1975) include *Cyperus rotundus* (C_4), *Cynodon dactylon* (C_4), *Echinochloa crus-galli* (C_4), *Echinochloa colonum* (C_4), *Eleusine indica* (C_4), *Sorghum halepense* (C_4), *Eichhornia crassipes* (C_3), *Imperata cylindrica* (C_4), and *Lantana camara* (C_3). The C_4 pathway is not invariably correlated with weediness or high competitive ability, however. Most agricultural systems represent attempts to maximize plant growth rates for at least a part of the year. Thus, crops are usually grown in natural or managed environments with relatively high temperatures and high levels of light available to the plant canopy. The microclimate of the period of crop growth is frequently characterized by the warmer and sunnier season of the year. Under these climatic conditions, warm-season or tropical plants can function in an environment which favors a phenotypic performance that includes C_4 carbon fixation.

On the broad geographic scale, many widespread weed species have the C_3 pathway. Good (1974) has classified the particularly widespread weed species of the earth according to the climates in which they originated. Six of the eight tropical species that are now widespread in temperate climates are C_4 (Table 15.8). In contrast, all 12 of the temperate-origin weeds that are widespread in tropical climates are C_3 (Table 15.8). Tropical weeds at temperate latitudes appear to be predominant in open sites, such as in early-stage successional communities, where they grow during the summer season. The distribution patterns of introduced C_4 dicot species in the United States were not found to be significantly correlated with

TABLE 15.8.

THE PROBABLE PHOTOSYNTHETIC PATHWAYS OF WIDESPREAD WEED SPECIES OF TEMPERATE AND TROPICAL ORIGINS

Temperate species widely adventive in tropical regions		Tropical species widely adventive in temperate regions	
Species	Probable pathway	Species	Probable pathway
Capsella bursa-pastoris	C_3	Amaranthus angustifolia	C_4
Chenopodium album	C_3	Asclepias curassavica	C_3
Erigeron canadensis	C_3	Cyndon dactylon	C_4
Euphorbia helioscopia	C_3	Echinochloa crus-galli	C_4
Plantago major	C_3	Gnaphalium luteo-album	C_3
Poa annua	C_3	Paspalum distichum	C_4
Polygonum aviculare	C_3	Portulaca oleracea	C_4
Solanum nigrum	C_3	Setaria verticillata	C_4
Sonchus oleraceus	C_3		
Stellaria media	C_3		
Taraxacum officinale	C_3		
Urtica dioica	C_3		

the geographic patterns in any of the measured climatic variables (Stowe and Teeri, 1978). The competitive ability of any species must be characterized with reference to the environment in which that phenotypic ability is expressed. In the case of potential interactions between C_3 and C_4 grass species in the Great Plains, it appears that the presence of the two types of species may actually diminish the total competitive interaction among all species composing the total grass flora. Thus, a portion of the grasses grow only during the cool season (Table 15.5), when there is likely to be far less competition with the warm-season species for such resources as nutrients and water. The optimal growth performance of most C_4 angiosperm species is generally restricted to a certain set of enviromental conditions. The ability of a weedy species to invade and colonize new or open sites should not be confused with its ability to be a competitor under other environmental circumstances. A large proportion of the climax communities on earth are predominantly composed of C_3 species. This fact alone provides a strong suggestion that the competitive abilities of C_4 species are best suited to only a limited set of environmental circumstances.

PALEOCLIMATOLOGY

There is little direct evidence concerning the precise kinds of climatic circumstances in which C_4 species arose. It is clear that C_4-pathway lineages have been derived from C_3 ancestors independently in a number of plant families that are not closely related. The climatic distributions of the C_3 species in the C_4 families provide indirect evidence as to the origins of the C_4 taxa. In the Gramineae, the C_4-containing subfamilies are predominantly tropical and subtropical. The climatic distributions of the C_3 species of eight C_4 dicot families in North America have been analyzed (Stowe and Teeri, 1978), and, in all eight families, the C_3 species exhibited the same geographic pattern as the C_4 species; that is, the C_3 species were most abundant in regions of greatest aridity. The climatic variable that had the strongest positive correlation with high abundance of the C_3 species was summer pan evaporation, the same variable that was most highly correlated with the abundance of C_4 species. This observation is not surprising in view of the functional advantages of C_4 metabolism that have been discussed above. It appears that C_4 taxa have evolved in climates similar to those in which they presently exist, precisely the kinds of climates in which the C_4 pathway is at the greatest functional advantage relative to the C_3 pathway.

The strong correlations between specific climatic variables and the abundance of C_4 taxa indicate that C_4 species may have considerable value as indicators of paleoclimates. The Gramineae appear to offer an unusually sensitive indicator of paleotemperature. The presently available evidence suggests that the strong relationship between summer temperature and the relative abundance of C_4 and C_3 grass species is a global phenomenon. The statistical relationship is strong, with very little scatter (Teeri and Stowe, 1976). The effects of other climatic variables (irradiance, humidity, moisture availability, and precipitation) are predominantly seen at the microclimatic level and do not appear to severely alter the broad geographic–climatic relationship with growing-season temperature. Thus, for much of North America, it is possible to estimate the July minimum temperature with an accuracy of about 1°C from the percentage of C_4 grass species in a geographic region. In Figure 15.5, the present latitudinal distribution of North American C_4 grasses is compared with a hypothetical paleodistribution. The paleodistribution is based on the assumption of a uniform cooling of summer temperature by 5°C over the entire latitudinal gradient. The data are taken from the regression in Teeri and Stowe (1976). The precision of the relationship gives an indication of the potential for estimating changes in summer temperature at a particular location through time. Further efforts should be made to develop morphological and anatomical criteria for the identification of genera, and perhaps species, of fossil grasses, in order to permit a computation of the relative abundance of C_3 and C_4 taxa.

SUMMARY

The distribution patterns of C_4 plant species among major classes of climates agree with the functional interpretation of the

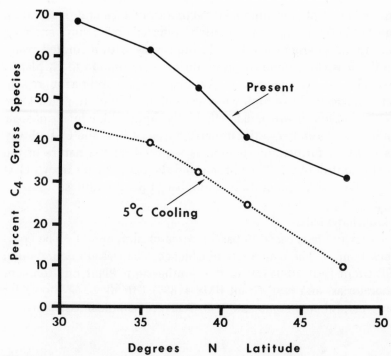

Figure 15.5. The present latitudinal gradient in abundance of C_4 grass species in the Great Plains compared with the hypothetical gradient that would result from a cooling of 5 °C in normal July minimum temperature over the entire range of latitudes.

adaptive significance of the C_4 pathway. The C_4 species are most abundant in climates characterized by high temperatures. In the cooler climates of temperate latitudes, the period of growth of C_4 species is restricted to the high-temperature part of the year. In the Gramineae, the distribution of C_4 species appears to be primarily a function of growing-season temperature. In the eight C_4 families analyzed in the Dicotyledonae, the abundance of C_4 species is greatest in regions of greatest aridity (as measured by summer pan evaporation or the annual dryness ratio). Thus, the C_4 pathway appears to have different ecological significance to species in these two separate groups. It is important to emphasize the quantitative, rather than qualitative, nature of the differences in climatic preference of C_4 and C_3 species. At the microscale, the growth performance of most individual C_4 plants shows a sensitivity to a number of climatic variables. Relative to most C_3 plants, C_4 plants are better able to utilize high levels of photosynthetically active irradiance and to maintain high temperature optima and high water-use efficiency, enabling them to gain carbon at relatively high rates with a limited water supply. However, none of these attributes is unique to C_4 plants. There are many C_3 species that have high thermal optima for growth. Some C_3 species have high water-use efficiency and, in at least one case (*Camissonia*), are able to efficiently capture high levels of solar irradiation. To increase the understanding of how the C_4 pathway has evolved, further information is required on the nature of the genetic control of the C_4 syndrome and the mechanisms by which it has evolved independently in a number of plant families.

Acknowledgments

　　This article is based in part on research supported by The Louis Block Fund of The University of Chicago, The DeKalb Foundation, NSF Grant DEB 76-04150 to the Southeastern Plant Environment Laboratories, and NSF Grant BMS-41837. I thank L. G. Stowe for many helpful comments and T. S. Teeri for technical assistance.

16 ON THE ADAPTIVE SIGNIFICANCE OF LEAF FORM

THOMAS GIVNISH

ALTHOUGH it is clear that leaves are central to a plant's adaptation for growth and competitive survival, the actual nature of the contribution of leaf form to plant adaptation is complex. For example, the size and shape of leaves can affect the rates at which leaves exchange heat, take up carbon dioxide, and lose water vapor. Leaf size and shape can also affect the efficiency with which the total photosynthetic surface can be arranged, supported, and supplied. Adaptations in leaf form thus touch on several aspects of plant form and function, with implications for thermoregulation (Gates et al., 1968), efficiency of water use (Parkhurst and Loucks, 1972), photosynthetic potential (Cunningham and Strain, 1969), branching and rooting strategies (Givnish and Vermeij, 1976), productivity (Tsunoda, 1972), and, presumably, competitive ability.

In view of the central importance of leaves and their intimate interrelation with many other aspects of plant form and resource allocation, it is important to understand the selective pressures that different environments place on leaf form. In this paper, I shall attempt to show how concepts based on natural selection and photosynthetic physiology can be used to understand several ecological patterns in leaf size, shape, and thickness. Many of these patterns are well known: Leaves tend to be small and thick in sunny or dry areas and large and thin in shadier or moister areas (Schimper, 1898). Leaves with ragged margins are frequent in the north temperate zone, but are largely replaced by leaves with entire margins in the tropics and the arctic (Bailey and Sinnott, 1916). Finely divided compound leaves are common in dry areas, like desert woodlands, that seem to favor effectively small leaves (Bews, 1925, 1927).

Although these patterns and their apparent selective control by climate and soil have been recognized for decades, they have yet to be explained satisfactorily. The distributions of certain leaf types are paradoxical. For example, plants growing on nutrient-poor soils often have small, thick leaves like those found in dry areas even though they have access to abundant supplies of water. Many tropical understory plants have leaves with ragged margins even though almost all the canopy species above them have entire leaves. Although compound leaves are frequent in certain dry areas, they are virtually absent in Mediterranean-type scrubs, even though effectively small leaves seem to be favored in both.

My aim here is to try to explain these patterns and paradoxes in terms of a few simple models. I will focus on two aspects of leaves that appear very strongly to be of adaptive significance. The first of these is the balance between carbon gain and concomitant water loss, with the implication that this balance has for the allocation of photosynthate between productive new leaves and unproductive, but necessary, roots. The second is the efficient support and supply of the total photosynthetic surface. These issues are handled in the sections below on the economics of gas exchange and the economics of support and supply, respectively. Questions that will be considered include: Why do plants in dry or nutrient-poor areas often show convergent morphological adaptations? What conditions favor entire leaf margins over nonentire ones? How can compound leaves be adaptive if they are equivalent to groups of smaller simple leaves but involve the seemingly wasteful shedding of rachises?

Since I have developed several of these points elsewhere (Givnish, 1976, 1978; Givnish and Vermeij, 1976), I will concentrate here on sketching the models briefly and commenting, where appropriate, on how empirical tests of these models and related ideas might proceed. Throughout, I will assume that natural selection tends to favor plants whose form and physiology maximize the net rate of carbon gain from a photosynthetic system of a given size, since such plants should have greater resources with which to compete for light, water, nutrients, and space (Givnish and Vermeij, 1976; Horn, article 2; but also see discussion on testing hypotheses below).

ECONOMICS OF GAS EXCHANGE

How can the size and shape of leaves possibly affect the net photosynthetic rate of a whole plant? Here we must avoid a common error [e.g., see Schimper (1898) and rebuttal by Thoday (1931)] and differentiate between the true effects of individual leaf geometry and the purely multiplicative effects of total leaf area. Bigger leaves are not better simply because they incorporate more photosynthetic tissue; the question is whether a collection of smaller leaves having the same total surface as a single large leaf might not be more effective. By considering an ideal plant whose total leaf area is fixed, we can analyze how different methods of subdividing this surface into smaller subunits affect the plant's balance of profit and loss. Let us first consider the effects that these methods of subdivision have on gas exchange and the implications that these effects have for optimal leaf size.

Effective Leaf Size

Leaf size affects gas exchange indirectly through its effect on the thickness of the leaf *boundary layer*. Generally, the larger and broader a leaf is, the more it interferes with the free flow of air around the photosynthetic surface. The stagnant film of air, or boundary layer, that forms next to the leaf surface is, on the average, thicker the larger the leaf is. However, large leaves that are deeply dissected can behave as if they were a collection of smaller leaves, with air passing around each leaf division rather independently (Raschke, 1960). The boundary layer of a sunlit leaf influences gas exchange through two main effects, one major and one relatively minor. The first effect is to impede convective heat loss. Large leaves with deep boundary layers tend to impede heat loss more than small

leaves do and so heat up more in sunlight (Gates and Papian, 1971).
Small leaves exchange heat rapidly and in this way remain close to
air temperature; broad sunlit leaves are frequently 3 to 10° C warmer
than the surrounding air (Gates et al., 1968; Taylor and Sexton,
1972).

The second effect of the leaf boundary layer is to increase the
length of the diffusive pathway in and out of the leaf, which creates
greater resistance to transpirational water loss and the uptake of
carbon dioxide. The resistance offered diffusion by the boundary
layer, however, is usually small compared with that found in the
stomata and mesophyll (Holmgren et al., 1965; Table 16.1). For
sunlit plants, the principal effect of individual leaf size on photosynthesis and transpiration is thus indirect, through the effects of leaf
size on heat loss and leaf temperature.

Higher leaf temperatures should increase the rate of transpiration rapidly. They exponentially increase the amount of water vapor
held at saturation in the intercellular air spaces, and hence greatly
enhance the concentration differential across the diffusion gradient
between leaf and atmosphere. To the extent that diffusion and the
temperature-dependent process of carboxylation limit photosyn-

TABLE 16.1.

DIFFUSIVE RESISTANCE OF BOUNDARY LAYER IN
CIRCULAR LEAVES AS A FUNCTION OF LEAF
DIAMETER AND WIND SPEED[a]

Leaf diameter (cm)	Wind speed (cm/sec)		
	10	100	1000
0.1	0.08	0.02	0.01
1	0.30	0.08	0.02
10	1.40	0.40	0.12
100	4.95	1.40	0.40

[a]Diffusive resistance with respect to water vapor (Gaastra, 1959)
was calculated using formula of Gates et al. (1968), with the correction
for field conditions suggested by Parkhurst and Loucks (1972). These
figures are for forced convection only; the free convection set up by
large, warm leaves would effectively reduce the values in the lower left-
hand corner of the table. Measured minimal values for stomatal resistance are typically in the range of 1 to 10 sec/cm (Holmgren et al., 1965;
Dunn, 1975).

thesis in sunlit leaves (see Rabinowitch, 1951; Fogg, 1972), higher leaf temperatures should also lead to greater photosynthetic rates. Although this is complicated by factors to be discussed presently, as leaf temperature increases, so should carboxylation, and the overall rate of photosynthesis it controls.

Although transpiration continues to increase with increasing leaf temperature, the photosynthetic response slows and eventually reaches a plateau as diffusion to the plastids limits carbon gain. Transpiration, being an outwardly moving process, faces no such limit (Givnish and Vermeij, 1976). With increased leaf size, leaf temperatures increase and diffusive resistance rises but slowly. We conclude that photosynthesis and transpiration should initially increase with increasing leaf size, with photosynthesis leveling off and falling away, as it becomes limited by the diffusive resistance of the boundary layer. These expected trends on photosynthesis and transpiration are displayed in Figure 16.1.

The real world is more complicated than this analysis. The relation between photosynthesis and leaf temperature is often diatonic, with carbon gain declining at high temperatures (Björkmann, 1973). The point at which photosynthesis plateaus and then declines, however, varies from species to species. Photosynthesis in a given species seems to peak at temperatures similar to those found in the natural environment of the species, with Arctic species peaking at lower temperatures and rates than many temperate or desert species (Mooney, 1963; Mooney and West, 1964; Strain and Chase, 1966; Mooney and Shropshire, 1967; Slatyer, 1978). Observations like those of Björkmann et al. (1972) on the gas exchange of summer-active plants in Death Valley suggest that the diatonic thermal response shown by most populations is not due to any absolute metabolic limitation but is instead the adaptive price paid for specializing on a particular thermal regime (Givnish and Vermeij, 1976).

As leaf size, and hence leaf temperature, is varied, it should not be expected that a plant will be "stuck" with a photosynthetic response adapted to a particular temperature. Instead, the photosynthetic system should coevolve with leaf size and temperature, perhaps so as to maximize the rate of carbon gain within the leaf at each

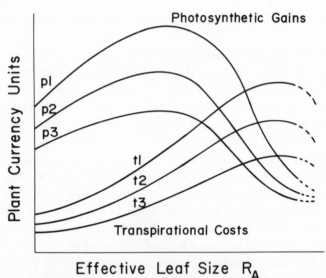

Figure 16.1. Benefit and cost curves of photosynthesis and transpiration for leaves in a sunny environment, as a function of effective leaf size. The different benefit curves represent expected photosynthesis in environments more (p1) or less (p3) sunny, warm, or rich in mineral nutrients. Similarly, the set of cost curves indicates the range of root costs associated with supplying transpirational losses in environments more (t1) or less (t3) dry or sunny. The optimal leaf size for a given habitat is at the point at which the benefit curve corresponding to those conditions most greatly exceeds the cost curve. Photosynthetic enhancement or depression of root costs favors larger leaves, whereas photosynthetic impairment or inflation of root costs favors smaller leaves.

temperature. If we are to assume that selection operates on both leaf size and photosynthetic response, then the appropriate photosynthetic model to consider is *not* one that accurately mimics the thermal response of any one particular species of physiological race, but rather one that models the envelope of responses adapted to various temperatures (see Figure 16.2 and legend). This response envelope should increase monotonically with light intensity and leaf temperature over a wide range—the greater the flux of photons and the more rapidly reacting the pool of CO_2 molecules, the easier it should be to trap either sort of particle. This rationale supports the analysis given previously for photosynthetic trends with leaf size.

As the rate at which a plant transpires increases, so should the amount of energy it devotes to the construction and maintenance of

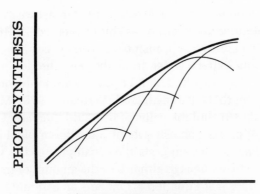

LEAF TEMPERATURE

Figure 16.2. Photosynthetic systems adapted to a particular leaf temperature should have the greatest photosynthetic rate available at that temperature and, hence, must be suboptimal at other temperatures. Therefore, the envelope (heavy line) of the thermal responses of species adapted to various temperatures indicates how the optimal photosynthetic rate at each temperature should vary with temperature.

roots and xylem in order to keep the leaves functioning (Givnish and Vermeij, 1976; Chabot and Bunce, article 14). Otherwise, the increased resistance to water uptake will cause the leaves to dry somewhat and may thereby reduce the rate of photosynthesis at the plastid level (Boyer and Bowen, 1970; Slavik, 1975). Thus, although an increase in individual leaf size brings a beneficial increase in photosynthesis, this must be weighed against the cost in unproductive roots and xylem associated with increased transpiration.

If we assume that, as the demand for water increases, the ability to supply that demand should increase proportionately, we can formulate a simple model to account for the above effects of leaf size on production. This model expresses the whole-plant rate of net carbon assimilation (N) in terms of the daily rates (per unit leaf area) of photosynthesis, night leaf respiration, and transpiration (P, R_1, and E, respectively), relating N to total leaf area A as (Givnish and Vermeij, 1976):

$$N = (P - R_1 - bE)A \tag{1}$$

The parameter b in this equation is a proportionality constant measuring the metabolic cost of arranging for the supply of a unit flow rate of water. The value of this parameter depends on factors that affect root metabolism and the availability of moisture in the soil. As soil water potential decreases or as soil oxygen concentration, temperature, permeability, or transmissivity increases, the ease with which the roots extract water from the soil should increase (see Slatyer, 1967). As a result, it should pay a plant to devote relatively more energy to roots (Givnish and Vermeij, 1976; Mooney and Gulmon, article 13), and the value of b should increase. Givnish and Vermeij (1976) present an estimate of b for the tropical tree *Anthocephalus cadamba* that suggests the sensitivity of the whole-plant carbon balance to changes in either b or the amount of transpiration.

If we assume that Equation (1) models an assembly of identical leaves in a uniform environment, we can express N (the whole-plant rate of net carbon assimilation) as a function of the characteristics of the individual leaves. The optimal combination of these characteristics can then be found by maximizing N with respect to them. Applying this approach to a plant whose leaves are arranged over a range of different environmental conditions becomes a complex procedure if we include the effects that one layer of leaves may have on another. A plant might optimize its total return by using at one level what would be, by itself, a suboptimal strategy, but which so improves the physical conditions at other levels that the gains at these levels are dramatically increased. For this to work, a plant must keep track of whose leaves it shades and is shaded by. Such perceptiveness seems unlikely in plants, so that the analysis for leaves in one set of conditions may apply directly to cases in which leaves are scattered over several sets of conditions. For now, I will assume that the analysis can be so extended, and proceed to develop the theory arising from Equation (1).

Let us assume that the rate of night leaf respiration (R_1), total leaf area (A), stomatal resistance, and leaf infrastructure (including thickness) remain constant as leaf size varies, so that we can separate the effects of leaf size from those of the other variables on plant growth. This reduces the problem of maximizing N in Equation (1) to maximizing

$$N' = P - bE \qquad (2)$$

Graphical means are perhaps the most suitable for analyzing the maxima of this equation (Givnish and Vermeij, 1976). This method has the advantage that a slightly nonlinear dependence of transpirational costs on the rate of water loss will not affect the qualitative predictions that result, except in detail, since such nonlinearity will tend only to steepen or flatten the cost curve as a whole, not change its basic shape. Similarly, to the extent that transpirational costs are not those associated with increased amounts of roots but instead are the decrease in photosynthesis in drier leaves (Mooney and Gulmon, article 13), these costs can be transferred to the bE curve and handled graphically.

In order to determine graphically where N' in Equation (2) reaches its maximum, we need to analyze where the difference between the plotted curves $P(s)$ and $bE(s)$ reaches its maximum, where s is effective leaf size and P and bE are the rates of photosynthesis and transpirational costs per unit area, respectively. The forms of these curves in a sunny environment are depicted in Figure 16.1. Let us consider how the supply of moisture in the soil and atmosphere and the supply of nutrients in the soil can affect the optimal leaf size at which the difference between photosynthetic gains and transpirational costs is maximized.

Soil Moisture

As the amount of water stored in the soil increases, its transmissivity increases, and it becomes progressively easier for the roots to remove water, at least at first (see Gardner, 1960). As a result, the value of b, and thus the steepness of the cost curve in Figure 16.1, should decrease. At the same time, the photosynthetic profits curve should steepen or remain roughly the same; leaves needn't dry out as much to draw water from moist soil, and so should have a greater potential for photosynthetic enhancement (see Boyer and Bowen, 1970; Slavik, 1975). These effects tend to favor larger leaves in moister environments by decreasing the effective leaf size at which the difference between photosynthetic profits and transpirational costs is maximized (Figure 16.1).

This trend should not be monotonic, however. As soil moisture increases, water will eventually begin to fill the larger soil pores and impede aeration. This tends to inhibit root function by causing a

buildup of carbon dioxide and a lack of oxygen in the soil. Consequently, the efficiency of water and nutrient absorption decreases,
and more carbon must be devoted to accomplish the same tasks. As a
result, the value of b and the steepness of the cost curve should
increase. Other means of coping with waterlogged soils, such as the
development of aerenchyma or sprawling, superficial root systems,
should also tend to increase root costs. Saturated soils also discourage aerobic nitrification and encourage anaerobic denitrification
(Kormondy, 1969). The resulting soil infertility can limit the production of carboxylating enzymes and impede carboxylation and photosynthesis (Medina, 1970, 1971); this tends to flatten the photosynthetic profits curve. Thus, leaf size should first increase and then
decrease with increasing soil moisture (Figure 16.3). The small size
of leaves in deserts, chaparral, woodlands, and other arid communities (e.g., see Volkens, 1887; Schimper, 1898; Clements, 1905; Maximov, 1929; Shields, 1950; Oppenheimer, 1960; Webb, 1968); and the
abundance of conifers in temperate swamp forests (e.g., *Taxodium,
Chamaecyparis, Picea, Larix*) are compatible with these predictions.

Atmospheric Humidity
        Increases in relative humidity tend to reduce transpiration and
evaporative cooling while causing leaf temperature to rise more

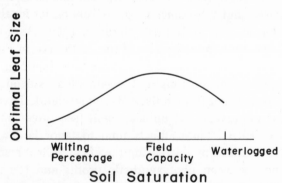

Soil Saturation

Figure 16.3. Expected trend in effective leaf size with increasing soil moisture. As
soil moisture increases from an initially low value, plants are able to extract water
more easily from the soil, and leaf size should increase. However, as soil moisture
increases beyond field capacity (where many of the larger soil pores are drained by
gravity), leaf size should decrease as soil aeration becomes adversely affected.

rapidly with effective leaf size. This tends to flatten the transpirational cost curve and steepen the photosynthetic profits curve (at least to the extent that photosynthesis does not become strongly limited by diffusion rather than carboxylation), and should thus favor larger leaves in more humid environments.

Wind Speed

The effect of wind speed on optimal leaf size is complex. On the one hand, stronger winds should reduce the thickness of the leaf boundary layer and cause the leaf to behave as if it were effectively smaller (Table 16.1). This would seem to favor larger leaves. On the other hand, stronger winds can create greater atmospheric mixing within a layer of leaves, perhaps causing a drop in relative humidity and an increase in air temperature next to the transpiring surfaces. This would favor smaller leaves. On balance, if leaves are strongly lit and arranged in a single layer, an increase in wind speed should favor larger leaves since convective cooling will reduce evaporation (Gates, 1968). If the leaves are arranged in a number of layers, so that many of them are in cool, moist shade, an increase in wind speed may favor smaller leaves by reducing humidity and increasing transpiration. Certainly the latter situation would cause a smaller increase in optimal leaf size than the former. A trend toward small, thick leaves in windswept habitats like lake and ocean shores has been noted (Chrysler, 1905).

Soil Nutrients

As the amount of nutrients stored in the soil decreases, it should become harder for the roots to take up those nutrients most limiting to photosynthesis and plant growth. As a result, the steepness of the profits curve in Figure 16.1 should decrease in response either to the directly deleterious effects of mineral deficiencies on current photosynthesis or to the indirect, deferred effects of a reduced rate of protein synthesis under these conditions (Givnish and Vermeij, 1976; Medina, 1970, 1971). At the same time, the transpirational cost curve should steepen or remain roughly the same, since the protein costs of the roots and xylem will be relatively greater. These effects tend to favor small leaves on poor soils and larger leaves on richer soils. Reduced leaf size should also occur on soils that contain minerals that poison the roots and inhibit their absorptive function or form complexes with leaf proteins once absorbed (Givnish, 1976).

Small leaves should thus be found where either the substrate originally lacked vital nutrients or had them leached by excessive precipitation relative to evaporative potential. Indeed, small leaves are found in bogs, boreal forests, montane tropical forests and elfin scrub, arctic and alpine tundra, and in a variety of other habitats having leached or nutrient-poor soils (Schimper, 1898; Beadle, 1954, 1962, 1966; Grubb et al., 1963; Howard, 1969; Leigh, 1975; Whitmore, 1975). Although plants in these areas do generally have small leaves, this may sometimes have as much to do with the excessively cold, drained, or sodden nature of the soil (all favoring high transpirational costs) as with lack of nutrients. Woodwell et al. (1975) provide data suggesting that nutrients actually play a role in determining leaf size. They studied the concentration of eight major nutrients in the growing tissues of seven plant species native to the Long Island Pine Barrens. Six of these species are deciduous and, growing in the open barrens, are often found in similar light and moisture regimes on relatively uniform sand. Two of the species, *Quercus alba* and *Q. ilicifolia*, had much greater concentrations of the four most important nutrients (nitrogen, phosphorus, potassium, and calcium) than did the other five species (Figure 16.4). It is perhaps significant that the first two species have effectively larger leaves than the others. The basis for the observed differences among species in ability to accumulate nutrients is not clear but may have to do with successional status and the ability to persist through fires that release many nutrients into the sterile sand.

Perhaps another important illustration is provided by the leaf form and distribution of the genus *Gunnera*. Most species of *Gunnera* occur in montane rain forests and drenched ravines in the southern hemisphere, in areas frequently dominated by small-leaved species of the Myrtaceae and Ericaceae (Carlquist, 1970). *Gunnera*, however, often has immense leaves with a diameter of up to 2 m! What is interesting about *Gunnera* is that, in an area having superabundant supplies of ground water and deficient levels of soil nitrogen, it has symbiotic blue-green algae (*Nostoc*) in its petioles that

Effective Leaf Size in Relation to Stored Nutrients in Plant Tissue

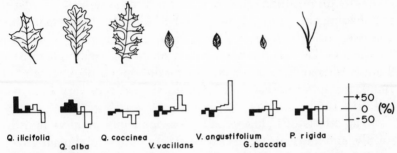

Q. ilicifolia Q. coccinea V. angustifolium P. rigida
Q. alba V. vacillans G. baccata

Figure 16.4. Effective leaf size of seven woody species from the Long Island Barrens in relation to stored nutrients in their tissues (nutrient data from Woodwell et al., 1975). The bar graphs below each leaf represent the concentrations of nitrogen, phosphorus, potassium, calcium, magnesium, sulfur, iron, and sodium in the leaf and twig tissues of that plant relative to the average for the seven species. The four major nutrients that are apt to be limiting in the sandy soil of the Pine Barrens (N, P, Ka, and C) are denoted by shaded bars.

provide it with nitrate. The remarkable leaf size of Gunnera may thus be an exception that proves our rule.

Leaf Thickness and the Duality of Xeromorphism and Oligomorphism

We see that the observed similarity of leaf size in dry and nutrient-poor habitats can be accounted for theoretically, and need not depend on infertile soils being effectively dry ones (Schimper, 1898; Beadle, 1962, 1966; Brunig, 1970, 1971). There is a natural duality between xeromorphism and *oligomorphism*, or morphological adaptations to mineral shortages. Decreased photosynthetic profits due to mineral poverty will mean relatively increased transpirational cost, so that plants facing shortages of either water or nutrients are really facing the same adaptive problem of balancing carbon gain with water loss. The duality between xeromorphism and oligomorphism should thus extend to other features of leaf form, such as stomatal resistance and leaf thickness, that have a major impact on gas exchange. An argument for this statement in the case of leaf thickness is given in the following model.

Consider first the effect of leaf thickness on the rate of photosynthesis per unit leaf area. Increased leaf thickness θ should yield gradually diminishing photosynthetic returns as a result of internal self-shading and competition for carbon dioxide. At the same time, increases in leaf thickness should incur proportionate increases in the costs of leaf construction and maintenance. Finally, because of the low resistance to water loss in the mesophyll (Holmgren et al., 1965), transpiration per unit leaf area should vary very little with leaf thickness.

The effects of light intensity, leaf longevity, relative moisture supply, and nutrients on the above functions and total plant growth can be combined if we assume that the gross photosynthesis of a plant with total leaf area A and leaf thickness θ can be adequately approximated as

$$\mathbf{P} = \frac{P_m \theta}{(\theta + k)} \qquad (3)$$

where P_m and k are constants. This form incorporates the initial linear response of photosynthesis to increased leaf thickness, as well as the ultimate leveling off with increased self-shading and competition for CO_2. Transpiration is modeled as $E = EA$, and leaf respiratory costs as $\mathbf{R_1} = R_1 \theta A$; the latter does not include within-leaf differences in photo-respiration due to differences in light intensity and CO_2 concentration. The problem is to maximize N, the whole-plant rate of net carbon gain:

$$N = \frac{P_m \theta}{(\theta + k)} - R_1 \theta A - bEA \qquad (4)$$

for plants with a given total effort in the photosynthetic system of leaves and roots.

In this case, we hold the quantity $Z = (R_1 \theta A + bEA)$ constant in Equation (4) and maximize N with respect to leaf thickness θ. This is the same as maximizing

$$N = \frac{P_m \theta Z}{(\theta + k)(bE + R_1 \theta)} - Z \qquad (5)$$

where Z is a constant since $A = Z/(bE + R_1\theta)$. Taking the derivative of N in Equation (5) with respect to θ and setting it equal to zero, we find that $\theta_{opt} = \sqrt{kbE/R_1}$. This result is relatively easy to interpret. The constant k is an inverse measure of plastid photosynthetic activity, since the smaller k is, the steeper is the initial rise in photosynthetic activity. The value R_1 is an inverse measure of leaf longevity, since the amortized rate of construction costs and hence total leaf costs decrease as the length of time the leaf is held increases. The term bE is a direct measure of transpirational costs. As k, b, or E increases, or as R_1 decreases, optimal leaf thickness should increase. Thick leaves are thus expected in sunny, dry, or effectively dry, habitats and nutrient-poor, habitats, as well as in foliage of long life. This accords with most observations (see literature cited above for patterns in leaf size).

Although adaptations to drought and soil poverty should parallel each other, there are certain aspects of plant function that ought not to show a similar response. Among these is the physiological tolerance of very low water potentials. Plants regularly exposed to water shortages will have to dry out considerably in order to extract moisture from the soil and remain active; an upper limit to their internal degree of wetness is set by that of the soil. Plants exposed to nutrient shortages, although under selective pressure to adopt gas-exchange strategies like those of plants in dry areas, need not encounter very low internal water potentials. This may be part of the reason that conifers are rarely found in the driest habitats, like deserts, even though they have a leaf morphology that would seem to suit them equally well to effectively dry or nutrient-poor situations. The gymnosperm line developed during the warm, humid conditions of the Carboniferous, Permian, Triassic, and Jurassic (Chamberlain, 1966). The presumed excess of rainfall to evaporative demand during this period would have favored acid, nutrient-poor soils, thus favoring effectively small leaves like those of most gymnosperms, lycopods, ferns, horsetails, lepidodendrons, and various arborescent fern allies. However, such conditions would not necessarily have selected for physiological tolerance of large negative water potentials, a needed adaptation for growth during harsh drought.

ECONOMICS OF SUPPORT AND SUPPLY

One cannot understand leaf shape on the basis of gas exchange alone, since leaves of different shapes can have the same properties with respect to water loss and carbon gain per unit area. Yet, of the myriad leaf forms that have the same effective size, most do not occur at all in nature, and most of those that do occur differ in their geographic distribution. Trees with entire leaves, for example, are most common in the tropics and south temperate zone; those with toothed or lobed leaves, in the north temperate zone; and those with compound leaves, in certain areas with warm, dry climates (Sinnott and Bailey, 1915; Bailey and Sinnott, 1916; Givnish, 1978). Why has natural selection favored so few leaf forms? What are the relative advantages of each?

Leaves of different shapes can require different amounts of support and supply tissue per unit area and can divert different amounts of support tissue from the permanent skeleton of the plant. In this section, I will briefly examine how the constraints of support and supply, efficient packing along branches, and partitioning of support tissues between the leaves and stem can affect the optimal shape of leaves in various environments.

Support and Supply

The *support* system of a leaf must ensure appropriate orientation of the leaf and prevent its collapse or untimely loss. At the same time, a leaf's *supply* system must provide flows of water and nutrients adequate to the demands created by transpiration and photosynthesis. Unfortunately, the optimally efficient means of support and supply are incompatible.

An optimal supply system has radial veins arranged about a central entry point and supplies a circular leaf, since this arrangement minimizes the length of supply lines needed for a given area. Ovate leaves with parallel venation can fulfill the same role if they are arranged in a rosette. These shapes and venations are found frequently among plants with floating or ground-hugging leaves, where support might be expected to be a less important constraint than in most self-supporting, aerial leaves. Perhaps the most familiar examples of this syndrome are found among the waterlilies (Nymphaeaceae). Round, floating leaves with radial or subradial venation are found in several waterlily genera, including *Nymphaea*, *Brasenia*, and *Victoria* (frequently, emergent *Nuphar* has pinnate venation). Convergent forms are found in *Nymphoides* (Gentianaceae). Parallel venation in dicots is particularly common among rosette-forming herbs and is found in families like the Ranunculaceae (*Ranunculus lingua*, *R. alismaefolius*), Umbelliferae (*Eryngium*), Plantaginaceae (*Plantago*), and Compositae (*Scorzonera*, *Tragopogon*) (Stebbins, 1974).

Whereas an optimal supply system would have parallel or subparallel radial veins, an optimal support system favors veins that are consolidated and arranged in a branching system. This is because the cost of any vein cross-section varies as the square of its diameter, but its ability to withstand stress varies as the cube of the diameter (Howland, 1962; Givnish and Vermeij, 1976). Since the strength of a vein increases more rapidly than its cost or cross-sectional area, two nearby, subparallel veins will cost more than a single, slightly larger vein.

This may partly be why most aerial leaves have a single, large midrib. The greatest mechanical stresses within a leaf should occur along the longest support arms, so the greatest tendency toward consolidation of supporting veins should be along a leaf's longitudinal axis. Those plants that do have multiple midribs should presumably have leaves so thick or so constantly turgid that they do not require much extra stiffening. Among the dicots, the Epacridaceae, Australian *Acacias*, and various species of *Hakea* (Proteaceae), *Melaleuca*, and *Callistemon* (Myrtaceae) have parallel venation and do indeed possess thick leaves or phyllodes (Carlquist, 1974; Stebbins, 1974).

Although many monocots have parallel venation, broad-leaved forest herbs like *Arisaema* and *Trillium*, with strong reinforcing midribs are notable exceptions. Thin, broad leaves with solitary midribs and pinnate secondaries are found among a wide variety of such plants in the Araceae, Cannaceae, Marantaceae, Musaceae, Zingiberaceae, Dioscoreaceae, and Liliaceae. Monocots other than the highly specialized grasses, sedges, rushes, and restionids tend to be plants of mesic or hydric habitats (Stebbins, 1974). It is difficult to say whether this implies that monocots tend to have greater turgidity than dicots and thus can have parallel venation. It is interesting to note, however, that parallel or radial venation is also common in the ephemeral petals and sepals of flowers, which are usually held erect by turgor pressure.

Economic considerations thus appear to favor the consolidation of longitudinal support tissue into a single midrib in most leaves. Increased consolidation of these and lesser veins imposes a requirement for the support of tissue lying between these veins. Design of efficient support for this surface should take into account the fact that it can support its own weight, at least over a lever arm of some length d. Auxiliary veins to support this tissue should thus, among other things, be arranged so as to maximize the area over which the lamina supports itself. Parallel, linear veins spaced roughly d units apart should cover this area efficiently, allow the leaf tissue to support itself over a maximal area, and have relatively short lever arms. Thus, the constraints of optimal support seem to favor the typical pinnate venation found in most leaves.

Support–Supply Model

Since the midrib in most leaves is particularly massive and appears largely responsible for both the lengthwise support of the leaf and its gross pattern of supply, it is important to know the optimal shape of the area this vein serves.

The midrib tapers from leaf base to leaf tip, which parallels the decreases in the stresses it must resist as it supports less and less of the leaf surface. If this tapering is coupled with an allometry between the support and supply tissues, it implies that some supply capacity is made available at each point along the midrib to serve the adjacent tissue. This means that leaf shape should depend directly on the taper of the midrib if the demands of each region along the midrib are proportional to its area and the supply capacity available at that point. Similarly, if midrib width is adjusted to the stresses it must bear at each point, midrib taper should depend directly on leaf shape. These two relations between midrib width and leaf shape can help determine both traits, as follows.

Imagine a flat, idealized leaf that is symmetric about a straight midrib of length L and rests horizontally at equilibrium, exerting no skewing torque on any part of the midrib. Let us assume that the midrib supports the entire weight of the leaf and that the weight of the midrib itself is negligible (cf. Howland, 1962). For mathematical simplicity, let us assume that the midrib cross section is of constant shape, so that the cross section can be characterized by a single number, its width or characteristic dimension.

Let $w(x)$ represent this characteristic dimension, measured in a cross section at distance x along the midrib from the leaf based, and let $f(x)$ be the width of each symmetric half-leaf at that point. Let us assume that the secondary veins are all parallel line segments perpendicular to the midrib, of length $f(x)$ at x. The strength of each cross section is proportional to $w(x)^3$ (Howland, 1962). In an optimally designed leaf, the midrib will be equally resilient to the stresses it faces along its entire length, lest energy spent in reinforcing certain sections be wasted by the failure of a "weak link." Making this assumption, we set the strength of a vein cross section proportional to the stress to which it is subjected:

$$w(x)^3 = 2\alpha \int_x^L f(z) \cdot (z - x)dz \tag{6}$$

where the right-hand side of the equation measures the total torque on the cross section at x caused by the weight of leaf strips of length $2f(z)$ applied through a lever arm of length $z - x$.

If the midrib were constructed of a constant proportion of xylem vessels of like diameter, then its supply capacity, or ability to deliver water at a given gradient of water potential, would scale like the area of the midrib cross section, or $w(x)^2$. If either the proportion of vessels or their diameters varied with midrib diameter, the supply capacity might scale like $w(x)^{2+2a}$, where a can be either positive or negative. The demand for water should scale like the area of each region along the midrib. If we assume that different regions of the leaf should have to dry out to the same extent to move water down a given length of midrib, then we should set the demand at each point along the midrib proportional to the supply capacity made available there by midrib taper:

$$f(x) = \frac{\beta}{2} \frac{dw(x)^{2+2a}}{dx} \tag{7}$$

where β is a proportionality constant.

Equations (6) and (7) are a system of functional relations involving both $f(x)$ and $w(x)$. This system can be solved using the fundamental theorem of calculus to obtain:

$$w(x) \propto (L - x)^{1/(1-2a)} \tag{8}$$

and

$$f(x) \propto (L - x)^{(1+4a)/(1-2a)} \tag{9}$$

This solution represents the leaf shape and midrib taper that best integrate the demands of support and supply. The predicted leaf shape is thus triangular if the supply capacity of the midrib varies linearly with its cross-sectional area ($a = 0$, Figure 16.5). If a is negative (support tissue increases, basipetally, faster than supply capacity), the leaf margin should be convex, since

$$(1 + 4a)/(1 - 2a) < 1$$

If a is positive and is less than 0.5, the leaf margin should be concave:

$$(1 + 4a)/(1 - 2a) > 1$$

In none of these cases, however, should leaf width ever decrease in moving from leaf tip to base (Figure 16.5). The inclusion of oblique

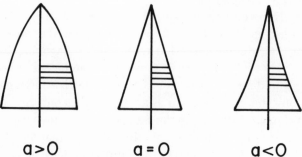

$$a > 0 \qquad\qquad a = 0 \qquad\qquad a < 0$$

Figure 16.5. Leaf shapes predicted by the support–supply model for leaves, with perpendicular secondaries and veins with a supply capacity proportional to the $2 + 2a$th power of their diameter.

rather than perpendicular secondaries does not affect this conclusion, except that leaf width should not decrease until the tip of the lowest secondary is reached, at which point the margin should then parallel the last supply line to the midrib and leaf base (Givnish, 1976).

Leaf Packing

The support–supply model predicts that simple leaves should be roughly triangular, or wedge-shaped, in outline. Only a relatively small group of plants have leaves of this shape, though a great many leaves can be decomposed into wedge-shaped subunits supported and supplied by major veins (e.g., see Figure 16.6). However, the majority of all leaves are roughly oval to lanceolate in form (like those of the beech, *Fagus grandifolia*), taper inward before the last secondary, and cannot be decomposed into triangular units. This is probably because the requirements of efficient leaf packing are at odds with those of efficient support and supply within the leaf.

The support–supply model contained the implicit assumption that a given leaf is essentially independent of, and relatively unaffected by, the shapes and positions of its neighbors. This assumption

Figure 16.6. Resolution of leaf (*Sida aggregata*, Malvaceae) into wedge-shaped regions supported and supplied by the secondary venation.

comes close to being realized when the leaves are spaced far enough apart on a branch so that they neither touch nor shade another, but this mode of leaf arrangement is often inefficient. It wastes space near the branch, so that greater lengths of branch are required to accomodate a given amount of leaf surface; it is also not an efficient means of shading competitors. Plants have evolved various ways in which they hold their leaves so as to avoid self-shading and improve the density of shade cast. Two basic kinds of leaf arrangement are especially common and involve the packing of leaves in spirals about erect twigs (typical of sun plants) or in planar arrays along branches (typical of shade plants) (Horn, 1971, 1975; Leigh, 1972, 1975).

Although a strictly wedge-shaped leaf may be most effective as a unit standing alone, it does not use space efficiently when packed, for example, in a planar array along a branch (Figure 16.7). Even when such leaves are packed so close together that they touch, much of the space adjacent to the branch is not utilized, although it has been paid for in terms of woody tissue. If these leaves were more closely packed and the amount of branch invested per leaf were reduced, the leaves would overlap. The area of overlap is a respiratory drag for the shaded leaves and a liability for all leaves in that such close proximity could induce wear and tear. On balance, plants without the region of overlap would be favored. If the overlap is removed, so that the leaves remain symmetric and the space between them is efficiently covered (leaving some space to permit the convective decoupling of adjacent leaves), the leaf arrangement in the lower part of Figure 16.7 results. The limit to the packing process comes either as the optimal effective leaf size is reached or as the leaves become too narrow and long to be supported efficiently as separate units. The optimal shape of leaves to be packed in planar arrays must be modified from a wedge-shaped form to one in which the leaf margin roughly parallels the midrib over much of the midleaf and tapers toward either end. The apical section should remain roughly wedge-shaped. Similar principles operate in the packing of leaves in spirals about erect twigs, although, because the leaf bases must be packed radially, the ovate shape that obtains is quite different. This is not the complete story, of course, since plants that arrange their leaves diffusely can hold several layers produc-

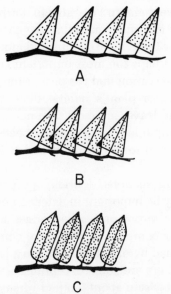

Figure 16.7. Packing of triangular leaves in a plane along a horizontal branch. (A) Arrangement of leaves so that they do not touch or overlap wastes much space along the branch. (B) Closer packing of similarly shaped leaves leads to overlap, shading, and potential wear and physical damage. (C) Efficient packing of triangular leaves in a plane along a horizontal branch entails the symmetric deletion of areas of overlap from each leaf in such a way that the area is covered efficiently, consistent with the convective decoupling of the leaves from one another and the overall efficiency of the leaf array as a cantilevered ribbon of photosynthetic tissue along the branch.

tively and grow rapidly in well-lit situations (Horn, 1971). However, the importance of the trade-off between leaf packing and individual support and supply is clear, should play a predominant role in shade-adapted species where there is no obvious advantage to packing leaves diffusely, and suggests costs not considered by Horn (1971, 1975) in his analysis of light-gathering strategies in intense sunlight.

Leaves, Leaflets, and Leaf Margins

Bailey and Sinnott (1916) documented a tendency for leaves to have entire margins in tropical, subtropical, and polar environments and for leaves to have nonentire margins in north temperate regions. They also noted that nonentire margins are generally more common in forest understories than in the canopy and more common in mesic areas than in dry. Perhaps the most important contribution of these students was the observation that the possession of entire margins is correlated more with a plant's habitat than with its phylogeny. Similarly, compound leaves are most common in certain warm, dry regions, whereas simple leaves are frequent elsewhere, particularly at high latitudes and altitudes (Sinnott and Bailey, 1915; Givnish, 1978).

In the following sections, I briefly propose some biological phenomena that may be important in determining whether leaves or leaflets should have entire or toothed edges and whether leaves should be simple or compound. The more complex case of lobed leaves is not discussed here, but because most lobed leaves are also toothed (although there are some exceptions—*Sassafras, Liriodendron*), much of what is said about toothed versus entire margins can be extended to the more general case of entire versus nonentire leaves.

Toothed versus Entire Leaves

In thin leaves, the areas served by the secondary veins should be more independent of each other than those in thicker leaves. The area near each vein derives progressively less support from the other veins as the leaf becomes thinner and flimsier. The resistance to the flow of water in the mesophyll also increases in thinner leaves and serves to isolate further the areas served by each secondary. In addition, the secondary veins themselves become relatively more pronounced in thin leaves as the support lent by unspecialized tissues decreases, which favors the consolidation of specialized support tissue. This suggests that the secondary veins act as several parallel midribs packed tightly together; the leaf shape predicted is one with wedge-shaped teeth centered on each secondary, with the depth and acuity of the toothing dependent on leaf thickness. Thick leaves, where the support and supply contributed by different secondary veins overlap strongly, should be entire.

Toothed leaves are most common among thin-leaved deciduous

species in the north temperate zone, fast-growing herbs, and shade-dwelling plants (Bailey and Sinnott, 1916; Dorf, 1969). Bailey and Sinnott state, after reviewing dozens of regional floras, that "among woody plants, well-developed non-entire margins occur commonly on comparatively thin, soft leaves with prominent veins. Entire margins, on the other hand, usually occur on thicker, stiffer, more leathery leaves with structures that seem to retard evaporation." They also note that "in regions with marked alternating periods of hot and cold or dry and wet seasons, the evergreen foliage may be entire when the deciduous types are strikingly non-entire."

Howard (1969) provides crucial quantitative data with which to test the idea that the thin leaves in a flora tend to be the toothed ones. He measured the leaf thickness of 40 species in elfin forest on El Yunque in Puerto Rico. All are evergreen, so that deciduosity is not a factor. The increasing trend in the entirety of leaf margins with increasing leaf thickness is shown in Figure 16.8. As expected, the percentage of species with nonentire leaves has a strong negative correlation with leaf thickness—indeed, 6 of the 7 thinnest-leaved species have nonentire leaves, whereas 17 of the 19 thickest-leaved species have entire leaves. The median thickness of nonentire leaves in this tropical locale is 218 μm. This compares with a mean of 154 μm for 80 species of deciduous trees with mostly nonentire leaves in the northeastern United States and a mean thickness of 406 μm for 38 species of evergreen trees with mostly entire leaves in New Zealand (Wylie, 1952, 1954). The high percentage of species in the El Yunque flora with nonentire leaves having thicknesses in the range characteristic of north temperate trees and the high percentage of species with entire leaves having thicknesses characteristic of tropical and south temperate trees support the predictions made above (Givnish, 1976).

These predictions do not explain the occurrence of thick leaves with spinose teeth, like those found in some hollies (e.g., *Ilex opaca*). Ehrlich and Raven (1967) have advanced a possible explanation for holly spines, that is, that they prevent caterpillars from feeding inwards from the leaf margin and devouring the entire leaf. This effect might be important in certain cases, but, generally, spinose teeth seem to be set too far apart to deter insects that could be met more effectively by closely set hairs and trichomes (Pillemer and

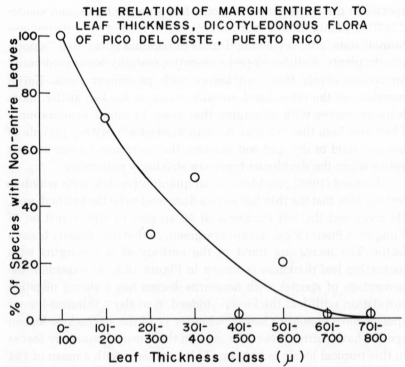

Figure 16.8. The relation of marginal entirety to leaf thickness in the dicotyledonous flora of an elfin forest on Pico del Oeste, Puerto Rico (after data from Howard, 1969, and personal communication).

Tingey, 1976). I propose an alternative hypothesis, that spinose teeth are a defense against mammalian herbivores, the analogs of thorns placed on stems and branches to protect foliage.

Either defense should be favored in semiarid areas where low rainfall keeps plants within the reach of large browsers. Thorns on stems and branches should be the more effective strategy for deciduous plants. Thorns are an investment that can protect several crops of leaves, whereas spines are not. Also, spines would be ineffective on soft, easily foldable deciduous leaves. On the other hand, spinose teeth would be advantageous on hard, evergreen leaves since they are more closely associated with the edible foliage and their cost can be amortized over a long period of time. We thus predict that spinose leaves should be relatively common in semiarid areas of winter rainfall that encourage short, evergreen woody plants (Mooney and Dunn, 1970), that is, in chaparral or similar Mediterranean-type ecosystems. Thorns should occur in semiarid areas that encourage the deciduous habit. Although it is beyond the scope of this paper to test these conclusions thoroughly, most of the spinose species with which I am familiar are evergreen shrubs and trees found in Mediterranean-type climates, like *Heteromeles*, *Quercus dumosa* and *Q. wizlizenii*, and *Prunus ilicifolia* in California (Munz, 1974), *Quercus* spp. in Mediterranean macchia, and the numerous saw-edged species of *Banksia* and *Hakea* in southwestern Australia. Brenner (1902) found that oaks with hollylike leaves are common only in areas of winter rainfall in California, the Mediterranean littoral, Anatolia, and the western Himalayan region. The apparent absence of spinose leaves in the Cape region of South Africa (E. Moll, personal communication, 1976) and in Arctic and Alpine tundra areas that support short, evergreen plants suggests the need for further investigation.

Compound Leaves

It may at first seem difficult to see any adaptive advantage in compound leaves: Why should plants throw away branchwork, in the form of leaf rachises, in which they have just invested a considerable amount of energy? Yet there are at least two advantages of such seemingly profligate behavior. First, in seasonally arid areas favoring the deciduous habit, shedding the highest order branches (or rachises) is the most efficient means of reducing residual water loss after leaf drop, since these branches have the highest surface-to-volume ratio and the lowest suberization. Second, among plants that invade light gaps (cf. Hartshorn, 1978) or other early-successional habitats, a premium is placed on height competition for light. Such plants should branch rarely or not at all, at least initially, since such branching diverts energy from the leader and may thus slow growth in height. For this reason, and because the side branches that are grown will have a short lifetime before being shaded by new and higher branches, these branches should be made as cheaply as possible. Compound leaves are the ideal throwaway branches because of the low cost of parenchyma in rachises relative to that of woody tissue in twigs. They may have an advantage over deeply lobed leaves of the same effective size because of the greater potential for movement of the leaf surface, perhaps to avoid desiccation or structural damage, when it is mechanically decoupled into leaflets.

These predictions are explored in greater detail elsewhere (Givnish, 1978). Compound leaves are indeed frequent in seasonally arid habitats, like deserts, savannas, tropical seasonal forests, and the upper stories of rain forests, which also encourage deciduosity. I have shown that the increase in the percentage of species with compound leaves along a gradient is, in certain instances, due to an increase in the percentage of species with *deciduous* compound leaves and a decrease in the percentage of those with evergreen simple leaves (Table 16.2). Certain semiarid habitats that favor evergreenness and not deciduosity, like the Californian chaparral, have relatively few species with compound leaves. Many chaparral species have evergreen simple leaves, even though they belong to families that typically have compound leaves; these include *Rhus integrifolia*, *R. laurina*, and *R. ovata* in the Anacardiaceae (*R. trilobata* is deciduous and has compound leaves); *Cneoridium* in the Rutaceae; and *Adenostoma*, *Cercocarpus*, and *Heteromeles* in the Rosaceae (Munz, 1974). Members of the Rutaceae in South African and Australian regions of winter rainfall mostly have simple evergreen leaves, even though the remainder of the family in the subtropics and tropics is characterized by compound leaves (Engler and Prantl, 1931).

Species of light-gap or early-successional habitats are often characterized by large compound leaves that seem to be used as throwaway branches. Many palms, for example, appear to be distributed in areas where unstable soils, battering winds, or repeated cycles of flooding and desiccation provide a chronically disturbed and open canopy (Givnish, 1978). Although unbranched trees are absent in the north temperate zone, several species of early succession have large, pithy compound leaves. These include sumacs (*Rhus* and *Toxicodendron*), Kentucky coffee tree (*Gymnocladus dioica*), devil's walking stick (*Aralia spinosa*), Hercules' club (*Zanthoxylum clavaherculis*), certain ashes (*Fraxinus*), and the introduced tree-of-heaven (*Ailanthus*). Walnuts (*Juglans*) are interesting

TABLE 16.2.
CROSS-TABULATION OF JAMAICAN DRY FOREST SPECIES BY
LEAF COMPLEXITY AND DECIDUOSITY[a]

Dry evergreen forest   *Evergreen bushland*

	E	D			E	D	
C	.00	.30	Msph (20)	C	.11	.15	Mcph (47)
S	.50	.20		S	.49	.25	

	E	D			E	D	
C	.09	.00	Mcph (33)	C	.08	.00	Nph (12)
S	.70	.21		S	.67	.25	

Δ^b Δ

	E	D		E	D
C	−.09	.30	C	.03	.15
S	−.20	−.01	S	−.18	.00

[a]E, evergreen; D, deciduous; C, compound; S, simple; Msph, trees 8 to 20 m; Mcph, trees 2 to 8 m; Nph, trees 0.25 to 2 m. Numbers in parentheses represent number of species found in each stratum. Each entry represents the fraction of species in each stratum with the leaf form and behavior cited.
[b]Represents the change in the fraction of species having the leaf form and behavior cited, as observed in moving from the lower to upper stratum. As expected, there is a large increase in the fraction of deciduous, compound-leaved species.

"early-successional" trees that create and maintain their own gaps in surrounding vegetation by poisoning their neighbors. Other compound-leaved species, like elderberry (*Sambucus*), box elder (*Acer negundo*), and ash (*Fraxinus*), are common in floodplain forests, where the frequently disturbed and open canopy favors opportunistic growth.

Other rapidly growing, sparsely branched trees of early succession, like the catalpas (Catalpa speciosa, C. bignoniodes) and introduced empress tree (Paulownia tomentosa), do not have compound leaves but may reap similar benefits by placing large, simple leaves on branchlike petioles. The same kind of advantages are probably also attained by the various monopodial trees of tropical succession, like Cecropia, Musanga, Macaranga, and Ochroma [see Whitmore (1975) for observations on huge-leaved trees in early tropical succession]. This can be seen as follows.

Howland (1962) rightly recognized that large leaves must contain proportionately more support tissue than smaller ones because they contain longer cantilevers and hence are subject to greater stresses. He concluded that it is therefore more costly to build large leaves than small. This view overlooks the fact that a plant must build support for these leaves both inside and outside the leaves themselves and that the woody tissue used to support groups of leaves is usually more expensive energetically than the parenchyma used for support within the leaf. Thus, for a plant with a fixed total leaf area, it may initially be cheaper to build one huge leaf, with woody twigs and branches being replaced by parenchymatous midribs and secondary veins. Such a strategy can be counterproductive in the long term, however, since all support tissue is shed when the leaf is shed. The plant would gain no benefit by consolidating new support tissue with old and would lose a considerable amount of height.

Thus, if a unit of leaf surface need not be replaced when it falls, and its support is not needed for further growth, big leaves are mechanically cheaper than small leaves. If the leaves must be replaced or if their external support can be used for further extensive growth, then small leaves are cheaper than bigger ones. As argued above, early-successional trees or species that invade under gaps in the canopy may present examples of plants where large units of leaf surface and ancillary support often need not be replaced. Generally, where height competition favors "large" leaves and gas exchange favors small leaves, compound leaves should be advantageous. Where both height competition and gas exchange economics favor either large or small leaves, simple leaves should be advantageous. Where the demands of extensive, slow growth favor small leaves and gas exchange economics favor large leaves, the optimal leaf should be simple to minimize area for a given effective size. Since the mechanical costs of leaves depend on their arrangement in time and space, and not in any simple way on individual leaf size alone, this factor was not included in the original leaf size model.

FUTURE TESTS OF HYPOTHESES

How might further tests of my models for leaf form proceed? If we first consider the predictions of gas exchange economics, it is clear that further confirmation requires *measurement* of the costs and benefits associated with various aspects of leaf form, rather than more comparative analysis of patterns in their distribution. Most of the patterns observed in leaf size, for example, are compatible not only with the gas exchange model but also with Walter's (1973) hypothesis that shortages of water or nutrients favor smaller leaves by slowing leaf development and expansion.

Measurements of the photosynthetic benefits and root and xylem costs associated with effective leaf size, thickness, and stomatal resistance are required to test the general theory in which all of these are allowed to vary simultaneously. Perhaps the theoretical construct that will prove most useful is not one which assumes that transpirational costs are proportional to the total rate of water loss, but instead incorporates these costs, partly as the observed investment in roots and xylem and partly in the photosynthetic depression caused by lower leaf water potentials. This notion is implicit in the marginal-profit model of Givnish and Vermeij (1976, p. 771) and explicit in that of Mooney and Gulman (article 13). The needed measurements include the following: the dependence of plastid photosynthesis on leaf water potential, light intensity, temperature, and CO_2 concentration; the calculated or measured effects of leaf thickness and stomatal resistance on these factors (e.g., see Lommen et al., 1975; Yocum and Lommen, 1975; Hall and Björkmann, 1975;

Alberte et al., 1975); and the respiratory cost, duration of activity, and associated resistance to water uptake in absorbing roots (Givnish and Vermeij, 1976; Caldwell, article 17).

One problem that some will see as particularly damaging to my models is the central notion that selection tends always to favor individuals that fix carbon at the greatest rate. Some of the difficulties with this notion have been dealt with elsewhere (see Horn, article 2). Selection probably no more actually favors maximal carbon gain in plants than economic competition actually favors maximal short-term profits in businesses and monopolies. Other factors always complicate the story. If a monopoly were to set prices so as to maximize profits in the short term, it would be more likely that a competitor would invade than if prices were set at a less than profit-maximizing level. Similarly, a plant might maximize its total short-term rate of carbon gain by maximizing the rate of increase of active meristems, since this maximizes the rate at which new carbon is converted into carbon-producing leaf tissue. However, such a plant might lose out in height competition because its lack of apical dominance diverts energy from the leader (Cook, 1976; Givnish, 1978). Here selection would not favor the plant with maximal carbon gain, because its strategy so constrains the use of its products that it would be overtopped by less productive individuals. Problems of this sort may be less severe for leaves than for more integrative aspects of plant form since leaves are usually relatively small, interchangeable units that can be deployed flexibly.

There are other, more important factors, however, that *can* prevent maximal carbon gain from being the optimal strategy. Competition for an exhaustible resource, e.g., water, is one such factor. Plants growing in dry areas generally have higher water-use efficiencies (ratio of photosynthesis to transpiration) than those in moist areas (Ting et al., 1972; H. Johnson, 1975), as expected on theoretical grounds (Givnish and Vermeij, 1976). As suggested by Cohen (1970), even though plants with high water-use efficiency may have the greatest carbon gain over a season, they are imperiled by plants with a lower water-use efficiency since these can "waste" the soil water needed by the efficient plants for photosynthesis late in the season. As a result of losing late-season photosynthesis, the highly efficient plants may not realize a greater gain than the less efficient ones. This may be part of the reason why cacti and other desert plants having crassulacean acid metabolism (CAM) are succulent. Plants with CAM open their stomata and fix carbon at night, when evaporative demand is low, and then photosynthesize during the day using carbon stored in organic acids (Ting et al., 1972). As a result, CAM plants have a very high water-use efficiency and are particularly abundant in arid habitats (e.g., see Mooney et al., 1974). One reason cacti are succulent may be to set aside the "winning margin" of water needed to extend their period of photosynthetic activity sufficiently to ensure coexistence with less efficient, water-wasting competitors.

Whatever the real reason behind CAM succulence, the point is that the criterion of maximal carbon gain is crude and may have to be replaced by the more general criterion of an "evolutionarily stable strategy" (Maynard Smith and Price, 1973; Maynard Smith, 1974, 1976). An evolutionarily stable strategy is one which, if most members of a population use it, cannot be bettered by other strategies under consideration. The predominant opinion has been that capacity for maximal carbon gain constitutes an evolutionarily stable egy for plants (e.g., see Mooney and Dunn, 1970; Horn, 1971). As quantitative data about the costs and gains associated with gas exchange become available, we may find evidence that the modes of conflict between plants are more subtle and involve complex strategies that balance growth with shading competitors, hoarding or wasting water and nutrients, mimicking plants protected against herbivores or diverging from those that are not (Gilbert, 1975; Barlow and Wiens, 1977), and the like. Meanwhile, even if these latter effects are important, they must be weighed against the advantages of maximal carbon gain, so that understanding how leaf form should vary with environment to maximize growth will give us insight into at least one component of the selective forces favoring leaves of different sizes and shapes.

17 ROOT STRUCTURE: THE CONSIDERABLE COST OF BELOWGROUND FUNCTION

MARTYN M. CALDWELL

SINCE ROOT SYSTEMS are not a conspicuous feature of the landscape, they have received relatively little attention in ecological studies—particularly in the context of ecosystem function. Root system biomass and, occasionally, the rooting volume and depth of dominant species have traditionally been the only parameters included in studies at the community level. Although no one would deny the important roles played by root systems, the energetic costs of root system development and maintenance in relation to the useful functions performed by these organs have only recently come into focus in ecosystem-level studies. Closely allied to this cost–benefit balance is the question of selective tactics in the structure and growth patterns of root systems.

A consideration of the magnitude of root system involvement in community energetics, albeit at a rather rudimentary level, was found to be a necessary component in ecosystem-level syntheses which resulted from the recent International Biological Program. The results of these studies in a variety of ecosystems suggest that the energetic costs of root system activity are very high and that the belowground system thus constitutes a rather large sink of carbon in many communities. These costs of root system activity will first be briefly examined, and consideration of the selective mandate for energy conservation as it relates to the effectiveness and longevity of individual roots and factors involved in the maintenance and replacement of root elements will follow. Finally, patterns of root growth will be discussed in relation to a hypothesis involving relocation of absorption microsites within the rooted soil profile.

ENERGY INVESTMENT IN ROOT SYSTEMS

Although no one would refute the primary role played by root systems in water and nutrient absorption, species which minimize the investment of energy in this heterotrophic system would seem to be at selective advantage. The close coordination of root and shoot growth is evident not only in studies of hormonal regulation of root and shoot development but also in the equilibrium that plants attempt to maintain between shoot and root tissues when subjected to pruning or grazing (Jameson, 1963). The regulation of root and shoot development and, indeed, the competition of these two systems for photosynthate (e.g., Webb, 1976) tend to minimize the production of excessive root tissues or other forms of superfluous energy investment in the root system. Nevertheless, a view of energy deployment to root systems at the level of the plant community reveals that these costs are remarkably high.

Some indication of the allocation of energy to root systems is gained from a survey of root/shoot ratios. Although these ratios may provide a reasonable picture of energetic costs for annual plants, for perennial vegetation they are less germane. In the case of established perennial plants, it is the total cost of maintaining the root and shoot systems that is of importance rather than the biomass at any point in time. The inability to distinguish the exact proportion of living biomass of root systems in most investigations further reduces the meaningfulness of root/shoot ratios. For established plants, the true energy costs would include: maintenance respiration (which is associated with basic cellular processes such as protein turnover and maintenance of ionic gradients), the energy associated with exudates and that relinquished to mycorrhizae and belowground predators, and of greatest consequence, the energy devoted to the periodic replacement of the more ephemeral elements of the root system, along with the associated biosynthetic respiratory costs. For a perennial plant in steady state, this replacement is equivalent to the annual belowground productivity.

An estimation of the magnitude of belowground productivity has been gleaned in the last several years, largely under the auspices of the International Biological Program (IBP), in which a holistic approach to energy and nutrient flow through ecosystems provided the impetus for considering the productivity or turnover of the belowground system in a variety of ecosystems. A summarization of belowground production from IBP and other related studies of the past decade (e.g., Coleman, 1976; Caldwell, 1975), reveals that the annual production or turnover of the belowground system in perennial vegetation is sizable and can account for 50 to 80 percent of the total net production for a wide variety of plant community types, including arid shrub steppe, shortgrass prairie, montane meadow, deciduous forest, and Arctic tundra.

Root system respiration represents an additional energy expenditure which is not included in the belowground production term. Root respiration has two basic components: biosynthetic respiration associated with root production and growth, and basic maintenance respiration (Penning de Vries et al., 1974; Penning de Vries, 1975a; Ledig et al., 1976). Biosynthetic respiratory activity should, in most cases, be proportional to root biomass production and usually amounts to 0.24 g of CO_2 for each gram of biomass produced (Penning de Vries et al., 1974). Penning de Vries suggests that biosynthetic respiration should not, in theory, vary substantially with biomass production in different plant species, since alternative biochemical pathways or compartmentalization of synthetic processes are not likely to result in a large change in this respiratory cost. Maintenance respiration, which he considers to be primarily associated with protein turnover and maintenance of ion balance, can vary considerably from species to species (Penning de Vries, 1975; McCree, 1974). Not surprisingly, maintenance respiration rates may also vary substantially between different plant organs. Hansen and Jensen (1977) reported that maintenance respiration rates of *Lolium multiflorum* roots are substantially greater than those of shoot tissues.

Ledig et al. (1976) assessed the carbon balance of *Pinus rigida* seedlings in a growth-chamber environment and concluded that root respiration, although lower than respiration of shoot tissues, still represented a major drain on the carbon budget of these seedlings. After 66 days, the respiration of roots accounted for 12 percent of the diurnal assimilation, and, by 185 days, the root respiration term was 69 percent of diurnal assimilation. Although these rates are probably much higher than would be encountered in a field situation, since both night and day temperatures in the growth cabinet were high (29–32°C), root respiration rates were still of the same order of magnitude as those reported in several studies in which CO_2 efflux was measured from excavated roots for a variety of woody species. Thus, the potential at least exists for root respiration to constitute a sizable carbon loss from plants. On the other hand, studies which employed different methods of assessment reported that root respiration constituted a rather small proportion of total plant respiration for a *Pinus taeda* plantation (Kinerson et al., 1977) and an *Agropyron–Koeleria* grassland (Warembourg and Paul, 1977).

Exudates and shedding of tissues from living roots have been observed and described in a number of studies. Head (1973) and Coleman (1976) have reviewed many of these studies and conclude that, this is certainly a significant pathway of potential carbon loss, but quantitative assessments of the magnitude of this term have seldom been undertaken.

Maintenance of root system symbionts can also represent a significant energy allocation. For example, Barnard and Jorgensen (1977) reported that roots of *Pinus taeda* with active, well-developed mycorrhizae had respiration rates twice those of roots with little or no mycorrhizal development.

Taken in total, the carbon (and thus energy) commitment to the belowground system is sizable. Annual root reconstruction costs alone can often exceed aboveground primary production, and, though not easily quantified in the field, root respiration and carbon loss through exudates, tissue shedding, and various symbiotic energy commitments can only inflate the magnitude of these costs.

MINIMIZING COSTS OF ROOT SYSTEMS: EVOLUTIONARY CONSIDERATIONS

The title of this section is quite presumptuous in view of our current state of knowledge. In fact, not only are the costs of plant root systems poorly quantified at present, but little can be stated, even with considerable equivocation, concerning adaptive evolutionary strategies of different plant species with regard to the structure and functional activities of roots. Nevertheless, the subject is of immense importance in view of the dedication of energy to the belowground system.

Although not often explicitly stated, a fundamental premise in plant population biology and competitive theory is that plants are basically energy limited. This is not to ignore other bottlenecks or constraints in the life cycle that limit the success or distribution of plant species. Nevertheless, discussions in this volume concerning the adaptive geometry of plant canopies, photosynthetic optimization in microenvironments, and, indeed, competition for most resources are based on the energy-limitation premise. This accepted, the selective pressures on plant species to economize costs associated with root systems cannot be ignored.

Genetic Resources

In contrast to the obvious structural and functional diversity of the aboveground plant world, a casual observation suggests that structural diversity of the root systems among plant species is rather limited. Apart from elaborated storage organs, the root systems of all plants consist of a dendritic structure with several orders of branching—commonly many more than in the shoot system. This, of course, maximizes the effective environmental volume which can be explored. In other words, this structure places the maximum amount of root surface within close proximity of the largest soil volume for the quantity of root biomass involved (Wiebe, 1978).

Despite the basic similarity of root system structure, a closer inspection certainly reveals variation among species in characteristics such as degree of branching, growth patterns, rooting density, and distribution of root hairs and secondary thickening, as well as physiological characteristics (Troughton and Whittington, 1969; Zobel, 1975). Intraspecific differences in such characteristics have also been clearly demonstrated in agricultural species (e.g., Hackett, 1969; Troughton and Whittington, 1969), though there is much less information at this level for nonagricultural plants. Zobel (1975) suggests that 30 percent of the genome of higher plants is associated with rooting characteristics and that genetic variability in plant populations, though little studied because of the obvious problems in sampling the large numbers of individual plants required to arrive at statistically significant differences, is quite prevalent. Troughton and Whittington (1969) also point out that many of the differences in root system characteristics between species can be linked to differences in the effectiveness of water and nutrient uptake and other characteristics of selective significance.

It is also likely that intraspecific variability in rooting characters is associated with differences in functional effectiveness which may provide options for natural selection at the plant population level.

Despite the genetic variability that apparently exists in root system characteristics of plant populations, it would also appear, from the meager evidence at hand, that the functions provided by root systems come at a rather high energetic cost and that evolution has not resulted in a marked circumvention of this expense.

Replacement of Root Elements

Production of new roots probably represents the single greatest expenditure of energy in the belowground system and appears to be a significant phenomenon in a variety of ecosystems. Although annual plants have little alternative but to undertake the production of an entire root system during their comparatively short growing season, new root production of the apparent magnitude which has been discovered for established perennial species is less immediately explicable. At least 25 percent of the root system is reconstructed annually in such diverse systems as an Arctic tundra (Shaver and Billings, 1975), a *Liriodendron* forest (Edwards and Harris, 1977), a tallgrass prairie (Kucera et al., 1967), a shortgrass prairie (Sims and Singh, 1971), and a shrub steppe (Caldwell et al., 1977). Some species such as the perennial Arctic tundra sedge

Eriophorum angustifolium apparently undertake a complete turn-over of the root system each year (Shaver and Billings, 1975).

The classicial depiction of root growth and development suggests that absorption of water and nutrients is confined principally to a zone of the new root immediately behind the region of cell elongation where root hairs are well developed. This "absorption" zone is, however, generally considered to be rather short-lived in its effectiveness since suberization and loss of root hairs are often observed within a couple of weeks. Certainly, if absorption of nutrients and water were entirely confined to this absorption zone of new roots and if this region had a strictly limited functional life, the necessity of continued root growth in perennial plant root systems would not come into question. The matter deserves closer scrutiny.

The necessity of new root growth hinges on several questions: Is absorption of water and nutrients confined to new, unsuberized roots? Are viable root hairs critical in the absorption process? If effective absorption is indeed dependent on new root tissues, is it the suberization of roots, which normally takes place within a few weeks of root formation, that effectively curtails the functional activity of roots? What is the adaptive significance of suberization? Is new root growth necessary simply to relocate centers of nutrient and water absorption in the soil profile so as to avoid localized depletion of water or nutrients? Unfortunately, these questions cannot be answered with much certainty at present. Nevertheless, each should be considered.

Effectiveness of new and old roots. There are few data concerning the permeability of older suberized roots to either water or nutrient uptake. Although unsuberized root tips exhibit greater permeability to water (Brouwer, 1953; Hansen, 1974), there is some suggestion that the older suberized portion of the root system is not impermeable to water. Kramer and Bullock (1966) confirmed older reports in the literature (e.g., Hayward et al., 1942) that suberized roots did have a significant permeability to water, although this might be only a fraction of the permeability of unsuberized new roots on a unit-area basis. Nevertheless, they suggest that, because of the predominance of older suberized root elements in the root systems of woody plants, water uptake by these older roots might account for at least 75 percent of the total water absorption. This surmise was based simply on the proportion of suberized and unsuberized root elements in the root system and their relative permeabilities. Nevertheless, it would appear that water absorption in perennial plant root systems is probably not confined to the suberized segments of new roots.

Older suberized root elements may be at least moderately effective in nutrient absorption. As is the case with most of the research on water permeability of roots, it is the new roots of young plants that have received most of the attention in studies of nutrient uptake (e.g., Bowen and Rovira, 1967; Bhat and Nye, 1973), and thus much less is known about nutrient absorption in older roots of perennial plants. Recent studies of Chapin (unpublished data), however, indicate that older root segments of Picea mariana are almost as effective in phosphate uptake as new roots. Since most of the root system is composed of older root elements, their contribution to the phosphorous uptake of this species may far outweigh that of the new growing root tips. Thus, though evidence is scarce, there is some basis for suggesting that new roots are not essential for either water or nutrient uptake by perennial plants.

Necessity of root hairs. The necessity of viable root hairs in the absorption of water and nutrients may also be questioned. Root hairs are usually considered as ephemeral appendages of epidermal cells concentrated in the absorption zone of new roots. There is little question that root hairs substantially enlarge the surface area of roots, and it might seem only logical that these small appendages greatly facilitate absorption of either water or nutrients.

Based on assumptions and calculations of the longitudinal hydraulic conductivity of root hairs, Newman (1974) argued that, even for a root with a high density of root hairs, the hydraulic conductivity of soil surrounding the root would be orders of magnitude greater than that of root hairs themselves under moist soil conditions. Even in soil with a matric potential of -15 bars, which would lead to a substantial lowering of hydraulic conductivity, he calculated conductivity of the root hair pathway to be only 5 percent that of the soil. Thus, the importance of root hairs in water uptake is not clearly established. Root hairs may be of importance when gaps develop between the root and adjacent soil owing to shrinkage of soil or, perhaps more frequently, shrinkage of roots themselves (e.g., Huck et al., 1970). Newman (1974) considered this possibility and calculated that water transport through root hairs across a root–soil gap would encounter less resistance and therefore require a smaller water potential difference than water-vapor movement across this gap, under some circumstances. In other situations, water-vapor diffusion across the gap can be calculated to be more efficacious.

Root hairs may be more important in facilitating nutrient absorption. Certainly, if gaps develop between the root and the bulk soil, root hairs would be essential for nutrient uptake. Even in the absence of gaps, there is convincing evidence that root hairs enhance nutrient absorption. Phosphorus uptake rates by onion roots completely devoid of root hairs were much lower than those of the roots of rape (Brassica napus), which are densely covered with root hairs (Bhat and Nye, 1974). The phosphorus depletion zones around the onion roots were also much narrower than those around the rape roots. The differences in phosphorus uptake were attributed to the presence of root hairs in these experiments. It appears that root hairs enhanced absorption in these studies. This might not, however, be only the result of better contact with a larger volume of soil. In a subsequent experiment, Bhat et al. (1976) emphasized the importance of phosphorus solubilization by root exudates in the facilitation of phosphorus uptake by roots; furthermore, they suggested that this may even be the result of exudates excreted by the root hairs themselves.

The role of root hairs in the absorption of water and nutrients, then, may not derive so much from the mere amplification of root surface area as from the bridging of soil–root gaps or chemical facilitation of nutrient absorption. The essentiality of root hairs and their quantitative impact on absorption have yet to be elucidated.

Even if the hypothesis that root hairs enhance water or nutrient absorption is accepted, the common assumption that these appendages are very ephemeral should also be made with caution. Persistence of intact root hairs for periods up to a year or two has been reported (e.g., McDougall, 1921; Dittmer, 1938; Fernandez and Caldwell, 1975). Although the viability of such roots hairs might be questioned, their physical persistence at least suggests that enhanced absorption by root hairs may not be limited to a period of a few days or weeks.

Although the newly produced root elements, particularly those with viable root hairs, appear to have higher absorption rates, it is by no means clear that the absorption capacity of older, suberized perennial roots is inadequate to supply the demands of perennial plants. Certainly, more discriminating studies are needed.

Suberization. Although it appears that older roots are to some degree permeable to water and nutrients, suberization of roots may be responsible for curtailment of absorption effectiveness. Field observations of growing root systems generally corroborate the common assumption that only the new growing root elements are free of the pervasive darkening in color usually considered to be the development of a suberin layer in older roots (Head, 1973). This darkening of the roots usually occurs within 2 or 3 weeks after formation, and even more quickly if the soil is dry or if roots are exposed to the atmosphere (e.g., Fernandez and Caldwell, 1975). Whether the darkening or browning of roots in field observations is always due to the process of suberization should, of course, be questioned. Since suberin is impermeable to water, deposition of suberin would seem to defeat the primary purpose of root elements as organs of absorption. Thus, though roots and even root hairs might persist for long periods of time, their permeability to water, and presumably nutrients, might be reduced and their effectiveness in absorption thereby curtailed.

As Newman (1974) has discussed, suberization is a rather general term and may refer to the deposition of suberin in the normal endodermal cells forming the Casperian strip, deposition in the exodermis immediately beneath the epidermis, or the formation of a periderm during secondary thickening. As he points out, even though suberin itself is impermeable to water, the deposition of suberin layers is not always continuous, particularly if they are deposited in the walls of the exodermis. Even in the case of periderm formation, water could still enter through lenticels, wounds, or other breaks or discontinuities in the suberin layer. Although older roots may exhibit much lower absorption rates, they are not necessarily impermeable to water or nutrients, and absorption is thus not restricted to unsuberized growing root tips.

Explanations of the adaptive import of suberization are not abundant. Restricted permeability of roots would reduce the loss of water from root tissues to dry soil. Presumably, during the night, when shoot water potentials approach those of the root system and there is not a prevailing water potential gradient for water movement to the shoot, water could migrate from root tissues into regions of the soil profile when the soil is dry. If this occurred near the soil surface, moisture might even be lost through soil surface evaporation during the subsequent day. Even during the day, when there is a water potential gradient from roots to the shoot, water potentials of dry, surface soil layers may be more negative than the shoot system, in which case water loss from roots to the soil might occur.

The suggestion that water might move from plant roots to the soil is hardly new (e.g., Breazeale et al., 1950; Borman, 1957; Jensen et al., 1961), and this movement has been demonstrated in several experiments. Nevertheless, the magnitude of this phenomenon is still in question. Molz and Peterson (1976) recently conducted experiments on 3-month-old potted cotton plants to determine the magnitude of potential water loss from roots to the soil. They reported that, for either live or dead root systems, the water loss was surprisingly low when compared to flux in the normal direction from soil to roots. (This low flux in the negative direction occurred at a greater water potential gradient than did the higher flux in the positive direction in their experiments.) They did not consider this phenomenon to be a property of the living roots but thought that the difference might be attributed to a hysteretic unsaturated flow phenomenon in the soil. They did not report the degree of suberization of the roots, but if the roots were suberized, reductions in permeability to water transport would be the same in either the positive or negative direction. As they point out, the picture is further complicated by the fact that many plants have growth-related water potentials of an approximate magnitude of -4 bars, which would thereby reduce the water potential gradient for water loss from roots to the soil. Thus, a combination of growth-related plant water potentials and this unexplained but apparent hysteretic soil phenomenon may result in water flux rates in the negative direction which are much lower than might otherwise be anticipated. This study should stimulate a reevaluation of the degree to which suberization restricts water loss from roots. Although suberin may reduce the permeability of roots, it is apparent that other factors play important roles.

Other explanations for suberization or development of a cuticle involve resistance to infection or herbivory by other soil organisms. Our concept of the adaptive import of suberization is hardly comprehensive. Although suberization probably results in some type of protective barrier for plant roots and reduces permeability, which would be of benefit for the portion of the root system that serves principally as a conduction system, this suberin deposition apparently is usually a leaky barrier, the ultimate biological significance of which is still in question.

Localized depletion and relocation of absorption sites. Another question presented in this section concerned the necessity of relocating centers of nutrient and water absorption within the soil profile in order to avoid localized depletion of water or nutrients. Such a question is necessarily based on two assumptions: that absorption of water and nutrients is not uniform within the rooted soil mass, and that diffusion of water or nutrients within the rooted zone of the soil is not sufficient to keep pace with absorption and therefore limits uptake of water or nutrients by the plant.

Most of the physical models of water absorption by plant root systems assume that water uptake within the rooted zone is reasonably uniform throughout the length of the root system and that roots are randomly distributed within the rooted zone. Thus, the geometrical basis of these models assumes water movement to take place from a cylinder surrounding each root element and the diameter of this cylinder to be a simple function of root density (Gardner, 1960; Cowan, 1965).

Several workers have employed the basic models of Gardner (1960) and Cowan (1965) and assumed uniform uptake within the rooted zone. If realistic root densities are employed, most have agreed that diffusion of moisture within this microcylinder to the root would indeed keep pace with water uptake. Therefore, what is termed a "rhizospheric resistance" to water uptake by the plant would not be an appreciable component of the overall resistance to water flux from the soil to the plant shoot (Newman, 1969, 1974; Lawlor, 1972; Williams, 1974, 1976). A detailed quantitative review and discussion of these models and their implications are not appropriate in this paper but have been covered elsewhere (Newman, 1974; Caldwell, 1976). Since an appreciable resistance to water uptake is often apparent in many plants, i.e., absorption of water often cannot keep pace with transpirational demand, many workers have concluded that resistance must be primarily within the root tissues rather than in the soil. Unfortunately, an experimental partitioning of root tissue and rhizospheric soil resistance to water movement is not easily accomplished, and the issue is therefore unresolved.

Based on the scant information available on permeability of new root as opposed to older root tissues and making assumptions concerning the nonuniform uptake of water along the length of older roots, Caldwell (1976) has presented an elaboration of the Cowan model. Even under moist soil conditions, this model suggests that rhizospheric resistances and hence localized depletion of water can easily develop. In this case, rhizospheric resistances would be much more important than root tissue resistance, and the need to relocate centers of water absorption by new root growth within the rooted zone would be emphasized.

Localized depletion of nutrients about actively absorbing roots may be an even more compelling reason for root growth. A case in point is phosphorus, the diffusion and availability of which is highly dependent on the buffering capacity of the soil (Cole et al., 1977). Soils with low buffering capacity have very slow solubilization and, consequently, slow diffusion of phosphorus. Elaborate experiments by Bhat and Nye (1973, 1974) have vividly demonstrated the localized depletion of phosphate around absorbing roots. This depletion zone developed quickly, and up to 60 percent of the exchangeable phosphorus depletion observed within the root hair cylinder took place in the first 3 days of uptake. Subsequent transfer of phosphorus to this cylinder was quite slow. Such experiments, though not easily subject to quantitative extrapolation to root systems under field conditions, do nevertheless lend support to the hypothesis of the necessity of relocating the centers of uptake within the rooted zone.

Although there is some support, both theoretical and experimental, for the hypothesis that root growth may be necessary simply to relocate centers of absorption in order to avoid localized depletion of water and nutrients, much is yet to be learned before this hypothesis can be solidly accepted. Most satisfying would be a demonstration of depletion zones around roots under field conditions. In addition, it would be necesary to demonstrate that the rate of new root growth is of the appropriate magnitude to satisfy the nutrient- and water-absorption requirements of the plant that are directly limited by diffusion. If the relocation hypothesis is indeed supported for at least some higher plant species, this would go far in explaining the magnitude of plant root system turnover.

In addition to relocation necessitated by development of water and nutrient depletion zones, if soil–root gaps are prominent, the consequent resistance to water and, in particular, nutrient uptake might be reason alone for new root growth.

Of particular interest in the study of interspecific competition and evolutionary progress is the extent to which plant species might differ in the degree to which relocation, and hence root growth, is necessary. For example, it is clear that species differ in permeability of root elements to water and nutrients. Edwards and Barber (1976) showed that phosphorus uptake declined as a function of root age in soybeans to a much greater extent than in maize. In a somewhat different vein, Addoms (1946) used dyes to demonstrate that the nature and regularity of sites of dye penetration through the periderm of older roots was dependent upon the species under consideration. Therefore, it is expected that the degree of nonuniformity in water uptake in root systems of different species might also vary substantially. The hypothesis of relocation is to a large extent dependent on the degree to which uptake of water and nutrients is nonuniform and localized in specific microsites within the rooted zone. The greater the degree of nonuniformity, the greater the degree to which localized depletion would occur, and a mandate for new root growth would result. Thus, anatomical, as well as physiological, characteristics of both young and old roots of perennial plants certainly deserve much more attention than they have received in the past 50 years. Understanding of root structural characteristics has lagged far behind physiological studies of ion flux across root membranes and the development of elaborate physical models of water and nutrient uptake.

Root Longevity

Longevity as related to relocation of absorption sites. Although there are undoubtedly basic biological limits for longevity, in the interest of energy and nutrient considerations, long-lived root elements would seem to be of advantage for a plant. However, if root–soil gaps or the necessity of relocating centers of absorption within the rooted zone are compelling reasons for root growth, the question of root-element longevity may be viewed in a different perspective.

Maintenance of long-lived, fine root elements may, in fact, be a poor evolutionary strategy. For example, it may be much less energetically expensive to form new root elements which systematically explore microsites of the rooted zone in the process of scavenging elements such as phosphorus than to maintain long-lived absorbing root elements in sufficiently large density to meet the nutrient demands of the plant. If diffusion of an element such as phosphorus is extremely slow, a completely static perennial root system might need to have a much greater density than a dynamic system, in which new roots are continually produced at random locations within the rooted zone and subject to a high attrition rate. Of course, the most resourceful mean root-element longevity would depend to a considerable extent on the environmental conditions. Soil nutrient concentrations, nutrient buffering capacity, average frequency of soil moisture recharge, and prevailing soil temperatures, as well as various physical attributes of the soil, would all be important determinants of the extent to which localized depletion zones would develop around microcells of absorption and root growth would be necessary to relocate centers of absorption. The degree to which soil physical conditions are conducive to the development of gaps around roots is also of considerable consequence. Naturally, the aerial environment and physiological status of the shoot system are also important since they determine nutrient and water demands of the plant, as well as photosynthate supply.

Analogous to the oft-discussed evolutionary strategies of ever-green and deciduous modes of leaf retention, it would seem likely that longevity of small root elements could differ strikingly between species. As leaf photosynthetic capacity appears to be inversely correlated with leaf longevity (e.g., Mooney and Dunn, 1970a; John-son and Tieszen, 1976), so might absorptive intensity of root ele-ments be inversely correlated with their normal longevity.

It might be particularly instructive to investigate neighboring species in which longevity of root elements and the extent of annual root system turnover differ to a considerable degree. For example, in the Arctic tundra studies of Shaver and Billings (1975), species varied greatly in the magnitude of root system turnover. *Eriophorum angustifolium* exhibited a complete turnover of the root system annually, whereas species such as *Carex aquatilis* and *Dupontia fischeri* were estimated to experience a complete root system renewal only after 6 to 8 years. Although this might be partly explained by differences in microsites, the differences in longevity and turnover may be linked with other characteristics of these root systems, such as patterns of growth and the degree of nonuniformity in nutrient absorption. Such striking differences in root longevity and new root production must have a heritable basis, and it seems only reasonable that such differences would be coupled with differ-ences in other root system characteristics, the whole of which would constitute a particular evolutionary strategy in the Arctic environment.

Energetic considerations. This discussion of root longevity should include a consideration of the trade-off between maintenance of root elements of considerable longevity and production of new root elements as needed. Or, in other terms, What is the ledger balance between maintenance respiration and carbon costs associ-ated with exudates and material relinquished to symbionts, para-sites, and predators versus the carbon costs of new root production, along with the associated biosynthetic respiratory expenditures? Although not directly involved in this discussion, considerations of nutrient costs in each case should not be overlooked.

Unfortunately, it is not possible to provide a simple statement or calculation of the results of this ledger balance. Reports of root respiration rates are, first of all, not abundant and, second, indicate a great variation between species and between measurement techniques (e.g., Ledig et al., 1976). Some of this variation probably reflects differences in root culture conditions and artifacts of measurement techniques. Not surprisingly, determinations of respiration rates of roots extracted from the field are even less abundant.

Ledig et al. (1976) have summarized results of root respiration rate studies for a variety of woody plant species. A value of 0.4 mg of $CO_2 \cdot g^{-1}$ dry weight $\cdot hr^{-1}$ falls close to the minimum of the range of respiration rates they reviewed and probably reflects a basal maintenance respiration rate. By extrapolating this rate to an annual carbon cost, it would appear that over twice as much carbon would be utilized in root maintenance as would be necessary to invest in new root production, even with the accompanying biosynthetic respiratory expenditures. Considerable caution however, is warranted in such hasty calculations. For example, some of our recent measurements of respiration rates of roots extracted from the desert shrub *Atriplex confertifolia* in the field (Holthausen and Caldwell, unpublished data) indicate that basic respiration rates vary considerably during the course of the year. Rates during that portion of the year when the shoot system is not particularly active can be quite low— in October, rates ranged from 0.01 to 0.1 mg of $CO_2 \cdot g^{-1}$ dry weight $\cdot hr^{-1}$ at 12°C. If this is representative of the average annual maintenance costs, an extrapolation suggests that maintenance of existing roots would be considerably less expensive than production of new roots. Again, this is a rather blatant extrapolation which also warrants caution. In any case, acclimation or adjustment of respiratory rates so as to conserve expenditure of carbon has been clearly demonstrated in this shrub, as well as in earlier studies using potted plants (Cox, 1975; Lister et al., 1967; Negisi, 1966; Shiroya et al., 1966).

The only conclusion that can be reached at this point is that the trade-off between maintenance and new root production could be in favor of either alternative, based on available information. A detailed energetic evaluation would be necessary to decide the final outcome for a particular situation.

Of course, a simple analysis as to which alternative should be more advantageous certainly cannot be based solely on energetic criteria.

Nonsynchrony of Root Growth

The patterns and extent of root growth reflect both the genetic control of the plant and characteristics of the soil environment. Degree of root branching and other growth characteristics exhibit both inter- and intraspecific differences in the same environmental setting, which indicates that root growth in part follows a genetic program (Hackett, 1969; Troughton and Whittington, 1969). Clearly, however, soil environmental factors such as mechanical strength and configuration, texture, moisture, nutrient concentration, and CO_2 concentration have a direct bearing on root growth and development. It is not particularly surprising that flushes of new growth in perennial plant root systems occur in response to significant wetting or fertilization of the soil (e.g., Levin et al., 1973; Ares, 1976; Russell, 1970). There also appears to be a certain coordination of root meristem activation and root extension in response to normal seasonal environmental changes (e.g., Ares, 1976; Reynolds, 1975).

Both the normal seasonal phenological changes of the root system and the response of root growth to changes in the soil environment such as wetting or nutrient additions follow our expectations of the proper adaptive course of action that root systems should pursue. Even more intriguing is the pattern of meristem activation and root growth that takes place apart from the general season trend or in specific response to changes in the soil environment. This is particularly germane in the context of the hypothesis of relocation discussed earlier.

When soil moisture and nutrient status are more or less static and some root growth is in process at a particular zone of the soil profile, a certain nonsynchronous growth of root tips may be evident. Reynolds (1975) has described such behavior from his extensive studies of Douglas fir *(Pseudotsuga menziesii).* He described a pronounced variability in the status of root apices with respect to shape and color and apparent differences in degree of activation within a seemingly uniform zone of the soil profile. Furthermore, he reported a similar heterogeneity in the localization of starch reserves and even root hair production. As he suggested, this is probably not a "simple response to the local soil environment, but may reflect tactical organization" of the root system. This nonsynchronous root growth agrees with the hypothesis that new root growth in perennial plant root systems is at least in part due to the need to relocate microcells of water or nutrient absorption.

We have observed a similar phenomenon in observations of root growth patterns in desert shrubs (Fernandez and Caldwell, 1975). Although there is an overriding pattern of new root growth activity in these shrub root systems, in which regions of root growth progress from the upper to lower zones of the rooted soil profile with progression of the growth season, when zone growth is underway within any particular depth zone, it is quite nonsynchronous. Furthermore, activity of any particular root tip is usually limited to 2 weeks or less. Thus, widely and apparently randomly spaced root meristems are activated, root extension occurs for a period of 10 to 14 days, after which there is a progressive darkening of the root tip, and there is no subsequent activity at a later date. Therefore, root growth activity exhibits a series of short pulses within any particular local zone of the profile. Such pulsing of root growth and the subsequent relocation of centers of moisture and nutrient uptake spaced throughout the soil may be a reasonably efficient mode of root growth, as opposed to a synchronous, slow, and continuous growth of all meristems within a particular zone.

The short-lived activity of small root meristems which has been observed in these and other studies may be under the regulation of plant hormones since they interact with growth substrate concentrations, as has been discussed by Street (1969). Under certain tissue culture conditions involving regular subculture, it has been possible to prolong individual meristem activity some 40 to 50 weeks. Under most tissue culture conditions, however, meristems undergo a pronounced aging and cease growth after a strictly limited period of time, as has been observed in the field. Although the exact mechanism by which chemical growth regulators are involved in the curtailment of root meristem activity has not yet been elucidated, the possibility that this regulation of meristem activity, and hence the pulsed pattern of root growth, is under an overriding genetic program should be given serious consideration and may, in fact, constitute part of the "tactical organization" of the root system which Reynolds suggested.

Although the phenomenon of pulsed and nonsynchronous root

growth coincides well with the hypothesis of relocation, it hardly constitutes proof. Much has yet to be demonstrated before this hypothesis can be endorsed.

CONCLUSIONS

Research of the past 5 to 10 years has indicated that perennial plants commit a surprisingly large quantity of energy to root systems. Much of this energy appears to be utilized in the continual replacement of short-lived root elements within the established rooted zone. Despite this seeming inefficiency, such tactics of growth and energetic deployment may represent the best evolutionary compromise. Either physiological limits or the need for relocation of microcells of water and nutrient absorption may dictate this need for new root production. Attrition of unused fine roots and their replacement when needed at a later time may be more efficient than maintenance of a sufficiently high density of long-lived roots to accomplish the same end.

A genetic program of pulsed, nonsynchronous patterns of new root growth may agree well with the hypothesis of the need to relocate microcells of water and nutrient absorption.

Intraspecific, as well as interspecific, genetic variability exists with respect to root system characteristics, and selective pressures of the belowground environment—particularly under the general mandate of energy conservation—are certainly no less severe than those of the aboveground system. There is no reason to believe that plant root systems are any less finely tuned to their environment than are aerial plant parts. Fine tuning is probably needed more in our understanding of root structure and dynamics than in root systems themselves.

18 CANOPY STRUCTURE AND ENVIRONMENTAL INTERACTIONS

P. C. MILLER AND W. A. STONER

PATTERNS OF vegetation have generally been recognized first on the basis of physiognomy, i.e., the structure of the vegetation canopy, and second by species composition. Several investigations have sought general relationships between physiognomy and climate; others have correlated physiognomy with soil properties. Raunkiaer (1934) emphasized the position (height above the ground surface) of the perennating bud and the protection of the buds during an unfavorable period. Clements (1916) proposed that with moderate, i.e., nonlimiting, conditions, the vegetation would tend toward the tallest height possible. Monk (1966), Beadle (1966), and Small (1972a,b) have emphasized the role of soil nutrients in selecting for evergreenness and sclerophylly, whereas others have suggested that drought leads to evergreenness and sclerophylly. Thus, a diversity of environmental factors has been suggested as important in selecting several elements of canopy structure.

The selection of plant properties can be direct, as when the condition of the physical–chemical environment exceeds the physiological limits of the species, or indirect, as when tolerable conditions of the physical–chemical environment are modified by the presence of coexisting species to exceed the physiological limits of a particular species. As examples, light intercepted by overtopping species is unavailable to understory species; water transpired by some species reduces the soil moisture available for other species; and nutrients taken up by one species are unavailable to other species. Discussion of canopy–environment interactions will be couched in terms of the interactions between canopy properties, availability of resources such as light, water, nitrogen, and phosphorus, and the control of temperature. All resources will be considered in terms of the rate at which the resource is made available rather than the amount present. Light is clearly a flux. Nutrients taken up are many times greater than the amounts present at any one time, and thus the rate of supply is important. These resources have different availabilities owing to different mobilities in the soil. Some crude rules of thumb are that to exploit soil water fully in a few days, roots should be 8 cm apart (Lambert and Penning de Vries, 1973); to exploit NO_3^-, roots should be 2 cm apart (Keulen et al., 1975); and to exploit phosphate, roots should be 0.2 cm apart (Bieleski, 1973). These values mean that roots must be more dense, and root turnover perhaps greater, in a phosphorus-limited system than in a nitrogen- or water-limited system. The investment by the plant in root growth should be greater in a phosphorus-limited system and less in a water-limited system.

For the purposes of this paper, the canopy is defined as the volume occupied by leaves and associated stems. With a well-developed understory, the canopy may extend to the ground surface. The canopy can be defined quantitatively in terms of certain properties. Since the work of Watson (1947), the leaf area index, defined as total leaf area per unit of ground area, has been used to define the canopy. Leaf area density, i.e., total leaf area per unit canopy volume, has also been used. In agricultural usage, stem area is usually ignored. But in forest and shrub communities, stem area is significant and must be considered. The stem area relevant to shading and radiation profiles is the planar area; the stem area relevant to heat absorption is the surface area. The foliage area index (Anderson, 1966; Monsi and Saeki, 1953) has been used to include both leaf and stem area indexes. The relative merits of using leaf area index or leaf area density are unclear, since theory does not usually distinguish between them. However, wind profiles may be affected differently by leaf area density than by leaf area itself (Uchichima, 1962).

In addition to leaf and stem areas, the canopy is defined in terms of height, which involves the vertical profiles of leaf and stem areas and the vertical position of the growing points for leaf production or the perennating buds. The density of growing points per unit of ground surface may also be important. Leaf properties such as duration and size are also recognized and used to distinguish canopy types such as evergreen or deciduous and broad-leaved or narrow-leaved. In addition, degree of sclerophylly and leaf inclination are recognized leaf properties.

The objective of this paper is to evaluate, from current understanding, what can be said quantitatively about the interactions between the canopy and the environment. One hypothesis to be considered is that the existing canopy is optimal for maximum production or resource-use efficiency. An alternative hypothesis is that the structure is not optimal but is constrained by genetic capabilities and evolutionary history.

The discussion will largely be illutrated with results from Mediterranean scrub vegetation, since the convergence of vegetation form in phylogenetically disjunct Mediterranean climatic regions has been recognized and documented (Axelrod, 1973; Raven, 1973; Mooney et al., 1974; Mooney, 1977; Kummerow and Fishbeck, 1977). Additional information will be drawn from the Arctic tundra, which is also a well-documented vegetation (Tieszen, 1978; Brown et al., 1970). The Mediterranean-type climate is thought to have developed since the Pleistocene, although the current species existed long before (Axelrod, 1973). Thus, the convergence of form in Mediterranean-type regions has probably been by selection of existing shrub species and the evolution of herbs (Specht, 1969; Raven, 1973). The common element in these regions is domination of the vegetation by sclerophyllous, broad-leaved, evergreen shrubs; thus, the species present have in common many of the properties defined above, i.e., leaf width, leaf duration, height, and location and density of growing plants. Leaf area indexes are below 4.0, and stem area indexes may be 1 to 6 times the leaf area index.

SIMULATION OF CANOPY PROCESSES

The processes which define the interaction of the vegetation and the aboveground environment are fairly well understood. These include radiation, turbulent transfer, plant water relations, and canopy photosynthesis. Radiation models have been developed from concepts of light penetration in continuous vegetation canopies and have had broad usage in studies of temperate-zone crops (Monsi and Saeki, 1953; Davidson and Philip, 1958; de Wit, 1965; Duncan et al., 1967; Anderson and Denmead, 1969; Yim et al., 1969; Acock et al., 1970; Idso and de Wit, 1970). Our model of radiation in the canopy calculates infrared radiation within the canopy, a component not found in previous models, and includes the effects of stems on the radiation distribution within the canopy, a factor which has rarely been considered. This model was developed for the red mangrove vegetation of south Florida (Miller, 1975) and has been applied to the canopies of the Arctic and Alpine tundra (Tieszen et al., 1976; Miller et al., 1976) and the chaparral (Lawrence, 1975; Miller and Mooney, 1974; Roberts and Miller, 1977).

Models of the turbulent transfer of heat, water vapor, and CO_2 in the canopy have been proposed by Denmead (1964), Waggoner and Reifsnyder (1968), Waggoner et al. (1969), Murphy and Knoerr (1970, 1972), and Stewart and Lemon (1972) and tested with data from red clover, millet, pine, and corn canopies, with reasonable success. In our model, profiles of air temperature and humidity between the top of the canopy and a reference height above the soil surface are calculated with the algorithms proposed by Waggoner and Reifsnyder (1968) and Waggoner et al. (1969). The model has been tested on a preliminary basis on chaparral, mangroves, and Arctic tundra, with fair success (Miller et al., 1974; Miller, 1975a; Miller et al., 1976; Tieszen et al., 1976). To estimate the partitioning of energy by the canopy and soil, the soil surface temperature and soil heat conduction are calculated with a surface heat exchange model.

The leaf and stem temperatures are modeled by adapting the energy budget equation for individual leaves (Gates, 1962, 1965).

The model calculates the physical processes of solar and infrared radiation absorption, infrared loss from the leaves, convectional exchange, and transpirational exchange. The leaf temperature is calculated by balancing the incoming and outgoing fluxes of energy. The leaf temperatures have been tested against measured leaf temperatures in mangroves (Miller, 1972) and chaparral (Lawrence, 1975). In the canopy model, energy exchange processes and leaf temperatures are calculated for sunlit and shaded leaves at different levels in the canopy.

Models of plant water relations have been proposed by Honert (1948) and Rawlins (1963) and have been described in nonmathematical terms by Jarvis and Jarvis (1963). Refinements in concepts have led to the formulation of more complex models (Cowan, 1965; Philip, 1966). These models, developed as electrical analogs, attempt to explain the regulation, by plant organs and physical variables, of water movement from the soil through the plant and into the atmosphere. The model of plant water relations allows for nonequal rates of water uptake and loss. The model of the development of water stress in the plant was first proposed for the mangrove species (Miller, 1975) and evaluated and used in the Alpine tundra (Ehleringer and Miller, 1975a,b), the Arctic tundra (Stoner and Miller, 1975), and the chaparral (Miller and Mooney, 1974; Miller and Poole, unpublished data).

Models of canopy photosynthesis are based on the photosynthetic relations of a single leaf within the canopy. Monsi and Saeki (1953) and Davidson and Philip (1958) used a simplified model to calculate stand photosynthesis. De Wit (1965) used a more realistic model to calculate the annual course of potential production for different regions. Monteith (1965), Anderson (1966), Duncan et al. (1967), and Lemon (1967) proposed a series of refinements to the basic concepts. Miller and co-workers developed a canopy photosynthesis model for mangrove species (Miller, 1972) and used it to indicate the daily and seasonal progression of production and water-use efficiency in chaparral, mangrove, and tundra species (Miller and Mooney, 1974; Miller, 1975b; Miller et al., 1976).

The belowground environment of the vegetation is controlled by the well-defined processes of heat and water movement and by chemical and microbiological processes that are less clearly understood.

Energy exchange processes at the soil surface have been modeled by Goudriaan and Waggoner (1972) and Denmead (1973) by using simplifying assumptions about the moisture status of the soil. A more complex version of the energy budget model has been used for work in chaparral, mangrove swamp, and Arctic tundra (Miller et al., 1974, Ng and Miller, 1975, 1977). Soil temperatures may be simulated from a known surface temperature (Goudriaan and Waggoner, 1972; de Wit and Goudriaan, 1974) or directly from air temperatures (Hasfurther and Burman, 1974).

The heritage of canopy simulation models, which have been tested and developed with experimental data from diverse vegetation types, gives an established mechanism to generalize data with explicit assumptions and to explore the interactions among canopy properties, canopy microclimate, and production and water use. The models require, as input, information on the seasonal and daily courses of solar and infrared irradiance, air temperature, air humidity, wind, soil temperature, and the physiological relationships defining the fluxes of water and carbon dioxide into and out of the plant. This information has been established by measurements on four Californian and six Chilean shrub species during the past 2 years and by measurements of macroclimate and microclimate made in southern California and central Chile during the past 7 years. In the tundra, the data exist for species representing a diversity of plant growth forms, i.e., evergreen shrub, deciduous shrub, tussock gramminoid, and single-shooted gramminoid.

RESULTS AND DISCUSSION

Implications of the Shrub Form
    Height, density of growing points, and regrowth. We now consider the primary production system at several scales of time resolution. Primary production, in terms of grams of carbon per unit area of ground, is considered in terms of photosynthesis per unit leaf area, leaf area per growing point, numbers of growing points per individ-

ual or genet, and number of individuals (genets) per unit area of ground. The different time scales are apparent: Photosynthesis is considered through the day at different times of the season; the development of leaf area at a growing point is considered through the course of a season, the number of growing points changes annually; and the number of individuals, or genets, changes through generations. With annuals, the annual and generational turnover times are the same.

The job of the plant is to display photosynthetic tissue and replace this tissue as rapidly as possible if lost or dropped. The replacement of leaf tissue depends on the growth rate at each growing point or bud and on the number of growing points. Leach (1968, 1969, 1970) showed that the regrowth after cutting in alfalfa was related to number of crown sprouts (growing points), not to carbohydrate levels in the crown. The increase in the population of growing points on an individual can be summarized as a population process. Shaver (personal communication) estimates for *Ledum procumbens*, a tundra evergreen shrub, an r of 0.075 per year, with a gross reproduction rate of 0.31 and a mortality rate of 0.24 per year. Early-successional plants with rapid growth potential might be expected to show high numbers of growing points. The advantage of the shrub form over the tree form, when each is composed of the same biochemical constituents in the wood, leaves, and roots, may be this increased number of growing points at the root crown after top removal. The disadvantage is the lack of supporting tissue for increasing height when light is limiting.

Following this thought, we have estimated the number of growing points in shrub tundra canopies over a range of canopy heights. In the tundra, observations indicate that, with increasing protection from wind and abrasion, the height of the vegetation increases. The protection is ultimately provided by topographic relief, and, in windblown sites, the height of the vegetation corresponds to the depth of the topographic depression, so that the surface of the landscape is smooth. Different species seem to have minimum heights of protection, below which mature individuals do not occur. *Salix pulchra* did not occur below 10 cm, *Betula nana* below 5 cm, *Vaccinium uliginosum* below 4 cm, *Ledum* below 2 cm, or *Vaccin-*

ium vitis-idaea below 1 cm. With decreasing protection, the taller species were not present, but the lower ones remained. Thus, in stands of low stature, evergreen shrubs were obvious, and, in stands of higher stature, evergreen shrubs occurred beneath a canopy of deciduous shrubs. A survey of the density of growing points in sites of varying topography near Eagle Summit, Alaska, showed little change in density as height varied between 2 and 40 cm. The meristem density was about 4,000 meristems·m^{-2}. In the coastal wet-meadow tundra, the density of tillers is about 2,000 to 25,000·m^{-2}. Each tiller develops from one stem-base or meristematic region. The densities of the growing points mean that each leaf growing point in the wet meadow can exploit the light energy, water, and nutrient resources of 4 to 5 cm^2 of ground surface area; with taller canopies, each leaf growing point can exploit the resources of 2.5 cm^2. In this tundra area, these resources consist of light energy (200–500 cal·cm^{-2}·day^{-1}), water (about 0.8 g per square centimeter of surface and centimeter of depth, through a 25-cm thawed depth), nitrogen (about 0.03 g of N released·cm^{-2}·day^{-1}), and phosphorus (about 0.003 g of P released·cm^{-2}·day^{-1}). The greater number of growing points should mean more rapid growth of photosynthetic tissue early in spring or after loss, consistent with the limitation of the resources available to the growing point. Evidence available indicates that, of the three stages in primary production (photosynthesis, growth per growing point, and density of growing points), decreasing nutrient availability affects density first, growth rate second, and photosynthetic rate third (Tieszen 1979; Chapin, 1978).

Effect of foliage area index on heat exchange. The foliage affects the processes of heat exchange by intercepting and absorbing solar and infrared radiation, by radiating and transpiring some of this energy, and by exchanging energy by convection (Figure 18.1; Table 18.1). A leaf area index of 1.0, randomly arranged, intercepts about 63 percent of the incoming direct-beam solar irradiance and about 55 percent of the incoming diffuse solar irradiance. The interception of the direct beam varies with solar altitude, being higher with lower solar altitudes, and depends on the inclination of the leaves. At latitudes of 32.5°, where the mean daily solar altitude varies through

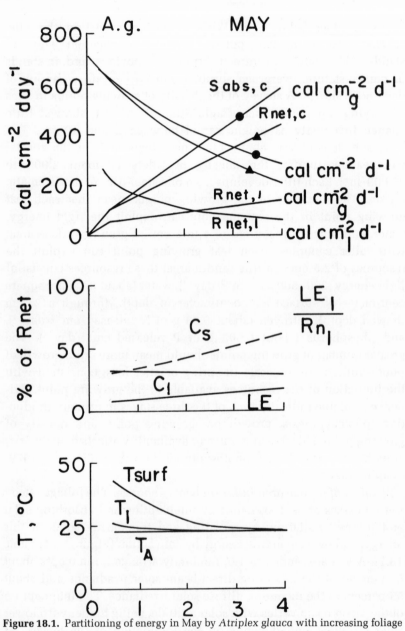

Figure 18.1. Partitioning of energy in May by *Atriplex glauca* with increasing foliage area index: *top*, energy absorbed; *middle*, percentage of total radiation, *bottom*, temperature.

TABLE 18.1.

SIMULATED PARTITIONING OF ENERGY OVER 3 MONTHS BY *Arctostaphylos glauca* FOR TWO FOLIAGE AREA INDICES

	May		September		January	
Foliage area index	3.8	0.5	3.8	0.5	3.8	0.5
Solar radiation absorbed by the canopy (cal·cm⁻²·day⁻¹)[a]						
Ground area basis	567	154	480	127	262	81
Foliage area basis	149	308	126	254	69	162
Net radiation absorbed by the canopy (cal·cm⁻²·day⁻¹)[a]						
Ground area basis	430	153	255	78	137	55
Foliage area basis	113	306	67	156	36	110
Fraction of net radiation absorbed by the canopy, lost as:						
Convection from stems	0.70	0.64	0.72	0.70	0.77	0.74
Convection from leaves	0.12	0.27	0.20	0.27	0.01	0.14
Transpiration	0.18	0.09	0.08	0.02	0.22	0.12
Fraction of net radiation absorbed by leaves, lost as:						
Convection	0.40	0.75	0.73	0.17	0.03	0.55
Transpiration	0.60	0.25	0.27	0.83	0.97	0.45
Temperature (°C)						
Air[a]	23.3	23.1	26.3	26.2	16.2	16.2
Leaf	23.3	26.2	27.4	30.2	16.3	17.7
Soil	24.2	43.5	26.4	43.0	13.0	24.1

[a]Data from Thrower and Bradbury (1977).

the season between 23 and 45°, interception with leaves inclined 60° varies between 78 and 50 percent, and with leaves inclined 30°, between 63 and 59 percent. This intercepted solar radiation is the major source of the energy required for photosynthesis. A fraction of the intercepted solar irradiance is reflected or transmitted. This portion is less effective in photosynthesis and is usually ignored, but it must be included in considering processes of heat exchange on plant and soil surface temperatures. In regions with clear skies, the diffuse radiation increases in the canopy as the direct beam is decreased.

As the foliage area index increases, the interception of solar radiation increases. At the ground surface, solar irradiance decreases, but the proportion of diffuse radiation increases. Infrared radiation from above also increases. Soil surface temperature decreases if the solar irradiance is moderately high, so that the infrared irradiance from the surface decreases. The net effect is that the total heat load on the foliage decreases as the foliage area increases, because of both increased shading and decreased infrared radiation from the surface.

The absorbed radiation is lost by reradiation, convection and evaporation. The heat lost by these processes increases as plant temperatures diverge from air temperature, but an increase in the rate of water loss is particularly significant for plants in arid environments. A plant conserves water if the total absorbed radiation is minimized by a reduction in leaf size or if the proportion of radiation absorbed by transpiring organs is reduced by the interception of radiation by low- or non-transpiring stems. Of the radiation absorbed in the chaparral canopy, 60 percent is lost by reradiation. Of the remaining radiation 30 to 40 percent is lost by convection. About 80 percent of the convectional loss is from stems. Thus, stems function to absorb solar radiation and dissipate this energy by convection, which reduces the evaporational loss of water. Leaves closely appressed to stems are shaded by the stems and should have lower transpiration demands.

Effect of foliage area index and height on microclimate profiles. Leaf temperature and air temperatures and humidities calculated by simulation models are close to measured values, which indicates that the models are valid. The effect of the total foliage area index and the vertical distribution of foliage area on the profiles of air temperature and humidities, as calculated in simulations, is slight. The mixing of the air is adequate to reduce air property gradients to within 1 to 2°C or 1 to 2 $g \cdot m^{-3}$ vertically or horizontally over a broad area, except close to the soil surface. The conclusion from simulations and measured profiles is that most natural canopies are not sufficiently dense to affect the profiles of air temperatures and humidity significantly, relative to our knowledge of biological processes. Leaf temperatures may vary because of the direct effect of shading. Because of our interest in how grasses would fare in the Mediterranean climate, we placed the foliage area index in a profile expected for a grass canopy. Such a configuration caused little effect on microclimate and leaf temperatures.

Effect of foliage area index on production and water balance. Early models of canopy production indicated that, as leaf area increased, production increased up to an optimum leaf area index and then decreased (Monsi and Saeki, 1953; Davidson and Phillip, 1958; Donald, 1961). The shape of the production—leaf area index curve was parabolic. An argument ensued (Donald, 1961) over the concept of optimum leaf area index, since many natural and agricultural communities had higher than optimal leaf area indexes. The argument was resolved by McCree and Troughton (1966), who showed that acclimation of leaves to shading results in a reduced respiration rate, which thus increases the net carbon dioxide exchange of leaves at the bottom of the canopy. The shape of the production—leaf area index curve then becomes more asymptotic, without a clear maximum. Duncan et al. (1967) showed a small effect on production by vertical distribution of leaf area index and leaf inclination. Production increased about 10 percent with leaves concentrated at the bottom of the canopy or with steeply inclined leaves. Other studies have shown increased production with leaves steeply inclined at the top of the canopy and becoming more horizontal within the canopy, as compared to other configurations. Miller (1972) showed that production in red mangroves was highest at the observed leaf area index, leaf inclination, and leaf absorptance. Miller and and Mooney (1974) concluded that the observed seasonal patterns of leaf area index in the chaparral were consistent with high production and water-use efficiency. Production and water-use efficiency were decreased at the bottom of the canopy because of high soil surface temperatures and leaf temperatures in summer and because of low solar irradiance in winter.

In simulation of the diurnal and seasonal progression of photosynthesis and transpiration of Mediterranean-climate shrubs from Chile and California (Figure 18.2), photosynthesis (net CO_2

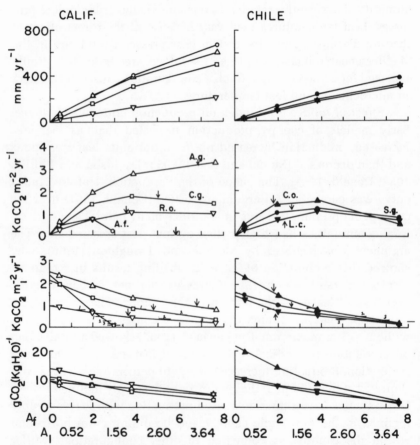

Figure 18.2. Transpiration (millimeters per year), production on a ground area and leaf area basis (kilograms of Co_2 per square meter per year), and water-use efficiency (grams of CO_2 per kilogram of H_2O) with increasing foliage (A_f) and leaf (A_l) area indexes for Mediterranean climate species from California and Chile.

exchange) was positive for all species throughout the year, except for *Adenostoma fasciculatum*. This species showed negative CO_2 exchange during the winter with a foliage area index of 1.92, and negative rates throughout the year with a foliage area index of 7.7. Photosynthesis is highest in the spring in both Chile and California. The suppression of photosynthesis in March in California is apparent. Increasing leaf area does not increase photosynthesis in winter; in fact, at the highest leaf area indexes, photosynthesis is decreased. In spring, when favorable conditions prevail, photosynthesis increases with leaf area index up to a leaf area index of 1 to 2 (foliage area index of 2 to 4). In summer and fall, photosynthesis is suppressed by a shortage of soil moisture. Photosynthesis and water-use efficiency were highest in *Arctostaphylos glauca*, followed by *Ceanothus greggii*, *Adenostoma fasciculatum*, and *Rhus ovata*, in that order. In the current simulations, light intensities are relatively low at the bottom of the canopy throughout the year. Consequently, photosynthesis and water-use efficiency are highest at the top of the canopy throughout the year.

In these simulations, the vertical profile of photosynthesis and water-use efficiency through the season followed a different pattern than that presented by Miller and Mooney (1974). The essential difference appears to be related to the temperature responses of photosynthesis and respiration. Miller and Mooney used data for *Heteromeles arbutifolia* which show a strong relation between photosynthesis and temperature (Harrison, 1971). The data used in the present simulations show a broad temperature optimum and positive photosynthesis at 0°C (Oechel and Lawrence, unpublished data).

In simulations that do not include an acclimation response to low light levels, canopy photosynthesis increased with increasing leaf area and reached a plateau for all the Mediterranean scrub species except *Adenostoma fasciculatum*. With *Artostaphylos glauca*, the highest canopy photosynthesis rates occur at a leaf area index of 3 to 4 (foliage area index of 7 to 9). With two others, *Ceanothus greggii* and *Rhus ovata*, the highest rates occur at leaf area indexes of 2 to 2.5. The rates are consistently high with leaf area indexes from 2 to about 5. With *A. fasciculatum*, the highest rates occur at a leaf area index of about 1.0. The leaf area index measured on the pole-facing slope at Echo Valley, California, where all these species occur, was 3.46 (Krause, 1975), which is equivalent to a foliage area index of 5 to 7. The leaf area index on the equator-facing slope at Echo Valley was 1.0 (Krause, 1975), which is equivalent to a foliage area index of about 2.0. Kummerow and Fishbeck (1977) summarized data on the leaf area indexes of these shrubs. The leaf area index estimated for the Echo Valley site was 2.65. The leaf area index estimated for each of these shrubs was appropriate for near-maximal photosynthetic rates, except for *A. fasciculatum*, for which the estimated leaf area index was much higher than the photosynthesis and respiration measurements indicate as being possible. The leaf area index estimated for the Fundo Santa Laura site was 2.02, which is near optimal for high production in the three species simulated, but the separate leaf area index of the shrubs was slightly below optimal.

The conclusion from these relations is that *A. fasciculatum* can survive in open communities, but not in closed communities, because of shading. Closed communities could contain *A. glauca* or *C. greggii*. These conclusions are consistent with general field observations. The high leaf area index of *A. fasciculatum* given by Kummerow and Fishbeck (1977) may be due to disparities between the leaf areas which can be maintained in open-grown shrubs and those maintained by a shrub in a continuous canopy. The three Chilean shrubs should be able to coexist, and do.

With an annual precipitation of 450 to 550 mm in both research sites (Miller et al., 1977) and a 10 percent interception loss (Hamilton and Rowe, 1949), maximum leaf area because of a water restriction should be 2 to 4, depending on the mix of *A. glauca* and *C. greggii*. *Rhus ovata* would require a large leaf area to use all the water. The highest water use would be by *A. fasciculatum*, but it cannot attain high leaf areas because of its unfavorable carbon balance. The three Chilean shrubs would use all the water if the foliage area were increased to the point of zero carbon balance, which allows no carbon surplus for maintaining the rest of the plant or for growth. The conclusion from the water relations is that *A. glauca* and *C. greggii*, singly or in a mixture, can use all the water which is available but that *A. fasciculatum* cannot, which is consistent with observations (Ng and Miller, unpublished data). *Adenostoma fasciculatum* is restricted to low leaf area indexes and open canopies because of its closely appressed and clustered leaves, which give it a high light compensation level. The other evergreen species have more widely spaced leaves and can tolerate higher leaf area indexes, thereby using more water. The ratios of water transpired to precipitation received are nearly 90 and 40 percent on the pole-facing and equator-facing slopes, respectively, in California and about 40 percent on the average, in Chile.

The water-use efficiency (grams of dry weight per kilogram of H_2O) decreases and the relative order of efficiencies of the species changes with leaf area index. At low leaf area indexes *A. glauca* has the lowest efficiency, although, at high leaf area indexes, it has the highest efficiency. *Adenostoma fasciculatum* has a low water-use efficiency. Of the Chilean species, *Colliguaya* has a high water-use efficiency.

Thus, the widespread species *A. fasciculatum* appears neither to be especially productive nor to use water efficiently.

Leaf Longevity

Leaf longevity and carbon balance. Leaf longevity may be related to the length of the period during which leaves must be maintained without showing a positive CO_2 uptake. Miller and Mooney (1974) assumed that if the glucose cost of maintaining leaves was greater than the cost of regrowing the leaves, deciduous forms should prevail, and that, if the converse occurred, evergreen forms should be favored. The cost of maintenance increases with the length of the unfavorable period. The break-even period was estimated to be about 100 days, which corresponds to the different lengths of soil drought in the drought-deciduous coastal sage and the evergreen chaparral.

Leaf longevity is also related to the length of the favorable period. Miller (1976) related the altitudinal pattern of deciduous and evergreen forms in the southern California chaparral to the altitudinal and seasonal patterns of precipitation, potential evapotranspiration, and temperature, and to the difference in the temperature responses of growth and photosynthesis. At temperatures below optimum, growth is more sensitive to temperature than is photosynthesis (Reuther, 1973; Tieszen, 1973; Warren Wilson, 1966; Miller, unpublished data). It was assumed that neither growth nor photosynthesis occurred during the drought period. Along the coast, at low altitudes, both growth and photosynthesis are possible through the winter, from the onset of rain to the onset of soil drought. The period is sufficiently long that growth costs could be regained with photosynthesis. With increasing altitude and lower temperatures, growth is suppressed throughout the winter, although photosynthesis is possible. At 1,000 m, after the rains begin in the fall, growth is possible for only a short period of time before temperatures are too cold. Photosynthesis is possible throughout the winter. The temperature suppression of growth means that deciduous forms show most of their growth in the short spring growing season. The carbon balance of evergreens is far more favorable than that of deciduous forms. At still higher elevations, i.e., 1,400 m, no growth is possible in fall and winter, but soil moisture lasts almost throughout the summer; the growing season is in late spring and summer, is longer than at 1,000 m, and has higher temperatures than any of the lower elevations. Here cold-deciduous, summer-growing forms have an advantage. At still higher altitudes, the summer growing season becomes shorter, and evergreen forms are favored.

Two conclusions can be drawn from these results. First, if the length of growing and photosynthesizing season is so short that the cost of leaf growth is not regained by photosynthesis, evergreen forms are favored because they can extend the carbon recovery period over several years. The evergreen leaves must be protected during the unfavorable period, as occurs in tundra snowbeds. Second, if the photosynthetic period precedes the growth period, as occurs in chaparral or in tundra fellfield, evergreen forms will be favored. Thus, the essential controls on the patterns of leaf duration are seen as the timing of periods favorable to growth and photosynthesis and the temperature responses of growth and photosynthesis.

The average rate of photosynthesis per unit leaf area decreases with increasing leaf area index because of increased shading. Decreased photosynthetic rates may be compensated for by increased leaf longevity. The assumption is that a leaf, over the period of its life, must produce at least as much as was involved in its production and support a fraction of root and stem growth. The balance of photosynthesis, leaf duration, and production and maintenance costs will be described after discussing the calculation of growth and maintenance respiration costs. This analysis is based on the work of Penning de Vries (1973, 1974, 1975a,b; Penning de Vries et al., 1974), and his calculations for compounds not given in these papers are gratefully acknowledged (Tables 18.2 and 18.3).

The total respiration of different plant parts is considered as coming from three sources: maintenance respiration, translocation respiration, and growth respiration. Maintenance respiration is associated with the replacement of proteins used in the metabolism of

TABLE 18.2.

BIOSYNTHETIC EFFICIENCIES OR CONVERSION RATIOS FOR VARIOUS PLANT PRODUCTS AND PROCESSES [a]

	Glucose (g)	CO_2 (g)	O_2 (g)	Other
Product				
Lignin	2.15	0.63	0.25	—
Lipid	3.03	1.61	0.35	—
Cellulose	1.17	0.08	0.06	—
Starch	1.17	0.08	0.06	—
Sucrose	1.09	0.06	0.04	—
Translocated organic N-containing compound (AA_t)	0.915	−0.015	—	+0.185 g of NH_3
Protein from AA_t	0.517	0.413	—	+1.174 g of AA_t
Process				
Mineral uptake	0.05	0.073	—	—
Sucrose translocation	0.05	0.073	—	—
AA_t translocation	0.04	0.059	—	—

[a]Based on Penning de Vries (1973, 1974) and de Wit (1973). Values are amounts of glucose required to synthesize 1 g of product, together with amounts of CO_2 evolved and O_2 required. See text for explanation of calculations.

TABLE 18.3.
PERCENTAGE COMPOSITION AND COSTS OF SYNTHESIS AND MAINTENANCE FOR DIFFERENT PLANT PARTS

	Percentage composition								Glucose required for 1 g of biomass	
	Lignin	Cellulose	Hemicellulose	Lipids	Proteins	Starch	Sucrose	Other	Synthesis ($g \cdot g^{-1}$)	Maintenance ($g \cdot g^{-1} \cdot day^{-1}$)
Dupontia fischerii										
Leaf	4	16	25	7	8	7	13	30[a]	1.31	
Rhizome	6	26	31	5	6	37	3	—	1.28	
Root	8	80[a]		4	6	0	2		1.27	
Dryas integrifolia										
Leaf-stem base	19	14	2			65[a]				
Inflorescence	27	21	0			54[a]				
Adenostoma fasciculatum										
New leaf	8	21	4	9	6		(10)	(42)[a]	1.71	0.004
Stem	12	47	12	4	4		(10)	(11)[a]	1.37	0.002
Root	12	54	14	0	2		(10)	(17)[a]	1.29	0.001
Arctostaphylos glauca										
Inflorescence	11	19	2	9	4		(10)	(39)[a]	1.72	0.002
New leaf	11	18	9	9	5		(10)	(44)[a]	1.78	0.003
Old leaf	12	15	4	11	4		(10)	(20)[a]		0.002
Stem	12	40	12	3	3		(10)		1.34	0.002
Root	14	53	12	0	1		(10)	(10)[a]	1.30	0.001

[a] Found by difference.

the plant. In leaves, maintenance respiration is measured as part of net photosynthesis. From studies of temperate-zone plants (McCree, 1973; Penning de Vries, 1973), maintenance respiration has been estimated at 1.5 percent of the biomass per day for tissues with 20 to 25 percent protein. Maintenance respiration is assumed to be proportional to protein content and turnover and can be adjusted downward for plants with lower protein contents. For *Dupontia fischerii*, an Arctic tundra grass, maintenance respiration was assumed to be 1.2 percent per day in leaves and 0.6 percent per day in roots. This latter value is consistent with measured respiration rates of intact roots (Billings et al., 1977).

Translocation respiration is involved in the loading and unloading of phloem, the transport of sugars across membranes, and is proportional to the translocation rate. Penning de Vries (1973) estimated that sugar translocation respiration is 5 percent of the amount of sucrose translocated. Nutrient uptake and translocation were estimated to cost 0.05 g of glucose per gram of nitrogen absorbed and 0.04 g of glucose per gram of amino acid translocated (Penning de Vries, 1973, 1974; Table 18.2).

Growth respiration provides the energy required for the synthesis of constituents of new tissues (e.g., nitrogenous compounds, carbohydrates, lipids). Energy costs of synthesizing various biochemical constituents from glucose have been calculated by Penning de Vries (1973, 1974; Table 18.3). Although plants contain a variety of sugars which function as metabolic intermediates, these will be viewed as glucose equivalents. To synthesize 1 g of lipid, 3.03 g of glucose is used, which involves a cost of 2.03 g of glucose and a carbon dioxide evolution of 1.16 g. Protein synthesis is regarded as two steps: (1) the formation of transportable organic nitrogenous compounds (AA_t) from glucose and NH_3 and (2) the interconversion of these compounds and the formation of proteins. Many natural systems are ammonium dominated. Ammonium is organically bound in the roots and transported to the sites of protein synthesis. We assume that 0.915 g of glucose and 1.74 g of AA_t are used in the synthesis of these intermediary compounds, since organic binding is the major energy cost in the synthesis of amino acids.

With these biosynthetic efficiencies, the growth cost of producing 1 g of new biomass in terms of grams of glucose used per gram of biomass produced is:

$$\text{Cost} = 2.15 \text{ (percentage lignin)} + 1.17 \text{ (percentage cellulose)}$$
$$+ 3.03 \text{ (percentage lipids)} + 1.483 \text{ (percentage protein)}$$
$$+ 1.17 \text{ (percentage starch)} + 1.09 \text{ (percentage sucrose)}$$

By using this measured percentage composition of various plant parts in the tundra, the cost of producing 1 g of mature tissue is calculated as 1.31 g of glucose for leaves, 1.27 g of glucose for rhizomes, and 1.28 g of glucose for roots. For chaparral leaves, the cost is about 1.7 g of glucose per gram of new leaf because of the abundance of secondary compounds with lipidlike structures.

During senescence, tissue structures are broken down, remobilized, and withdrawn. The remobilization is not completely energy or carbon efficient, and energy losses in terms of glucose are estimated as follows. When 1 g of polysaccharide is broken down, 1.10 g of glucose is formed. Since the polysaccharide was originally formed from 1.17 g of glucose, a loss of 0.07 g of glucose is associated with polysaccharide turnover. When 1 g of lipid is broken down, 1.39 g of glucose is formed. However, 3.03 g of glucose was used to form 1 g of lipid, so a net loss of 1.64 g of glucose is associated with each gram of lipid turnover. When 1 g of protein is broken down, 0.485 g of glucose and 0.762 g of amino acids are formed; 0.39 g of glucose is used in the turnover of 1 g of protein.

Secondary compounds, which are becoming recognized as herbivore defenses, probably have high synthesis costs like those of lipids or lignin, since their chemical structure is similar.

In anerobic conditions of pond margins, bogs, and many tundras, respiration energy can be expected to increase, since biosynthetic processes become less efficient. Processes performed under anaerobic conditions can take 9 to 17 times more glucose equivalents than under aerobic conditions. In addition, toxic end products must be excreted, which thus requires additional energy expenditures.

With these estimates, the carbon balance of leaf tissue can be calculated. The CO_2 cost of producing an evergreen leaf is about 560 g of $CO_2 \cdot m^{-2}$ leaf, derived from an area:weight ratio of 200 $g \cdot m^{-2}$; a CO_2:dry weight ratio of 0.6, and a growth cost of 1.7 g of glucose per gram of dry weight (Miller, 1977). The growth cost of roots is 1.3 g of glucose per gram of dry weight (Penning de Vries, 1972). The calculated photosynthesis rates then imply that a leaf area index up to 3.5 can be supported by *A. glauca* if the leaves have a 1-year longevity (somewhat longer since the calculation did not allow for maintenance or growth in the rest of the plant), but the same leaf area index would require about 2.5- and 3.0-year longevities in *C. greggii* and *R. ovata*, respectively. Leaf longevities in *A. fasciculatum* must increase rapidly to increase leaf area indexes. Species differences depend on the photosynthetic rates, the biochemical costs of producing new leaf material, and the time of year of leaf growth.

Higher leaf area indexes require higher leaf longevities in all species. For the species in California to coexist with a leaf area index of 3.46, as they do on the pole-facing slope, leaf longevities must be higher in *C. greggii*, *A. fasciculatum*, and *R. ovata* than in *A. glauca*. On the equator-facing slope, the vegetation is comprised almost wholly of *A. fasciculatum* with a leaf area index of 1.0 (Krause and Kummerow, 1977) and should show leaf longevities of at least 1 to 1.5 years. In Chile, the species can coexist with a leaf area index of 1.0 if leaf longevity in *Colliguaga odorifera* is at least 1.0; in *Satureja gilliesii*, at least 0.15; and in *Lithraea caustica*, at least 1.6. A leaf area index of 2.0 can be associated with a leaf longevity of 1.0 in *C. odorifera*, 1.5 in *L. caustica*, and less than 1.0 in *S. gilliesii*. With the leaf area indexes given in Kummerow and Fishbeck (1977) and these calculations, leaf durations should be at least 0.9, 1.2, 1.5, 0.5, 0.3, and 1.0 for *A. glauca*, *C. greggii*, *R. ovata*, *C. odorifera*, *S. gilliesii*, and *L. caustica* and are calculated from Kummerow and Fishbeck as 1.85, 1.50, 2.31, 2.18, 0.41, and 1.23, respectively.

Deciduous shrubs should support a higher leaf area index than evergreens, first because of the lower cost of producing leaves (about 1.3 instead of 1.7 g of glucose per gram of dry weight) and second because of the higher photosynthetic rates. However, the deciduous form requires a period of at least 80 days in each growing season to regain the carbon costs of leaf growth.

If the minimal leaf longevities are calculated on the basis of the photosynthetic rates at the top of the canopy rather than the average rates in the canopy, the required leaf longevities become shorter. *Adenostoma fasciculatum* requires the least time to recover the carbon cost of a leaf, because of the relatively low costs of leaf construction related to its low leaf density. *Adenostoma fasciculatum* leaves, if initiated between October and early June, could recover the cost of their construction by September. Leaves initiated in May and June could recover their costs only by the following spring, because of the negative carbon balance during the late summer. The other Californian species require longer periods to recover the costs of constructing new leaves, the longest periods occurring in *R. ovata*. In *R. ovata*, leaves initiated in spring or summer will not recover their construction costs until almost a year later. This relatively long construction-cost recovery period may be related to the irregular growth of *R. ovata* branches. In this species, only a few branches on an individual produce new growth each year (Kummerow, personal communication). Of the Chilean species, *S. gilliesii* has the shortest recovery time—shorter than *A. fasciculatum*— which is appropriate for a drought-deciduous species. In this plant, leaves brown and dry in January but persist until the following fall (April). After the first rain, these persistent leaves turn green and then drop (Montenegro, personal communication). Leaves initiated in December, if maintained active, could not recover the growth costs until early fall.

Leaf growth costs need not be recovered within a short period of time, unless a mortality factor operates in a pulsed manner (Fig. 18.3). Fire-adapted species should recover construction costs before the fire season, which is September through October in California. Frequent fires should therefore tend to shift the growth–temperature relationship toward lower temperatures, allowing earlier leaf growth. *Adenostoma fasciculatum* has early leaf growth and rapid recovery of construction costs, which is consistent with the idea that it is a fire-adapted species. In contrast, *R. ovata* appears to be the least fire adapted of the Californian species studied, although it is somewhat fire resistant. The growth period in Chilean shrubs is not

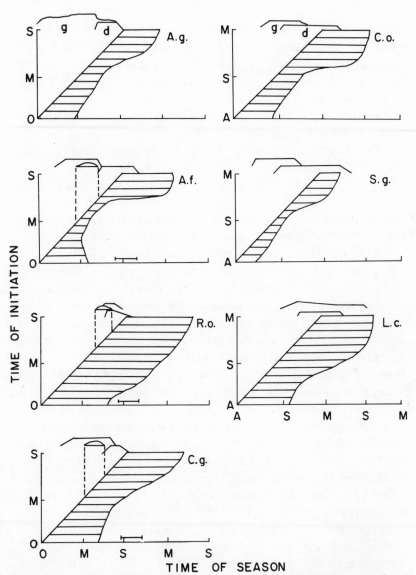

Figure 18.3. Period of year required to recover the costs of leaf and associated root growth if leaves are initiated at different times of the year and the observed period of growth (g) and litterfall (d). The observed periods are from Kummerow and Fishbeck (1977).

as pulsed as in the Californian shrubs, which is consistent with the view that fire is not as important in Chile as in California (Kummerow and Fishbeck, 1977). *Lithraea caustica*, the Chilean analog of *R. ovata*, is widespread in Chile, in contrast to the limited distribution of *R. ovata* here. *Lithraea caustica* is almost as poorly adapted to fire as *R. ovata*, and the different distributions of these species may be due to the absence of fire in Chile.

Leaf duration and nutrient balance. Stoner et al. (1978) calculated the increased advantage of a plant with low leaf turnover relative to a plant with high leaf turnover as nutrients become increasingly limited. The assumptions were that nutrients affect production by reducing the total leaf biomass rather than photosynthesis directly (Brown et al., 1978), that the turnover pattern is inherent, that there is a growth cost associated with new leaf biomass, that total photosynthesis depends on total leaf biomass, and that withdrawal of nutrients for new leaf growth hastens senescence of older leaves. With these assumptions, plants with high leaf turnover, e.g., grasses, cannot gain more leaf biomass because of nutrient limitations and do lose carbon in the growth costs. Plants with lower leaf turnover, e.g., evergreens, maintain the same leaf biomass and lose less carbon in growth. Thus, with low nutrients, the carbon balance becomes more favorable for the evergreens than for the others.

The production of deciduous and evergreen forms along a gradient of nutrient uptake by the total vegetation was estimated for the tundra. The production of different species or growth forms can be considered in relation to the amount of each inorganic nutrient taken up by the vegetation as a whole. The latter is an indication of nitrogen, phosphorus, or calcium made available. The production was estimated from the density of growing points, the number of leaves produced per growing point, the area per leaf, and a weight:area ratio. The nutrient uptake was estimated by multiplying production by the nitrogen, phosphorus, and calcium requirements. The nitrogen, phosphorus, and calcium relations were obtained from Stoner et al. (1978). For the deciduous shrub, 1 g dry weight of new tissue requires the uptake of 25 mg of N, 2.5 mg of P, and 1 mg of Ca. For the evergreen shrub, these requirements are 12 mg of N,

1.1 mg of P, and 1 mg of Ca. These amounts are the required uptake from the soil and allow for the retentions of nutrients with leaf senescence.

According to these simulations, if the nitrogen taken up is less than about 4 g of $N \cdot m^{-2} \cdot yr^{-1}$, evergreen shrubs predominate, and, if uptake is greater than this amount, deciduous shrubs predominate (Figure 18.4). With phosphorus, the dividing amount is about 0.25 g of $P \cdot m^{-2} \cdot yr^{-1}$. Within the evergreen form, fertilizing with nitrogen can increase leaf longevity, whereas fertilizing with phosphorus can decrease leaf longevity (Shaver, personal communication).

Interestingly, production in the Mediterranean scrub communities seems to follow the same relationships even though the data and logic are entirely different. Specht (1969a) gave the nitrogen and phosphorus release rates (in grams per square meter per year) for the garrigue in southern France as 2.9 and 0.18, respectively; for the chamise chaparral in southern California as 0.6 and 0.08; and for the heath and malle broombush in South Australia as 0.72 and 0.016 (for heath) and 0.62 and 0.024 (for broombrush). Thus, all these evergreen communities lie within the range of nitrogen and phosphorus uptake in evergreen-dominated communities in the tundra. In another paper, Specht (1963) gave nitrogen and phosphorus contents of the soil under heath, dry sclerophyll forest, and savannah woodland. Taking the release rates given in the 1969 paper and assuming that release is proportional to the nutrient content, the nitrogen and phosphorus release rates (grams per square meter per year) under the three types are 0.72 and 0.016, 2.6 and 0.14, and 4.5 and 0.52, respectively. The change from evergreen-dominated type to savannah woodland is associated with release rates of 4.0 g of $N \cdot m^{-2} \cdot yr^{-1}$ and 0.4 g of $P \cdot m^{-2} \cdot yr^{-1}$.

Nitrogen:phosphorus ratios in plants are usually about 10, except for nitrogen fixers (Miller, 1977). The ratios of N and P release found by Specht indicate that the chamise chaparral is nitrogen limited and the garrique and South Australian communities are phosphorus limited. Aboveground production given by Specht (1969b) is higher than expected from the tundra values, and the curves apply only to leaf production, because of nitrogen fixation and stem production. International Biological Program (IBP) data for

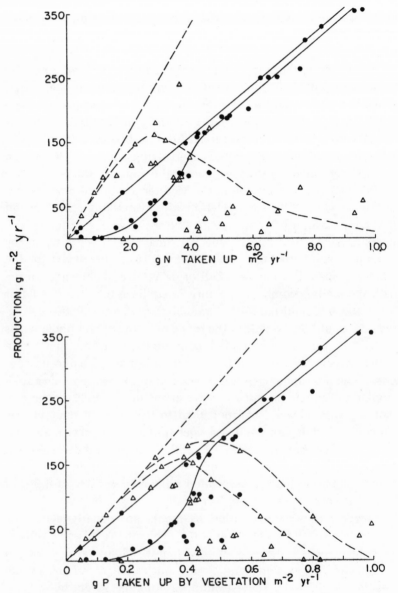

Figure 18.4. Production of (– – –) evergreen and (——) deciduous shrubs in relation to nitrogen and phosphorus taken up by the vegetation.

the chaparral indicate a nitrogen uptake of about 1.2 $g \cdot m^{-2} \cdot yr^{-1}$, combined with leaf production of about 130 g of dry weight $\cdot m^{-2} \cdot yr^{-1}$ in California.

Small (1972b) suggested that the ratio of photosynthesis accumulated through the life of the leaf to leaf nitrogen should be optimized in nitrogen-limited sites. A convergence toward similar values is seen in tundra plants for nitrogen, but not for phosphorus. With chaparral plants, the ratios were more similar for phosphorus.

Thus, as nutrients become deficient, the production system can be expected to shift from forms with high turnover to forms that have low turnover. Within the evergreen type, this can mean increased leaf longevities. As nutrients become deficient, the competing plants can be expected to converge to similar ratios of photosynthesis to leaf nutrient level throughout the life of the leaf. It must be recognized, however, that the rate of nutrient availability may be the result of vegetation type, not necessarily the cause. Evergreen leaves drop from the plant with high C/N ratios (≈ 100), which inhibit the rates of decomposition and release of nitrogen. In addition, evergreen leaves often have secondary compounds which inhibit decomposition and may reduce herbivory. Thus, a site with favorable nutrient relations, but on which the climatic conditions such as drought or the timing of favorable and unfavorable conditions favor evergreenness, may develop low rates of nutrient turnover and appear nutrient limited as the vegetation develops. Indeed, it is hard to escape the conclusion that competition, measured by the ratio of resource incorporated to resource made available, will always be more intense for nitrogen and phosphorus than for water, with competition for light being least intense, unless peculiar edaphic conditions override the normal relation for nitrogen and phosphorus.

Leaf Width and Inclination

Leaf width affects the potential for convectional exchange of energy; leaf inclination affects the absorption of solar energy. With incoming solar irradiance, leaf temperatures and transpiration rates will tend to increase with leaf width (Gates, 1962, 1965, 1968). This increase may influence photosynthesis directly, depending on temperatures and the temperature-response curve, and indirectly via the effect on leaf water balance and stomatal closure. In simulation, Miller and Mooney (1974) found slightly increased production and water-use efficiency in winter with wider leaves and found the same response in summer with narrower leaves. The overall effect of wider leaves was to decrease production, more so in California than in Chile, and decrease water-use efficiency. Winter-active plants should have broad leaves, and summer-active plants, especially in California, should have narrow leaves in order to increase photosynthesis and water-use efficiency slightly. California, as compared to Chile, should have plants with narrower leaves and of a more consistent size, since production is more sensitive to leaf size in California. Measurements of leaf size support these predictions (Parsons, 1973).

The increased width of leaves raises leaf temperatures higher above air temperatures. The increased rise in temperature in Chile, owing to increased leaf width, is almost equal to the decreased air temperatures in Chile as compared to California. Thus leaf temperatures may be more similar between countries than are air temperatures.

Leaf inclination was related to production and water-use efficiency only in the summer in California. Summer-active plants in California should increase production and water-use efficiency by having moderately inclined leaves.

Both leaf width and leaf inclination have less effect on production and water-use efficiency than position in the canopy, location with respect to the hot ground surface, and leaf duration–carbon balance considerations. Temperatures are usually not critical, and modification in width and inclination constitute fine-tuning of an already surviving system.

Interaction

Climatic and nutrient conditions clearly interact and confuse the picture. Combining the above discussions, leaf duration should be greater than 1 year with low rates of nutrient release (< 4.0 g of $N \cdot m^{-2} \cdot yr^{-1}$ or 0.4 g of $P \cdot m^{-2} \cdot yr^{-1}$) and a period favorable for growth and photosynthesis lasting more than about 265 days or less than about 50 days. As the period favorable for growth and photosynthesis shortens and as nutrient release rates decrease, leaf duration should increase. Leaf duration should be less than 1 year with high nutrient release rates and a favorable season of between 50 and 265 days. Longer leaf duration can be expected to persist over shorter leaf duration where either form is favored.

CONCLUSIONS

Canopy structure is constrained by the physical and chemical environment and can, within limits, modify the physical and chemical environment. The height of the canopy is a balance among competition for light (via the photosynthesis–light response curve), the cost of producing and maintaining supporting structures, the glucose (carbohydrate) reserves and the relation of growth to reserve level, and the severity of the aerial environment. Above 10 cm, increasing canopy height does not reduce temperatures. High growing-point densities aid in rapid regrowth of successional species and increase as a population process. Varying foliage area indexes hardly affect air and leaf temperatures but do affect soil surface temperatures. The absolute amount of transpiration and the percentage of net radiation lost in transpiration increases with foliage area index. Stems function to intercept and convert solar radiation, reducing the heat load on leaves and perhaps photosynthesis. Within the mixed chaparral community, *Adenostoma* is limited by light, *Arctostaphylos* by light, and *Rhus* by temperature. Considering several species in an array along one environmental axis may be misleading. Optimal foliage area indexes for production are observed in several, but not all, species. Leaf duration may increase with foliage area index. Evergreen leaves can be expected where the length of the growing and photosynthetic season is too short to recover leaf construction costs in one season, where the length of the period unfavorable for photosynthesis is short enough that maintenance costs are less than construction costs, or where a period favorable for photosynthesis precedes the period favorable for leaf growth. Leaf duration is increased as nutrient availability is decreased. Plants active when temperatures are above optimal should show narrow, steeply inclined leaves; those active when temperatures are below optimal should show wide, more horizontally inclined leaves. An evergreen vegetation which is initially not limited by nutrients may be selected by the climatic conditions and show nutrient limitations as rates of nutrient release are modified by leaf drop. Mature communities should almost always show the greatest competition for nutrients and lower competition for water and light. We can learn things by taking the negative approach. The direct selection of plants by the environment can be quantified and rigorously defined. After this definition, selection for optimizing processes can be considered.

Acknowledgments

This research was supported by grants from the NSF for study of resource utilization by vegetation in Mediterranean-type ecosystems and builds on the work of the IBP Origin of Ecosystem Structure Program, Mediterranean Scrub Subproject. We thank Dr. W. C. Oechel, Dr. Gaius Shaver, and Dr. J. Kummerow for use of unpublished data, Ms. Susan Richards for help with the computer programs and Ms. S. A. Barkely, Ms. C. Murray, and Ms. P. Miller for their help in preparing this manuscript. We are indebted to Dr. H. Mooney for many of the ideas developed in this paper.

4 SUMMARY AND CONCLUSIONS

19 FUTURE DIRECTIONS IN PLANT POPULATION BIOLOGY

Peter H. Raven

WHAT APPROACHES to the study of plant populations will be most fruitful during the closing decades of the twentieth century? Our complete dependence on plants as a source of food, the explosive growth of the human population, and the consequent imminent destruction of many kinds of ecosystems—these factors invest the question with an urgency that it has not had before.

Solbrig (1977, p. 202) has defined population biology as "a synthetic discipline with the aim of understanding the mechanisms that govern the growth and reproduction of individuals and populations in order to be able to make predictions regarding future states under normal or abnormal environmental conditions." Taken in its broadest possible extension, this definition is equivalent to that of Ehrlich and Holm (1962); population biology, in this sense, includes taxonomy, ecology, population genetics, and all other disciplines that concern themselves with the properties of populations or the population level of biological organization (Raven, 1977b). The restatement of the basic questions of plant population biology in non-taxonomic terms (Ehrlich, 1964), as well as the rigorous application of new and more objective methodologies to these fields, holds great promise for the resolution of many traditional questions, such as those concerning the origin of reproductive barriers and their genetic consequences, the means by which new species are produced, and the nature of adaptation.

The principal methodologies that have been applied to a study of plant populations in the past have been grouped under headings such as ecology, genetics, physiology, and, especially, taxonomy. It has become increasingly clear, however, that each of these approaches has limitations which are becoming more evident as additional information continues to accumulate. For further progress and new insights into the properties of populations, we need new conceptual frameworks (e.g., Ehrlich and Holm, 1962; Solbrig, 1977). In order to consider what these might be, it is first necessary to review some traditional and newly emerging areas of investigation.

THE SYSTEMATIC VIEWPOINT

Twentieth-century efforts to codify the definition of the category species have not been especially productive in helping us to learn about processes at the population level in nature (Raven, 1977a). Indeed, there now appears to be no definition that applies to plant species in general, other than one derived from purely logical and hierarchical principles (for a useful review, see Slobodchikoff, 1976). Although about a quarter of a million recognizable species of vascular plants exist in nature, they can be defined only in relational terms; there is nothing intrinsic to the taxon *Quercus alba* L. that is also common to the taxon *Clarkia lingulata* Lewis & Lewis except their position in the taxonomic hierarchy.

The kinds of distinct units that we conventionally regard as species cannot coexist in nature unless there are barriers to their free intercrossing or to the establishment of hybrid zygotes (Mayr, 1942). Nevertheless, species of woody plants, and perennials in general, are often, and perhaps usually, interfertile, and there is no evidence in most cases that they are evolving reproductive isolation or ever will evolve it (cf. Dobzhansky et al., 1977, p. 223). The interfertility of the morphologically and ecologically distinct units in such groups appears to be an important part of their adaptive system, allowing adjustment to the demands of an ever-changing environment by what might be termed interspecific recombination (summaries in Raven, 1977a,b). In contrast, short-lived plants often exhibit strongly developed genetic barriers of various kinds, but these often separate individual populations or portions of populations, as well as morphologically and ecologically definable units that are called species (Mosquin, 1966).

The development of genetic barriers, either prezygotic or postzygotic, in plants seems more and more clearly to accompany genetic differentiation. There is little evidence in plants for Lewontin's (1974, p. 161–2) "second stage in speciation," i.e., selection for the enhancement of differences between formerly isolated populations that come into contact. Such enhancement is what Verne Grant (1971, Chapter 10) has called the "Wallace effect." On the other hand, P. R. Grant (1975) has pointed out that the evidence for character displacement for ecological factors is much better than that leading to reproductive isolation in animals, and few botanists have looked for such ecological displacement, despite the suggestion of Levin (1970).

In recent years, the role of the taxonomic system as one that incorporates as many features as possible and is consequently of the most utility for studies of many kinds and of the highest predictive value is becoming more and more widely accepted (summaries in Raven, 1977a,b). Such a system of naming plants is a compromise between knowledge, tradition, and convenience (Ehrendorfer, 1976). The urge to classify has been an almost overwhelming one since the days of the encyclopedists in the seventeenth century, who seem to have left us with an intellectual legacy that regards the accumulation of a little information about a lot of organisms as better than a lot of information about a few. In a sense, population biology is seeking to counter this historical trend, a point to which I shall refer again later.

THE NATURE OF PLANT POPULATIONS

Careful studies of gene flow in plants (Levin and Kerster, 1974, 1975; Moore, 1976; Raven, 1977a,b) have made it clear that effective population size in plants is usually to be measured in areas only a few meters across. The unit of adaptation and evolution is the local population, usually consisting of a very few individuals (Bradshaw, 1971, 1972; Frankel and Galun, 1977; Raven, 1977a). In plants at least (Bradshaw, 1972), population differences are often maintained even in the face of strong gene flow between adjacent populations (Jain and Bradshaw, 1966; Ehrlich and Raven, 1969; Endler, 1973). Extremely local patterns of differentiation have been clearly demonstrated, for example, in a series of studies of local races adapted for survival in soils with toxic levels of heavy metals (Bradshaw, 1971; Antonovics et al., 1971); the differentiation of such races has often been rapid (e.g., Wu and Antonovics, 1976). To mention a few examples, populations of *Eucalyptus viminalis* Labill. on basalt and granite soils differed in growth rate and drought resistance over distances of a few hundred meters (Ladiges, 1976). Highly distinctive ecotypes of the wind-pollinated, sexually reproducing tree *Thuja occidentalis* L. in Wisconsin occurred within 0.7 km of each other (Musselman et al., 1975). In Northumberland, England, six populations of *Dactylis glomerata* L., a wind-pollinated grass, taken along a transect 130 m long over sand dunes, differed greatly in relative growth rate in relation to the relative dryness of their respective habitats (Ashenden et al., 1975). In the annual *Veronica peregrina* L. growing around vernally moist pools in California, plants growing at the margins of the pools were genetically strongly differentiated from those growing nearer the center, even though the populations were continuous (Linhart, 1974).

Self-pollination in many plant groups leads to the production of distinctive biotypes that may differ ecologically and occur in various mixes in local populations, as shown for *Avena* in California (Jain, 1969; Hamrick and Allard, 1972, 1974), for *Trifolium subterraneum* L. in Australia (Morley et al., 1962; Morley and Katznelson, 1965), and experimentally for *Hordeum vulgare* L. (Harlan and Martini, 1938). Where vegetative reproduction is well developed, clones that are well suited to a particular habitat may come to occupy wide areas, as in *Populus* (Kemperman and Barnes, 1976), *Taraxacum* (Solbrig and Simpson, 1974; Janzen, 1977b), *Apios americana* Medic. (Seabrook and Dionne, 1976), and *Festuca rubra* L. (Harberd, 1971). In such cases, the clones may or may not be capable of sexual reproduction (Raven, 1977a).

Interesting in terms of local adaptation are a number of examples in which the morphological and chemical characteristics of the
populations are not in agreement. Evidently, the selective demands
of the environment for different sets of characteristics differ. Such a
lack of concordance is evident in studies of the chemical races of
Ambrosia chamissonis (Less.) Greene (Payne et al., 1973), *A. confer-
tiflora* DC. (Payne, 1976), and in natural hybrids between *Cupressus
sargentii* Jeps. and *C. macnabiana* A. Murr. (Lawrence et al., 1975),
to mention but three examples. Analogous is the well-studied lack of
concordance between morphological features and chromosomal
ones in the transitional area between *Clarkia speciosa* Lewis &
Lewis ssp. *polyantha* Lewis & Lewis and *C. nitens* Lewis & Lewis
across a transitional area of some 100 km in the southern Sierra
Nevada of California (Bloom and Lewis, 1972; Bloom, 1976).

Often the adaptive meaning of apparently simple characteristics
is deceptive. For example, in *Anemone coronaria* L., plants with
scarlet flowers that contain pelargonidin pigments are universal on
all soil types, whereas plants with purple and violet cyanidin- and
delphindin-pigmented flowers are restricted to unleached metal-
rich soils (Horovitz, 1976). Pelargonidin derivatives cannot undergo
chelation with metals, whereas cyanidin and delphinidin derivatives can; the scarlet flower may have a lower metal content, as was
found for different phenotypes of *Centaurea* (Bayer, 1958) and *Lupinus* (Bayer, 1959). Plants succeeding on metal-rich soils may have
chelating pigments as an integral part of the genetic system, and
these pigments may enhance the plants' tolerance by removing
metal ions from sites of active metabolism (cf. Antonovics et al.,
1971) and immobilizing them in complexes, as suggested by Horovitz (1976).

In summary, although much more needs to be learned about
pollen and seed disperal in plants and about the effect of the occasional to rare introduction of genetic material from one population
to another (e.g., Hamrick, 1976), the general pattern—one of extreme
local differentiation—has become clear. Long-distance pollen transport (Janzen, 1971b; Heinrich and Raven, 1972) does occur but is
apparently very rare. Statements such as the following need to be
evaluated carefully in the light of these observations: "A sexual
species is a reproductive community, all members of which are

connected by ties of mating, parentage, and common descent. The reproductive community has a common gene pool. . . . Possession of a common gene pool makes a sexual outbreeding species an inclusive Mendelian population. More precisely, it is an array of subordinate Mendelian populations interconnected by regular or occasional gene flow. The Mendelian population is a form of supraindividual integration" (Dobzhansky et al., 1977, pp. 131–132).

As Stebbins (1950) implied and many others have pointed out subsequently, it would be best for the progress of population biology if the term "biological species," as a unit of classification, be abandoned and emphasis placed squarely on the processes occurring in natural populations. There are simply too many exceptions in plants, at least (Grant, 1971, p. 59), for the concept to be operational (cf. Sokal and Crovello, 1970). We do, however, need far more information than is presently available about the interfertility of and relationships between plant populations, especially in the tropics, and the reasons for lowered fertility in hybrids when it is encountered.

If traditional explanations, such as "gene flow" and "the possession of a common gene pool," are insufficient to account for the formation and maintenance of the often wide-ranging units that we recognize as species (Ehrlich and Raven, 1969), what does account for them? The same might be asked with equal logic about any morphologically or ecologically distinguishable race that consists of more than a single local population. It is clear that some traditional questions, such as the one concerning the frequency and importance of sympatric speciation (Dickinson and Antonovics, 1973), have an interpretation at variance with the traditional one when viewed in this new context, and the situation clearly is one in which new ways of thinking about traditional situations will be necessary for future progress.

THE CONTROL OF CHANGE

In ways that can be perceived only dimly at present, populations tend to remain constant because of the integration of populations based on genetic linkage and physiological interrelationships which lead to the kind of resistance to selectional change that Lerner (1954) termed "genetic homeostasis" and Mayr (1963, Chapter 10) discussed under the heading "the unity of the genotype." In this connection, Mayr (1963) has stressed that genes must combine well in populations; they must also act together to produce phenotypically harmonious and functional results (Dobzhansky et al., 1977). The kinds of interactions discussed by Johnson in this volume certainly lie at the base of such relationships. They help to provide a basis for the quantification of the resistance to change that is observed in integrated natural populations. The adoption of multilocus thinking may lead to a further clarification of many of the problems that retarded the reconciliation of Mendelian genetics with Darwinism in the early years of this century (cf. Provine, 1971). At a genetic level, models adequate to deal with the question of linkage are just beginning to emerge in theoretical population biology (cf. Lewontin, 1974, Chapter 6; Christiansen and Feldman, 1975; Feldman and Christiansen, 1975; Slatkin, 1975; Felsenstein, 1976). Furthermore, if populations are as local, integrated, and resistant to change as seems to be the case, then it is appropriate to consider the conditions under which adaptive genes or gene complexes will spread in natural populations, and to carry out appropriate experiments on this question both in outcrossing and inbreeding species. When our models and experimentation in this area are better developed, we shall be in a position to answer important questions such as those posed by Mather (1953) and Waddington (1957) concerning the ways in which populations differentiate (Antonovics, 1976) or are held constant.

Valuable clues to the way populations function and change are being provided by studies concerning the nature of variation in enzymes in natural populations. Polymorphism among enzyme loci appears to be related to their metabolic function (e.g., Hamrick and Allard, 1975; G. Johnson, 1973, 1974, 1976a,b), and plants are ideal for testing this hypothesis through the study of individual adaptation to a particular habitat. To what extent does the genotype of an individual oak tree relate to its exact ecological position (cf. Hedrick et al., 1976)? This question is of fundamental importance but is

468   P. H. RAVEN

difficult to address, both because of the complexity of the processes involved and because of the potential indeterminacy in the position of a single individual. Nevertheless, the demonstrated levels of polymorphism and heterozygosity reviewed by Hamrick (in this volume, article 4) would seem to reinforce the idea that this is an extremely promising field for investigation. In a weedy population of pears in Ontario (Waldron et al., 1976), what determined the limits of variability as the population became more diverse? More generally, what are the genetic and physiological bases of "weediness" in plants (Baker and Stebbins, 1965; Baker, 1974; Zimmerman, 1976)? These and other questions now appear to be amenable to direct investigation.

An important technical problem in allozyme studies should be pointed out, however. Because of the extreme difficulty of purifying enzyme prepartions from preparations of mature plant tissues, many of the analyses that have been carried out thus far have been concerned with germinating seeds. It is often the case, however, that progenies of outcrossing plant species grown in cultivation are more variable than populations of mature plants of the same species in nature, but also less heterozygous. For example, in certain species of genera such as *Clarkia* and *Gaura*, chromosomally homozygous plants are very rare in nature but relatively frequent in experimental progenies, which are morphologically more variable. For outcrossing species, then, estimates of variability based on allozyme frequencies in seedlings are apt to be too high and not to reflect accurately the representation of genotypes among mature plants in nature, whereas estimates of heterozygosity are apt to be too low. This becomes an important problem in attempts to analyze the adaptive role of such alleles in wild populations and should be considered in more detail with appropriate materials.

The role of chromosomal structural and numerical changes in plant populations is of great importance in setting the rate and character of change (Grant, 1964, pp. 207–12; Stebbins, 1976b; Dobzhansky et al., 1977, p. 217) and may be seen in a new light in view of what theoretical population genetics is revealing about the nature of linkage. Translocations, inversions, duplications, and deletions all play an important role in setting the genetic structure of plant populations, as do changes in DNA content. A better understanding of chromsome pairing and meiosis (Sybenga, 1975; Sears, 1976; B. S. Baker et al., 1976; Jones and Brandham, 1976) will be most helpful. The role of complex heterozygosity is being investigated with the tools of enzyme polymorphism (e.g., Levin, 1975a; Levy and Levin, 1975), and there are many other problems in this area that appear to be amenable to similar analysis.

In the well-studied annual genus *Clarkia*, for example, it appears logical that relatively autogamous species are chromosomally more or less homozygous. On the other hand, the reason that such an outcrossing, morphologically polymorphic species as *C. rubicunda* (Lindl.) Lewis & Lewis should also be chromosomally homozygous is obscure (Bartholomew et al., 1973). Among the chromosomally heterozygous species there is also a rich diversity of patterns that presumably has an adaptive basis. Thus, *C. dudleyana* (Abrams) Macbr. consists of a series of populations characterized by different chromosomal end arrangements (Snow, 1960), whereas *C. amoena* (Lehm.) Nels. & Macbr. (Snow, 1963) and *C. unguiculata* Lindl. (Mooring, 1958), without any obvious differences in population structure or habitat, each have a very widespread chromosomal end arrangement with more local variants. Chromosomally heterozygous individuals are frequent in naturally occurring populations of all three species. The kinds of differences between them not only reflect their evolutionary history but also must play a role in determining their patterns of adaptation in the future.

Polyploidy, an evolutionary process of paramount importance in plants, has been explored in increasing depth (e.g., by Harlan and de Wet, 1975; Jackson, 1976). In an analysis of recently formed polyploids in *Tragopogon*, Roose and Gottlieb (1977) have demonstrated that polyploids exhibit substantial enzyme multiplicity over the diploids from which they were derived and also have a number of novel heteromeric enzymes. This promising approach is likely to enhance our understanding of the functioning of polyploids when the enzymes involved are better understood functionally.

Finally, a word should be said about breeding systems in higher plants (Solbrig, 1976, and article 5). The evolutionary significance of inbreeding has been explored in a recent review by Jain (1976b), who offers a number of excellent suggestions for enhancing our understanding of this crucial phenomenon. Enzyme polymorphism already has proved a useful tool in understanding the genetics of inbreeding in, for example, *Phlox drummondii* Hook. (Levin, 1976b), *Clarkia* (Gottlieb, 1974b), and *Stephanomeria* (Gottlieb, 1973a), and such applications will doubtless become more frequent in the future. What appears to be needed is an increased understanding of the genetics of inbred populations, together with more actual measurements in the field of the degree of inbreeding and the genetics of the populations involved. That the level of inbreeding cannot accurately be predicted from floral structure is illustrated in studies of derivative species of *Clarkia*; in *C. exilis* Lewis & Vasek, outcrossing varied from 0.38 to 0.89 in different populations without any visible difference in either floral structure or ecological parameters (Vasek and Harding, 1976). On the other hand, the heterostylous *Amsinckia grandiflora* Kleeb. (Weller and Ornduff, 1977) and *A. vernicosa* Hook. & Arn. var. *furcata* (Suksd.) Hoov. (Ganders, 1976) display cryptic self-incompatibility, a phenomenon that may well be widespread.

DEMOGRAPHY

In recent years, a meaningful beginning has been made in understanding the regulation of numbers in plant populations (Harper, 1967; Bradshaw, 1972; Harper and White, 1974; Schaeffer and Gadgil, 1975; P. A. Werner, 1977a). The vigor of the field is suggested by the series of meetings and symposia that have been held, as well as by the formation of an International Society of Plant Demography. The key role of seedling recruitment (e.g., Clapham, 1956; Holt, 1972; Sarukhán, 1976), the various factors that make germination possible or lead to its inhibition (e.g., Harper, 1967; Thompson, 1973), and the implications of a genetically diverse population of seeds in the soil (e.g., Epling et al., 1960; Bartholomew et al., 1973; Gottlieb, 1974a; Raven, 1977a) have been stressed by many authors recently and are under active investigation. Despite the difficulties of assessing the genetic composition of ungerminated

seeds on the parent plants or in the soil, the importance of understanding their role in natural populations is so great that strenuous
efforts should be made to evaluate this factor more completely. Both
the size and shape of seeds are under selective control (e.g., Harper
et al., 1970; H. G. Baker, 1972). A recent article of interest that
compares plants with vertebrates in this respect is that of Wilbur
(1977).

Changes in the genetic composition of populations through
time in response to seasonal changes or the ages of the population
(cf. Schaal, 1975; Schaal and Levin, 1976), the limits to and meaning
of phenotypic plasticity in individual plants, and the changes that
occur in individual plants throughout their life cycle and in relationship to environmental changes are all critical factors that control the
characteristics of plant populations. In *Agave deserti* Engelm., what
are the selective factors involved in determining the one seed out of
1.2×10^6 (Nobel, 1977) that will survive to produce a mature
flowering individual? The theme recurs: We need accurate estimates, for a few well-chosen plant populations, of the role of specified genotypes in nature.

DEVELOPMENT

It has become almost a platitude to say that certain developmental problems could better be studied in plants, with their fixed
position and open system of development, than in animals. The
genetic information contained in the integrated genotypes of plant
populations is closely linked to the demands of a complex and
highly heterogeneous environment, and development is the process
by which the relationship between genetics and ecology is established. The kinds of strong directional selection that are often operative in plant populations appear to make the modulation of integration in them even more important than in animals.

In this volume (article 7), Jain has rightly stressed appropriate
measurement as the key to understanding the phenomenon of phenotypic plasticity as an adaptive strategy. Plasticity has proved a
very complex phenomenon, rarely investigated with much success;

an extensive effort was made to evaluate the ecotypes of *Potentilla glandulosa* Lindl. in terms of their performance in different environments by Clausen and Hiesey (1958), and many other extraordinarily complex situations have been discussed in the literature (e.g., by Wooten, 1970; Mitchell, 1976). The importance of assessing the genotypically controlled limits of phenotypic plasticity through a wide range of environmental conditions has been outlined well by Heslop-Harrison (1964) and emphasized again by Hamrick in this volume (article 4).

In considering the role of development in setting the limits to adaptation for plant populations, we soon discover that not much information is available. Just as it is impossible, in the light of present knowledge, to evaluate changes in chromosomes or the patterns of association between them in a meaningful way, because we do not yet adequately understand the structure of normal eukaryotic chromosomes and their functioning, so it is impossible to provide satisfactory answers to many problems concerning plant development, because we simply do not understand the fundamentals. No one can explain satisfactorily at the molecular level what stimulates plants to flower, what causes shade leaves to be broader and thinner than sun leaves, why ivy and eucalyptus have distinctive mature and juvenile leaves, or how any plant hormone works in detail at the molecular level. How then are we to explain the meaning of vegetative plasticity or its role in natural populations?

At a fundamental level, the problems of development converge on those of genetics. G. Bulledo's concept of homeotic genes controlling the developmental pathways taken by specific groups of cells as they divide and construct portions of the mature organism provides a glimpse of the kinds of mechanisms that are operating (review by Morata and Lawrence, 1977). The kind of developmental compartmentalization that has been demonstrated in *Drosophila* and *Oncopeltus* might also be characteristic of the higher plants, with their much simpler plan of construction. The linkage of genes into functional groups is barely understood, and the theoretical considerations bearing on linkage, and the role of linkage in natural populations, are being developed now. Cell culture methods, fast becoming routine, offer a powerful new means of investigation (Levin, 1975b;

Day et al., 1977; Langridge, 1977). When the organization of DNA and histones in the eukaryotic chromosome is better understood and the numerous feedback and control mechanisms that function between this level and that of a mature cell are explored satisfactorily, then we may begin to understand the genetic and developmental reasons that flowers are stable in structure under most conditions (a significant study is that of Huether, 1966, 1968, 1969) and how seeds retain their constancy of dimension and weight even under drastic, and often adaptive, environmental change (cf. Hickman, 1975, and article 10).

ECOLOGY

Antonovics (1977) has stated that the explanation of the distribution and abundance of organisms—the essence of ecology—is to be found in genetics. In the same symposium, Mooney (1976) declared that, for the future progress of plant physiological ecology and its integration with population biology, individual variation in populations will have to be considered carefully. Several of the chapters in this volume have reminded us of the dangers of interpreting evolutionary events chiefly or solely from the products of evolution rather than from an experimental approach to the process of evolution. Despite these excellent indications, however, this volume seems to me to provide more evidence that a true synthesis between ecology and genetics is a very difficult one, made across a gap that is bridged all too rarely. I believe that most of the difficulty lies in the enormous complexity of the problems involved—consider the analysis of the ecological role of photosynthesis presented in this volume by Mooney and Gulmon—and the individualistic approach that characterizes the field, a theme to which I shall return later. In any case, plant ecologists have almost all pragmatically continued to treat species as if they were genetically invariate, despite the obvious inherent problems this assumption creates at several levels and the limitations it creates for us in our attempts to advance theory.

In the more traditional areas of ecology, an explosion of experimentation and thought has profoundly deepened our understanding of the adaptive role and functioning of the characteristics of plants during the past decade. Numerous examples are presented in this volume, and I shall confine myself here to a few supplementary observations. In recent years, there has been rapid advance in our understanding the adaptive functions of many of the morphological features of plants (e.g., Stebbins, 1970, 1971, 1974; H. G. Baker, 1972; Levin, 1973; Ehrendorfer, 1973; Carlquist, 1975; Givnish and Vermeij, 1976; Cruden, 1976, 1977). For example, hooked trichomes (Gilbert, 1971; Rathcke and Poole, 1975; Pillemer and Tingey, 1976) and glandular trichomes (Gibson, 1971) have been found to provide defenses against insect herbivores and to vary from population to population within some species, with differing effects on predation. In addition, the papers in this volume have provided ample evidence for the role of such characters as leaf size and shape, root structure, and canopy structure in natural populations. An analysis of the adaptive significance of the various structures by which the major and minor groups of plants are separated has provided no evidence that the evolutionary processes involved in the origin of genera and families differ from those operative at the population level.

Recent studies of the low shrub *Encelia* have demonstrated that the thick, white pubescence in *E. farinosa* A. Gray, a desert species, reduces the absorptance of photosynthetically active radiation by as much as 56 percent more than a closely related but nonpubescent species, *E. californica* Nutt., which occurs in coastal regions (Ehleringer, 1976; Ehleringer et al., 1976). Under developmental control, the amount of pubescence in *E. farinosa* increases through the growing season and modifies the leaf energy balance by dramatically reducing the photosynthetic rate as the weather becomes hotter and drier. In this one example, therefore, we have experimental verification of the ecological role of a particular morphological characteristic that is also adjusted developmentally to the requirements of the changing season. The integration of genetics, development, and ecology is clearer in plants than in the higher animals, and the observed selection forces are often much stronger (Bradshaw, 1972; Antonovics, 1976). Many questions of general biological interest, therefore, may better be asked in plants than in animals, as this example from *Encelia* illustrates.

At another level, there are numerous examples in the literature of the ways in which individual plant species coexist, compete, and in general structure the environment by their differentiated activities. Many contributions in this field have been made by Harper, Bradshaw, Antonovics, and their students and associates and have been mentioned above. In addition, the literature now contains documentation for hundreds of examples of ecotypic differentiation in plants. In recent years, the methods of physiological ecology have begun to reveal the bases of such differences and the ways in which plants function in relation to the environment (Mooney, 1977; Mooney and Gulmon, article 13; Miller and Stoner, article 18). Studies that have looked at species composition, interactions, and the limits to adaptation, without regard to intraspecific variability, doubtless will continue to form the mainstay of ecological generalizations and predictions, but only the input from population genetics will allow proper analysis of the dynamic interactions occurring in ecosystems at the population level (Antonovics, 1977; Mooney, 1977). A detailed examination of carefully selected populations, involving demographic, genetic, and physiological input, appears to provide the best promise of future progress in the field of plant population biology.

A special word should be said about the field of ecological study that has come to be known as coevolution (Ehrlich and Raven, 1965). Evolutionary interactions between different organisms are now seen as key factors in studies of the functioning and evolution of ecosystems, and their importance has been marked by the appearance of numerous symposium and review volumes on this topic (e.g., Harborne, 1972; van Emden, 1973; Heywood, 1973; Gilbert and Raven, 1975; Wallace and Mansell, 1976; Jermy, 1976). Although much has been learned in general and descriptive terms (McKey, 1974; Levin, 1976a), and despite the fact that many productive hypotheses have been constructed that promise to illuminate our understanding of the broad picture of evolution and adaptation at the community level, few studies have yet been carried out at the population level stressed by Mooney (1977) and Antonovics (1977) as offering the greatest promise for future progress. A few studies of individual variation have been made, such as those of danaine butterflies and Asclepiadaceae (e.g., Brower, 1970; Brower et al., 1972; Brower and Moffitt, 1974; Roeske et al., 1976), but many more are needed to provide a solid basis for advance in the field.

To mention a few examples, in *Lotus corniculatus* L., preferential feeding by voles on acyanogenic individuals has been demonstrated by D. A. Jones (1966), this and similar pressures tending to maintain strongly cyanogenic forms in the populations. For instance, Jones (in Lindley, 1977) found that the greater tolerance of salt spray by acyanogenic plants of this species growing on a seacliff was balanced by the higher rate of reproduction of cyanogenic plants. The complexity of the situation is obvious from studies of both *Lotus corniculatus* and *Trifolium repens* L. (cf. D. A. Jones, 1973), as well as *Pteridium aquilinum* L., also polymorphic for cyanogenesis (Cooper-Driver and Swain, 1976). Populations of the woodland herb *Asarum caudatum* Lindl. in western Washington where the native slug *Ariolimax columbianus* (Gould) was relatively rare were dominated by clones allocating more energy to growth rate and seed production (Cates, 1975). Where the slug was frequent, they were polymorphic for growth rate, seed production, and palatability to the slug. Such relationships will certainly prove to be the rule rather than the exception. How, for example, are the bitter-tasting sesquiterpene lactones in *Vernonia* (Burnett et al., 1974) balanced with other features of the individual plants in making up variable populations in nature? And how is the alkaloid polymorphism in populations of the self-incompatible Californian annual *Lupinus nanus* Dougl. to be interpreted (Mainkinen et al., 1975)?

Plant–plant interactions may now be illustrated by a few examples. The importance of inter- and intraspecific competition in plants is enormous and is directly linked to phenotypic plasticity; it has been explored in an outstanding series of contributions by J. L. Harper and his associates. Other kinds of plant–plant interactions involve chemical interactions. For instance, studies of *Artemisia herba-alba* Asso in the Negev Desert of Israel, in which seedlings rarely emerged beneath the parent plants (presumably as a result of autoinhibition), showed that if plants emerged just beyond the canopy, 10 to 50 cm from the crown, they suffered high mortality, probably because of competition for water (Friedman and Orshan, 1975). Similar complex allelopathic interactions, for example in the vigorously cloning *Pteridium aquilinum* (L.) Kuhn (Glass, 1976), are being explored at present (e.g., Muller and Chou, 1972; Newman and Rovira, 1975). The ways in which relationships of this kind set the structure and diversity of communities have been explored by Janzen and others (e.g., Fox, 1977; Ricklefs, 1977) and need to be studied in much more detail.

Without the insight that might be provided by studies at the population level, the field of coevolution will be liable to stagnate and become largely descriptive—a broad field of implied evolutionary relationships that are difficult to test and of low predictive value. Nevertheless, the input from studies of animal interactions is certainly indispensable to a proper appreciation of the structure and functioning of plant populations. Despite the frightening complexity of studies at the population level, involving considerations of genetic diversity of both the plants and the animals with which they are interacting, they should be undertaken in particular groups. Until now, they have been most frequent, not surprisingly, in plant groups of agronomic importance [e.g., cotton: (Hedin et al., 1976); *Phalaris arundinacea* L. (Kendall and Sherwood, 1975)]. There appear, for example, to be very few studies of pollination systems that take into account variability in the plant and animal populations involved (e.g., Cumber, 1949; Levin, 1972a,b; Cruden, 1976). Without such studies, how can the functioning or evolutionary dynamics of the systems be built into a predictive theory?

We know just a little about the complex interorganismal interactions that set the place of individual plants in a community, but, considering the interactions that involve mycorrhiza and soil microorganisms in general (e.g., Zak, 1971; Lewis and Crotty, 1977), not to mention such presumably special cases as root grafting and parasitism in angiosperms (cf. Atsatt, 1973, 1977), it is clear that the concept of the individual has severe limitations in assessing natural situations. One of the more bizarre recent examples concerns the cryptic mimicry of a number of Australian trees and shrubs by their loranthaceous parasites (Barlow and Wiens, 1977). Another concerns the relationship between *Drosophila*, yeasts, and cacti in the Sonoran Desert (Heed et al., 1976). The numerous interacting factors that play a role in determining the patterns of distribution are suggested by the alternation of tree species in natural communities (Fox, 1977). Seedlings tend not to become established under a conspecific adult tree in undisturbed forest. They condition their own habitats in terms of such factors as light, moisture, mineral composition, allelopathic chemicals, and fungal floras in the soil (some of which are pathogenic). Much more should be added by careful physiological ecological studies. First suggested by Went and Stark (1968), direct nutrient cycling from the fallen leaves of one species to the mycorrhiza of another has recently been confirmed (Herrera et al., 1978). Janzen (1970, 1972a,b) has stressed herbivore pressure in considering the role of such interactions in determining the structure of tropical forests.

DISCUSSION

Over the past 75 years, single-gene Mendelian genetics has made a profound contribution to our understanding of the process of evolution in natural populations and helped to resolve the dilemma that faced Darwin in his attempts to understand the process of evolution. As our methods and concepts have become more sophisticated, however, it has become increasingly evident that the concepts provided by the geneticists of the first half of the twentieth century are far from adequate to describe the functioning of populations adequately. This is why Lewontin (1974, p. 159) has stated that evolutionary genetics has made no direct contribution to the solution of the problem of the origin of species. On the one hand, we have come to a growing realization that the gene is meaningful only in its context, and on the other, we have become aware that concepts such as biological species, ecotype, and introgressive hybridization provide only crude and at times misleading yardsticks by which to monitor the behavior of populations in nature and make predictions about them.

Many areas of investigation in the field of population biology are pushing the limits of time-honored concepts. In systematics, for example, taxonomic categories are being seen as useful ways to summarize natural variation and not as operationally definable units that have evolutionary, genetic, and ecological validity (Raven, 1977a,b). Scientists are starting to measure gene flow rather than talk about it, assessing it quantitatively, using both theoretical and experimental methodology in an effort to understand the behavior of populations. The ways in which populations are maintained as distinct units in nature are of great interest but should be studied for their own significance and not as ways of "defining" preconceived taxonomic categories which then are judged to have great significance in other fields.

As the nature of the eukaryotic chromosome and its functioning are elucidated, it should be possible to learn more about the kinds of functional blocks of genes that are predicted on theoretical bases. The interactions between these, and the ways in which they can be altered in producing variant phenotypes, will, when properly understood, provide a basis for the phenomena that have loosely been termed genetic homeostasis or genotypic integration in the past. A more profound understanding of the variability of enzymes significant in the adaptation of natural populations and the adaptive significance of such variation, coupled with an analysis of the ways in which the individual enzymes are synthesized, should provide another key to the structure of natural populations. When the methods of physiological ecology are applied at the population level, we shall be well on the way to understanding the means by which the limits of variability in natural populations are set and maintained.

Although many areas in more or less traditional fields remain to be explored—and I might mention the nature of species in woody tropical angiosperms as one example, in addition to the many ways of dealing with populations discussed in this volume—many lines of reasoning indicate that a multidimensional approach to a few well-selected populations, combining the methods of theoretical and experimental population genetics, physiology, development, and ecology, would be most fruitful (Gottlieb, 1976b). We need to know in detail how a few populations really function, not as abstractions in some broad theoretical hierarchy but in fact. We need to know at a biochemical level why some individual plants grow where they do within a population and why some survive while others do not. In order to know this, we need to be able to measure the habitat of each individual accurately. We need to understand by experiment and model the kinds of genetic changes that take place in populations as their age-structure changes and the meaning of the resultant reservoirs of genetic variability for the ability of the populations to change.

Two difficulties in achieving the kinds of information I have attempted to define as of central importance to effective future progress in plant population biology should be mentioned explicitly so that we may be better able to overcome them. First, the structure of research is highly individual, and no one individual is likely to be able to master the combination of sophistication in many distinct fields, depth, and sustained effort that will be necessary. Despite the outstanding success of Clausen, Keck, Hiesey, and their more physiologically inclined successors, plant population biologists in general have tended to be unwilling or unable to form teams sufficiently diverse and focused to attack significant problems in depth. Second, we operate within the boundaries of a 300-year-old encyclopedic mentality which leads us to believe intuitively that it is best to study many kinds of organisms descriptively and relatively superficially rather than a few in depth as systems. Perhaps we already have a fairly realistic picture of the diversity of organisms on a world scale, and the further application of traditional methods may not be capable of leading to conceptual advance. *Drosophila* is an obvious exception, but there are no comparable organisms among plants. In my opinion, we should work to educate granting agencies, editors of journals, and ourselves to the view that to have several or many people working on a particular organism or series of populations is not bad but highly desirable for future progress in the field.

It is of critical importance in a rapidly changing world to learn more about the functioning of populations in nature and to understand their adaptive and evolutionary significance. In attempting to do this, we must not be bound by traditional concepts or methodologies. The task is a very difficult one, but the increased understanding that will result warrants the effort and expense. The organisms to be studied in such exhaustive depth ought certainly to be selected with the greatest care. Perhaps even major funding, on an IBP (International Biological Program)-type model, could be found for a concentrated attack on the ecological genetics of a few well-selected plants. When we know how some populations function in detail, we shall then be in a position to know why species remain as coherent units even in the absence of gene flow, what a "genetic revolution" means at the population level (cf. Lewis, 1966; Carson, 1975, 1976; Gottlieb, 1976a, 1977), and how and under what conditions morphologically and ecologically distinct units are produced. In my opinion, not a single one of these questions will ever be answered by the continued separate application of the traditional methods of genetics, taxonomy, or ecology, although it is still extremely valuable to construct models and generalities at many levels of biological integration. For accelerated progress in understanding, however, we must resolve to create and to nurture the new field of plant population biology, a field that will have profound implications for human welfare and survival. To this end, the present volume may have made useful contribution, but it has also made it abundantly clear that the task has barely begun.

LITERATURE CITED

Abrahamson, W. G. 1975. Reproductive strategies in dewberries. *Ecology* 56: 721–26.

Abrahamson, W. G. and M. Gadgil. 1973. Growth form and reproductive effort in goldenrods (*Solidago*, Compositae). *Amer. Nat.* 107: 651–61.

Acock, B., J. H. M. Thornley, and J. Warren Wilson. 1970. Spatial variation of light in the canopy. In *Prediction and Measurement of Photosynthetic Productivity*. Wageningen: Center for Agricultural Publishing and Documentation.

Adams, C. D. 1972. *Flowering Plants of Jamaica*. Mona, Jamaica: University of the West Indies.

Addoms, R. M. 1946. Entrance of water into suberized roots of trees. *Plant Physiol.* 21: 109–11.

Adhya, S. L. and J. A. Shapiro. 1969. The galactose operon of *E. coli* K-12. I. Structural and pleiotropic mutations of the operon. *Genetics* 62: 231–47.

Al-Ani, H. and B. R. Strain. 1972. The physiological ecology of diverse populations of the desert shrub, *Simmondsia chinensis*. *J. Ecol.* 60: 41–57.

Alberte, R. S., J. D. Hesketh, and D. N. Baker. 1975. Aspects of predicting gross photosynthesis (net photosynthesis plus light and dark respiration) for an energy-metabolic balance in the plant. In D. M. Gates and R. B. Schmerl (eds.), *Perspectives of Biophysical Ecology*. New York: Springer.

Allard, R. W., S. K. Jain, and P. L. Workman. 1968. The genetics of inbreeding species. *Adv. Genet.* 14: 55–131.

Allard, R. W. and L. W. Kannenberg. 1968. Population studies in predominantly self-pollinated species. XI. Genetic divergence among the members of the *Festuca microstachys* complex. *Evolution* 22: 517–28.

Allard, R. W., G. R. Babbel, M. T. Clegg, and A. L. Kahler. 1972.

Evidence for coadaptation in *Avena barbata*.   *Proc. Nat. Acad. Sci. USA* 69: 3043–48.

Amen, R. D. 1966. The extent and role of seed dormancy in alpine plants. *Quart. Rev. Biol.* 41: 271–81.

Amen, R. D. and E. K. Bonde. 1964. Dormancy and germination in alpine *Carex* from the Colorado Front Range. *Ecology* 45: 881–84.

Andersen, R. N. 1968. *Germination and Establishment of Weeds for Experimental Purposes.* Geneva, N.Y.: W. F. Humphrey Press.

Anderson, D. J. 1967. Studies on structure in plant communities. III. Data on pattern in colonizing species. *J. Ecol.* 55: 397–404.

Anderson, D. J. 1971. Pattern in desert perennials. *J. Ecol.* 60: 555–60.

Anderson, E. 1929. Variation in *Aster anomalus*.   *Ann. Mo. Bot. Gard.* 16: 129–44.

Anderson, E. 1936a. An experimental study of hybridization in the genus *Apocynum*.   *Ann. Mo. Bot. Gard.* 23: 159–67.

Anderson, E. 1936b. The species problem in *Iris*.   *Ann. Mo. Bot. Gard.* 23: 457–509.

Anderson, E. 1948. Hybridization of the habitat. *Evolution* 2: 1–9.

Anderson, E. 1949. *Introgressive Hybridization*. New York: Wiley.

Anderson, E. and G. L. Stebbins. 1954. Hybridization as an evolutionary stimulus. *Evolution* 8: 378–89.

Anderson, M. C. 1966. Stand structure and light penetration. II. A theoretical analysis. *J. Appl. Ecol.* 3: 41–54.

Anderson, M. C. and O. T. Denmead. 1969. Shortwave radiation on inclined surfaces in model plant communities. *Agron. J.* 61: 867–72.

Antonovics, J. 1972. Population dynamics of the grass *Anthoxanthum odoratum* on a zinc mine. *J. Ecol.* 60: 351–65.

Antonovics, J. 1976a. The nature of limits to natural selection. *Ann. Mo. Bot Gard.* 63: 224–47.

Antonovics, J. 1976b. The input from population genetics: "The new ecological genetics." *Syst. Bot.* 1: 233–45.

Antonovics, J. and A. D. Bradshaw. 1970. Evolution in closely adjacent plant populations. VII. Clinal patterns at a mine boundary. *Heredity* 23: 507–24.

Antonovics, J., A. D. Bradshaw, and R. G. Turner. 1971. Heavy metal tolerance in plants. *Adv. Ecol. Res.* 7: 1–85.

Anxolabehere, D., P. Girard, L. Palabost, and G. Periquet. 1976.

Stabilité des polymorphismes morphologiques et enzymatiques d'une population naturelle de *Drosophila melanogaster*. *Arch. Zool. Exp. Genet.* 117: 169–79.

Ares, J. 1976. Dynamics of the root system of blue grama. *J. Range Manag.* 29: 208–13.

Armstrong, R. A. and M. E. Gilpin. 1977. Evolution in a time-varying environment. *Science* 195: 591–92.

Armstrong, W. 1967. The relationship between redox potentials and oxygen diffusion levels in some water-logged organic soils. *J. Soil Sci.* 18: 27–34.

Armstrong, W. 1968. Oxygen diffusion from the roots of woody species. *Physiol. Plant.* 21: 539–43.

Arouet de Voltaire, F. M. 1759. *Candide, ou l'optimisme.* Geneva: Cramer.

Arroyo, M. T. K. de. 1975. Electrophoretic studies of genetic variation in natural populations of allogamous *Limnanthes alba* and autogamous *Limnanthes floccosa* (Limnanthaceae). *Heredity* 35: 153–64.

Ashenden, T. W., W. S. Stewart, and W. Williams. 1975. Growth responses of sand dune population of *Dactylis glomerata* L. to different levels of water stress. *J. Ecol.* 63: 97–107.

Asprey, G. G. and A. R. Loveless. 1958. The dry evergreen formations of Jamaica. II. The raised coral beaches of the north coast. *J. Ecol.* 46: 547–70.

Atsatt, P. R. 1973. Parasitic flowering plants: How did they evolve? *Amer. Nat.* 107: 502–10.

Atsatt, P. R. 1977. The insect herbivore as a predictive model in parasitic seed plant biology. *Amer. Nat.* 111: 579–86.

Avdulov, N. P. 1931. Karyo-systematische Untersuchung der Familie Gramineen. *Bul. Appl. Bot., Suppl.* 44.

Avery, A. G., S. Satina, and J. Rietsema. 1959. Blakeslee: The Genus Datura. New York: Ronald Press.

Axelrod, D. I. 1973. History of the Mediterranean ecosystem in California. In F. diCastri and H. A. Mooney (eds.), *Mediterranean Type Ecosystems: Origin and Structure.* New York: Springer-Verlag.

Ayala, F. J. 1969. An evolutionary dilemma: Fitness of genotypes versus fitness of populations. *Canad. J. Cytol. Genet.* 11. 439–56.

Ayala, F. J. 1975. Genetic differentiation during the speciation process. *Evol. Biol.* 8: 1–78.

Ayala, F. J., ed. 1976. *Molecular Evolution.* Sunderland, Mass: Sinauer.

Babbel, G. R. and R. K. Selander. 1974. Genetic variability in edaphically restricted and widespread plant species. *Evolution* 28: 619–30.

Babbel, G. R. and R. P. Wain. 1977. Genetic structure of *Hordeum jubatum.* I. Outcrossing rates and heterozgosity levels. *Canad. J. Genet. Cytol.* 19: 143–52.

Babcock, E. B. 1947. *The Genus Crepis.* Univ. Calif. Publ. Bot., vols. 1 and 2. Berkeley and Los Angeles: University of California Press.

Babcock, E. B. and D. R. Cameron. 1934. Chromosomes and phylogeny in *Crepis.* II. The relationships of 108 species. *Univ. Calif. Publ. Agr. Sci.* 6: 287–324.

Babcock, E. B. and M. S. Cave. 1938. A study of intra- and interspecific relationships of *Crepis foetida* L. *Zeitschr. ind. Abst.-u. Vererbungsl.* 75: 124–60.

Babcock, E. B. and S. L. Emsweller. 1938. Meiosis in certain interspecific hybrids in *Crepis* and its bearing on taxonomic relationship. *Univ. Calif. Publ. Agr. Sci.* 6: 325–68.

Bailey, I. W. and E. W. Sinnott. 1916. The climatic distribution of certain types of angiosperm leaves. *Amer. J. Bot.* 3: 24–39.

Baker, B. S., A. T. C. Carpenter, M. S. Esposito, R. E. Esposito, and L. Sandler. 1976. The genetic control of meiosis. *Ann. Rev. Genet.* 10: 53–134.

Baker, H. G. 1955. Self-compatibility and establishment after "long distance" dispersal. *Evolution* 9: 347–48.

Baker, H. G. 1959. Reproductive methods as factors in speciation in flowering plants. *Cold Spring Harbor Symp. Quant. Biol.* 24: 177–91.

Baker, H. G. 1961. The adaptation of flowering plants to nocturnal and crepuscular pollinators. *Quart. Rev. Biol.* 36: 64–73.

Baker, H. G. 1965. Characteristics and modes of origin of weeds. In H. G. Baker and G. L. Stebbins, Jr. (eds.), *The Genetics of Colonizing Species.* New York: Academic Press.

Baker, H. G. 1970. Evolution in the tropics. *Biotropica* 2: 101–10.

Baker, H. G. 1972. Seed weight in relation to environmental conditions in California. *Ecology* 53: 997–1010.

Baker, H. G. 1973. Evolutionary relationships between flowering plants and animals in American and African tropical forests. In

B. J. Meggers, E. S. Ayensu, and W. D. Duckworth (eds.), *Tropical Forest Ecosystems in Africa and South America: A Comparative Review*. Washington, D.C.: Smithsonian Institution.

Baker, H. G. 1974. The evolution of weeds. *Ann. Rev. Eco. Syst.* 5: 1–24.

Baker, H. G. 1975. Sugar concentrations in nectars from hummingbird flowers. *Biotropica* 7: 37–41.

Baker, H. G. 1977. Chemical aspects of the pollination biology of woody plants in the tropics. In P. B. Tomlinson and M. Zimmermann (eds.), *Tropical Trees as Living Systems*. New York: Cambridge University Press.

Baker, H. G. and I. Baker. 1973. Some autecological aspects of the evolution of nectar-producing flowers, particularly amino acid production in nectar. In V. H. Heywood (ed.), *Taxonomy and Ecology*. New York: Academic Press.

Baker, H. G. and I. Baker. 1975. Studies of nectar constitution and nectar-plant coevolution. In L. Gilbert and P. H. Raven (eds.), *Animal–Plant Coevolution*. Austin, Texas: University of Texas Press.

Baker, H. G. and P. D. Hurd. 1968. Intrafloral ecology. *Ann. Rev. Entomol.* 13: 385–414.

Baker, H. G. and G. L. Stebbins, eds. 1965. *The Genetics of Colonizing Species*. New York and London: Academic Press.

Baker, I. and H. G. Baker. 1976. Variation in an introduced *Lythrum* species in California vernal pools. In S. Jain (ed.), *Vernal Pools: Their Ecology and Conservation*. Institute of Ecology Publications, no. 9. Davis, Calif.: Institute of Ecology.

Baker, J., J. Maynard Smith, and C. Strobeck. 1975. Genetic polymorphism in the bladder campion, *Silene maritima*. *Biochem. Genet.* 13: 393–410.

Baker, M. S. 1974. Floral ecology of coastal scrub in southeast Jamaica. *Biotropica* 6: 104–29.

Bammi, R. K. 1965. "Complement fractionation" in a natural hybrid between *Rubus procerus* Muell. and *R. lacinatus* Willd. *Nature* (London) 208: 608.

Bank, B. G. and S. Cobb. 1968. The size of the olfactory bulb in 108 species of birds. *Auk* 85: 55–61.

Bannister, P. 1976. *Introduction to Physiological Plant Ecology*. New York: Wiley.

Bänsch, R. 1966. On prey-seeking behavior of aphidophagous insects. In I. Hodek (ed.), *Ecology of Aphidophagous Insects.* Prague: Academia.

Barber, H. N. 1965. Selection in natural populations. *Heredity* 20: 551–72.

Barbier, M. 1970. Chemistry and biochemistry of pollens. *Prog. Phytochem.* 2: 1–34.

Barbour, M. G. 1968. Germination requirements of the desert shrub *Larrea divaricata. Ecology* 49: 915–23.

Barker, J., M. Khan, and T. Solomons. 1964. Mechanism of the Pasteur effect. *Nature (London)* 201: 1126–27.

Barker, J., M. Khan, and T. Solomons. 1966. Mechanism of the Pasteur effect. *Nature (London)* 211: 547–48.

Barlow, B. A. and D. Wiens. 1977. Host-parasite resemblance in Australian mistletoes: The case for cryptic mimicry. *Evolution* 31: 69–84.

Barnard, E. L. and J. R. Jorgensen. 1977. Respiration of field-grown loblolly pine roots as influenced by temperature and root type. *Canad. J. Bot.* 55: 740–743.

Bartholomew, B., L. C. Eaton, and P. H. Raven. 1973. *Clarkia rubicunda:* A model of plant evolution in semi-arid regions. *Evolution* 27: 505–17.

Baskin, J. M. and C. C. Baskin. 1971a. Germination ecology and adaptation to habitat in *Leavenworthia* spp. (Cruciferae). *Amer. Midl. Nat.* 85: 22–35.

Baskin, J. M. and C. C. Baskin. 1971b. Germination ecology of *Phacelia dubia* var. *dubia* in Tennessee glades. *Amer. J. Bot.* 58: 98–104.

Baskin, J. M. and C. C. Baskin. 1971c. Germination of *Cyperus inflexus* Muhl. *Bot. Gaz.* 132: 3–9.

Baskin, J. M. and C. C. Baskin. 1972a. Ecological life cycle and physiological ecology of germination of *Arabidopsis thaliana. Canad. J. Bot.* 50: 353–60.

Baskin, J. M. and C. C. Baskin. 1972b. Germination characteristics of *Dimorphia cymosa* seeds and an ecological interpretation. *Oecologia* 10: 17–28.

Baskin, J. M. and C. C. Baskin. 1972c. Physiological ecology of germination of *Viola rafinesquii. Amer. J. Bot.* 59: 981–88.

Baskin, J. M. and C. C. Baskin. 1972d. Influence of germination date on survival and seed production in a natural population of *Leavenworthia stylosa*. *Amer. Midl. Nat.* 88: 318–23.

Baskin, J. M. and C. C. Baskin. 1973a. Ecological life cycle of *Helenium amarum* in central Tennessee. *Bull. Torrey Bot. Club* 100: 117–24.

Baskin, J. M. and C. C. Baskin. 1973b. Delayed germination in seeds of *Phacelia dubia* var. *dubia*. *Canad. J. Bot.* 51: 2481–86.

Baskin, J. M. and C. C. Baskin. 1974a. Some eco-physiological aspects of seed dormancy in *Geranium carolinianum* L. from central Tennessee. *Oecologia* 16: 209–19.

Baskin, J. M. and C. C. Baskin. 1974b. Germination and survival in a population of the winter annual *Alyssum alyssoides*. *Canad. J. Bot.* 52: 2439–45.

Baskin, J. M. and C. C. Baskin. 1975a. Seed dormancy in *Isanthus brachiatus* (Labiatae). *Amer. J. Bot.* 62: 623–27.

Baskin, J. M. and C. C. Baskin. 1975b. Ecophysiology of seed dormancy and germination in *Torilis japonica* in relation to its life cycle strategy. *Bull. Torrey Bot. Club* 102: 67–72.

Baskin, J. M. and C. C. Baskin. 1976a. Effect of photoperiod on germination of *Cyperus inflexus* seeds. *Bot. Gaz.* 137: 269–73.

Baskin, J. M. and C. C. Baskin. 1976b. Some aspects of the autecology and population biology of *Phacelia purshii*. *Amer. Midl. Nat.* 96: 431–43.

Bawa, K. S. 1974. Breeding systems of tree species of a lowland tropical community. *Evolution* 28: 85–92.

Bawa, K. S. and P. A. Opler. 1975. Dioecism in tropical forest trees. *Evolution* 29: 167–79.

Bayer, E. 1959. Über Anthocyan Komplexe. II. Farbstoffe der roten violetten und blauen Lupinenbluten. *Chem. Ber.* 92: 1062–71.

Bayer, E. 1959. Über Anthocyan Komplexe. II. Farbstoffe der roten violetten und blauen Lupinenbluten. *Chem. Ber.* 92: 1062–71.

Beadle, N. C. W. 1954. Soil phosphate and the delimitation of plant communities in Eastern Australia. *Ecology* 35: 370–75.

Beadle, N. C. W. 1962. Soil phosphate and the delimitation of plant communities in Eastern Australia. II. *Ecology* 43: 281–88.

Beadle, N. C. W. 1966. Soil phosphate and its role in molding segments of the Australian flora and vegetation with special reference to xeromorphy and scleromorphy. *Ecology* 47: 992–1007.

Beatley, J. C. 1967. Survival of winter annuals in the northern Mohave desert. *Ecology* 48: 745–750.

Beatley, J. C. 1974. Phenological events and their environmental triggers in Mohave desert ecosystems. *Ecology* 55: 856–63.

Beattie, A. J. and N. Lyons. 1975. Seed dispersal in *Viola* (Violaceae): Adaptations and strategies. *Amer. J. Bot.* 62: 714–22.

Beimborn, W. A. 1973. Physical factors affecting establishment of *Solidago nemoralis* on the New Jersey Piedmont. Ph.D. thesis, Rutgers University, New Brunswick, N.J.

Bell, C. R. 1970. Seed distribution and germination experiment. In H. T. Odum (ed.), *A Tropical Rain Forest: A Study of Irradiation and Ecology at El Verde, Puerto Rico. Rio Piedras, P.R.: Nuclear Center.*

Bell, G. 1976. On breeding more than once. *Amer. Nat.* 110: 57–77.

Belling, J. and A. F. Blakeslee. 1926. On the attachment of non-homologous chromosomes at the reduction division in certain 25-chromosome Daturas. *Proc. Nat. Acad. Sci. USA* 12: 7–11.

Bennett, E. 1964. Historical perspective in genecology. *Rec. Scottish Plant Breed. Sta.* 1964: 49–115.

Bergmann, V. F. 1975. Identification of forest seed origin on the basis of isoenzyme gene frequencies. *Allg. Forst Jagdzeit.* 146: 191–95.

Berry, J. 1975. Adaptation of photosynthetic processes to stress. *Science* 188: 644–50.

Bews, J. W. 1925. *Plant Forms and their Evolution in South Africa.* London: Longmans, Green.

Bews, J. W. 1927. Studies in the ecological evolution of the angiosperms. I. *New Phytol.* 26: 1–21.

Bhat, K. K. S. and P. H. Nye. 1973. Diffusion of phosphate to plant roots in soil. I. Quantitative autoradiography of the depletion zone. *Plant Soil* 38: 161–75.

Bhat, K. K. S. and P. H. Nye. 1974. Diffusion of phosphate to plant roots in soil. II. Uptake along the roots at different times and the effect of different levels of phosphorus. *Plant Soil* 41: 365–82.

Bhat, K. K. S., P. H. Nye, and J. P. Baldwin. 1976. Diffusion of phosphate to plant roots in soil. IV. The concentration distance profile in the rhizosphere of roots with root hairs in a low-P soil. *Plant Soil* 44: 63–72.

Bibbey, R. O. 1948. Physiological studies on weed seed germination. *Plant Physiol.* 23: 467–84.

Bieleski, R. L. 1973. Phosphate pools, phosphate transport, and phosphate availability. Ann. Rev. Plant Physiol. 24: 225–52.

Billings, W. D., K. M. Peterson, G. R. Shaver, and A. Trent. 1977. Root growth, respiration, and carbon dioxide evolution in an Arctic tundra soil. Arct. Alp. Res. 9: 129–37.

Björkman, O. 1968. Carboxydismutase activity in shade-adapted and sun-adapted species of higher plants. Physiol. Plant. 21: 1–10.

Björkman, O. 1973. Comparative studies on photosynthesis in higher plants. Photophysiology 8: 1–63.

Björkman, O. 1975. Environmental and biological control of photosynthesis: Inaugural address. In R. Marcelle (ed.), Environmental and Biological Control of Photosynthesis. The Hague: Dr. W. Junk.

Björkman, O. 1976. Adaptive and genetic aspects of C_4 photosynthesis. In R. H. Burris and C. C. Black (eds.), CO_2 Metabolism and Plant Productivity. Baltimore: University Park Press.

Björkman, O., M. Nobs, H. Mooney, J. Troughton, J. Berry, F. Nicholson, and W. Ward. 1972. Growth responses of plants from habitats with contrasting thermal environments. Carnegie Inst. Washington Yearb. 71: 748–67.

Björkman, O., R. W. Pearcy, T. A. Harrison, and H. A. Mooney. 1972. Photosynthetic adaptation to high temperatures: A field study in Death Valley. Science 175: 786–89.

Black, E. C., T. M. Chen, and R. H. Brown. 1969. Biochemical basis for plant competition. Weed Sci. 17: 338–44.

Black, J. N. 1958. Competition between plants of different initial seed sizes in swards of subterranean clover (T. subterraneum L.) with reference to leaf area and light microclimates. Aust. J. Arid. Res. 9: 299–318.

Blackman, F. F. 1905. Optima and limiting factors. Ann. Bot. 19: 281–95.

Blakeslee, A. F. 1939. The present and potential service of chemistry to plant breeding. Amer. J. Bot. 26: 163–72.

Blakeslee, A. F. and R. E. Cleland. 1930. Circle formation in Datura and Oenothera. Proc. Nat. Acad. Sci. USA 16: 177–83.

Bliss, L. C. 1958. Seed germination in arctic and alpine species. Arctic 11: 180–88.

Bliss, L. C. 1971. Arctic and alpine plant life cycles. Ann. Rev. Ecol. Syst. 2: 405–38.

Bloom, W. L. 1976. Multivariate analysis of the introgressive replacement of *Clarkia nitens* by *Clarkia speciosa polyantha* (Onagraceae). *Evolution* 30: 412–24.

Bloom, W. L. and H. Lewis. 1972. Interchanges and interpopulational gene exchange in *Clarkia speciosa*. In C. D. Darlington and K. Lewis (eds.), *Chromosomes Today*, vol. III. Edinburgh: Constable.

Boardman, N. K. 1977. Comparative photosynthesis of sun and shade plants. *Ann. Rev. Plant Physiol.* 28: 355–77.

Böcher, T. W., K. Holmen, and K. Jakobsen. 1968. *The Flora of Greenland*. Copenhagen: P. Haase & Son.

Borman, R. H. 1957. Moisture transfer between plants through intertwined root systems. *Plant Physiol.* 32: 48–55.

Bossert, W. H. 1963. Simulation of Character Displacement in Animals. Ph.D. dissertation, Harvard University.

Botkin, D. B. 1977. Bits, bytes, and IBP. *BioScience* 27: 385.

Bourdeau, P. 1954. Oak seedling ecology determining segregation of species in Piedmont oak–hickory forests. *Eco. Monogr.* 24: 297–320.

Bourne, D. and S. Ranson. 1965. Respiratory metabolism in detached *Rhododendron* leaves. *Plant Physiol.* 40: 1178–90.

Bourque, D. P. and A. W. Naylor. 1971. Large effects of small water deficits on chlorophyll accumulation and ribonucleic acid synthesis in etiolated leaves of jackbean (*Canavalia ensiformis* (L.) DC.). *Plant Physiol.* 47: 591–94.

Bowen, G. D. and A. D. Rovira. 1967. Phosphate uptake along attached and excised wheat roots measured by an automatic scanning method. *Aust. J. Biol. Sci.* 20: 369–78.

Bowman, R. I. 1966. *The Galápagos*. Berkeley: University of California Press.

Boyer, J. S. 1970. Differing sensitivity of photosynthesis to low leaf water potentials in corn and soybean. *Plant Physiol.* 46: 236–39.

Boyer, J. S. 1971. Non-stomatal inhibition of photosynthesis in sunflower at low leaf water potentials and high light intensities. *Plant Physiol.* 48: 532–36.

Boyer, J. S. 1976. Water deficits and photosynthesis. In T. T. Kozlowski (ed.), *Water Deficits and Plant Growth*, vol. IV. New York: Academic Press.

Boyer, J. S. and B. L. Bowen. 1970. Inhibition of oxygen evolution in chloroplasts isolated from leaves with low water potentials. *Plant Physiol.* 45: 612–15.

Bradshaw, A. D. 1965. Evolutionary significance of phenotypic plasticity in plants. *Adv. Genet.* 13: 115–55.

Bradshaw, A. D. 1971. Plant evolution in extreme environments. In R. Creed (ed.), *Ecological Genetics and Evolution.* Oxford and Edinburgh: Blackwell Scientific Publ.

Bradshaw, A. D. 1972. Some of the evolutionary consequences of being a plant. *Evol. Biol.* 5: 25–47.

Bradshaw, A. D. 1974. Environment and phenotypic plasticity. *Brookhaven Symp. Biol.* 25: 75–94.

Bratton, S. P. 1976. Resource division in an understory herb community: Responses to temporal and microtopographic gradients. *Amer. Nat.* 110: 679–93.

Breazeale, E. L., W. T. McGeorge, and J. F. Breazeale. 1950. Moisture absorption by plants from an atmosphere of high humidity. *Plant Physiol.* 25: 413–19.

Brenchley, W. E. and K. Warrington. 1930. The weedseed population of arable soil. I. Numerical estimation of viable seeds and observations on their natural dormancy. *J. Ecol.* 18: 235–72.

Brenner, W. 1902. Klima und Blatt bei der Gattung Quercus. *Flora* 90: 114–60.

Brereton, A. J. 1971. The structure of the species populations in the initial stages of salt marsh succession. *J. Ecol.* 59: 321–39.

Brinkman, K. A. 1974. *Betula L.* In *Seeds of Woody Plants in the United States.* U.S. Dept. Agr. Handbook no. 450: 252–57.

Brinkman, K. A. and E. I. Roe. 1975. *Quaking Aspen Silvics and Management in the Lake States.* U.S. Dept. Agr. Handbook no. 486.

British Ecological Society. 1941. Biological Flora of the British Isles, Foreword and Schedule for Contributors. *J. Ecol.* 29: 356–60.

Brix, H. 1962. The effect of water stress on the rates of photosynthesis and respiration in tomato plants and loblolly pine seedlings. *Physiol. Plant.* 15: 10–20.

Brouwer, R. 1953. Water absorption by the roots of *Vicia faba* at various transpiration strengths. II. Causal relation between suction tension, resistance and uptake. *Proc. Kon. Ned. Akad. Wetensch.* (C) 56: 129–36.

Brower, L. P. 1970. Plant poisons in a terrestrial food chain and implications for mimicry theory. In K. L. Chambers (ed.), *Biochemical Coevolution.* Corvallis, Ore.: Oregon State University Press.

Brower, L. P. and C. M. Moffitt. 1974. Palatibility dynamics of cardenolides in the monarch butterfly. *Nature (London)* 249: 280–83.

Brower, L. P., P. B. McEvoy, K. L. Williamson, and M. A. Flannery. 1972. Variation in cardiac glycoside content of monarch butterflies from natural populations in eastern North America. *Science* 177: 426–29.

Brown, A. H. D. 1971. Isozyme variation under selection in *Zea mays*. *Nature* (London) 232: 570.

Brown, A. H. D. and R. W. Allard. 1971. Effect of reciprocal recurrent selection for yield on isozyme polymorphisms in maize (*Zea mays* L.). *Crop Sci.* 11: 888–93.

Brown, A. H. D., D. R. Marshall, and L. Albrecht. 1974. The maintenance of alcohol dehydrogenase polymorphism in *Bromus mollis* L. *Aust. J. Biol. Sci.* 27: 545–59.

Brown, A. H. D., A. C. Matheson, and K. G. Eldridge. 1975. Estimation of the mating system of *Eucalptus obliqua* L'Herit. by using allozyme polymorphisms. *Aust. J. Bot.* 23: 931–49.

Brown, C. R. 1977. A comparison of patterns of variability in *Limnanthes alba* Benth. and *L. floccosa* Howell. Ph.D. dissertation, University of California, Davis.

Brown, J., F. L. Bunnell, P. C. Miller, S. F. McLean, Jr., and L. L. Tiezen (eds.). 1979. *An Arctic Ecosystem: The Coastal Tundra of Northern Alaska*. Stroudsburg, Pa: Dowden, Hutchinson and Ross (in press).

Brown, W. V. 1958. Leaf anatomy in grass systematics. *Bot. Gaz.* 119: 170–78.

Brunig, E. F. 1970. Stand structure, physiognomy and environmental factors in some lowland forests in Sarawak. *Trop. Ecol.* 11: 26–43.

Brunig, E. F. 1971. On the ecological significance of drought in the equatorial wet evergreen (rain) forest of Sarawak (Borneo). In J. R. Flendley (ed.), *The Water Relations of Malesian Forests*. Hull: University of Hull Press.

Bryant, E. H. 1971. Life history consequences of natural selection: Cole's result. *Amer. Nat.* 105: 75–76.

Bunce, J. A. 1977. Non-stomatal inhibition of photosynthesis at low water potentials in intact leaves of species from a variety of habitats. *Plant Physiol.* 59: 348–50.

Bunce, J. A. and L. N. Miller. 1976. Differential effects of water stress on respiration in the light in woody plants from wet and dry habitats. *Canad. J. Bot.* 54: 2451–64.

Bunce, J. A., L. N. Miller, and B. F. Chabot. 1977. Competition for water among five eastern North American tree species. *Bot. Gaz.* 138: 168–73.

Burdon, J. J. and G. A. Chilvers. 1975. Epidemiology of damping-off disease (*Pythium irregulare*) in relation to density of *Lepidium sativum* seedlings. *Ann. Appl. Biol.* 83: 135–43.

Burger, H. 1941. Fichten und Fähren verschiedener Herkunft auf verschiedenen Kulturorten. *Mitt. Schweiz. Anstalt förstliche Versuchswesen* 22: 10–60.

Burger, W. 1975. The species problem in oaks. *Taxon* 24: 45–50.

Burnett, W. C., Jr., S. B. Jones, Jr., T. J. Mabry, and W. G. Padolina. 1974. Sesquiterpene lactones—insect feeding deterrents in Vernonia. *Biochem. Syst. Ecol.* 2: 25–29.

Bush, G. L. 1975. Modes of animal speciation. *Ann. Rev. Ecol. System.* 6: 339–64.

Butler, C. G. 1945. The influence of various physical and biological factors of the environment on honey bee activity. An examination of the relationship between activity and nectar concentration and abundance. *J. Exp. Biol.* 21: 5–12.

Caldwell, M. M. 1975. Primary production of grazing lands. In J. P. Cooper (ed.), *Photosynthesis and Production in Different Environments*. Cambridge: Cambridge University Press.

Caldwell, M. M. 1976. Root extension and water absorption. In O. L. Lange, L. Kappen, and E. D. Schulze (eds.), *Water and Plant Life*. Heidelberg: Springer-Verlag.

Caldwell, M. M., R. S. White, R. T. Moore, and L. B. Camp. 1977. Carbon balance, productivity, and water use of cold-winter desert shrub communities dominated by C_3 and C_4 species. *Oecologia* 29: 275–300.

Camacho-B, S. E., A. E. Hall, and M. R. Kaufmann. 1974. Efficiency and regulation of water transport in some woody and herbaceous species. *Plant Physiol.* 54: 169–72.

Cantlon, J. E. 1969. The stability of natural populations and their sensitivity to technology. In *Diversity and Stability in Ecological Systems; Brookhaven Symp. Biol.* 22: 243–51.

Carlquist, S. 1970. *Hawaii: A Natural History*. New York: Natural History Press.

Carlquist, S. 1974. *Island Biology*. New York: Columbia University Press.

Carlquist, S. 1975. *Ecological Strategies of Xylem Evolution*. Berkeley, Los Angeles, and London: University of California Press.

Carson, H. L. 1957. The species as a field for gene recombination. In E. Mayr (ed.), *The Species Problem*. Washington, D.C.: American Association for the Advancement of Science.

Carson, H. L. 1975. The genetics of speciation at the diploid level. *Amer. Nat.* 109: 83–92.

Carson, H. L. 1976. The unit of genetic change in adaptation and speciation. *Ann. Mo. Bot. Gard.* 63: 210–23.

Case, M. E. and N. H. Giles. 1971. Partial enzyme aggregates formed by pleiotropic mutants in the arons gene cluster of *Neurospora crassa*. *Proc. Nat. Acad. Sci. USA.* 68: 58–62.

Cates, R. G. 1975. The interface between slugs and wild ginger: Some evolutionary aspects. *Ecology* 56: 391–400.

Cavers, P. B. 1974. Germination polymorphism in *Rumex crispus*: The effects of different storage conditions on germination responses of seeds collected from different plants. *Canad. J. Bot.* 52: 575–83.

Cavers, P. B. and J. L. Harper. 1967. Studies in dynamics of plant populations. I. The fate of seed and transplants introduced into various habitats. *J. Ecol.* 55: 59–71.

Chabot, B. F. and W. D. Billings. 1972. Origins and ecology of the Sierran alpine flora and vegetation. *Ecol. Monogr.* 42: 163–99.

Chamberlain, C. J. 1966. *Gymnosperms: Structure and Evolution*. New York: Dover Press.

Chancellor, R. J. 1965. Emergence of weed seedlings in the field and the effects of different frequencies of cultivation. *Rept. Seventh Brit. Weed Control Conf.*, pp. 599–606.

Chandler, A. E. F. 1969. Locomotory behavior of first instar larvae of aphidophagous Syrphidae (Diptera) after contact with aphids. *Anim. Behav.* 17: 673–78.

Chapin, F. S. 1974. Phosphate absorption capacity and accumulation potential in plants along a latitudinal gradient. *Science* 183: 521–23.

Chapin, F. S. 1978. Nutrient uptake and utilization by tundra vegetation. In L. L. Tieszen (ed.), *Vegetation and Production Ecology of an Alaskan Arctic Tundra*. New York: Springer-Verlag (in press).

Charnov, E. L. and W. M. Schaffer. 1973. Life-history consequences of natural selection: Cole's result revisited. *Amer. Nat.* 107: 791–93.

Charnov, E. L., G. H. Orians, and K. Hyatt. 1976. Ecological implications of resource depression. *Amer. Nat.* 110: 247–59.

Cherry, J. P. and R. L. Ory. 1973. Electrophoretic characterization of six selected enzymes of peanut cultivars. *Phytochemistry* 12: 283–89.

Chippindale, H. G. 1948. Resistance to inanition in grass seedlings. *Nature* (London) 161: 65.

Christensen, N. L. and C. H. Muller. 1975. Relative importance of factors controlling germination and seedling survival in *Adenostoma* chaparral. *Amer. Midl. Nat.* 93: 71–78.

Christiansen, F. and M. W. Feldman. 1975. Subdivided populations: a review of the one- and two-locus deterministic theory. *Theor. Popul. Biol.* 7: 13–38.

Christiansen, F. and O. Frydenberg. 1973. Selection component analysis of natural polymorphisms using population samples including mother–offspring combinations. *Theor. Popul. Biol.* 4: 425–45.

Christiansen, F. and O. Frydenberg. 1976. Selection component analysis of natural polymorphisms using mother–offspring samples of successive cohorts. In Karlin and Nevo (eds.), *Population Genetics and Ecology.* New York: Academic Press.

Christiansen, F., O. Frydenberg, and V. Simonsen. 1978. Genetics of *Zoarces* populations, X. Selection component analysis of *EST III* polymorphism using samples of successive cohorts. *Hereditas* (in press).

Chrysler, M. A. 1905. Anatomical notes on certain strand plants. *Bot. Gaz.* 37: 461–64.

Chu, C. 1976. Physiological aspects of competition between redroot pigweed (C_4) and common lambsquarters (C_3). Ph.D. dissertation, Cornell University.

Clapham, A. R. 1956. Autecological studies and the "Biological Flora of the British Isles." *J. Ecol.* 44: 1–11.

Clark, R. B. and Levitt, J. 1956. The basis of drought resistance in the soybean plant. *Physiol. Plant.* 9: 598–606.

Clarke, B. 1975. The contribution of ecological genetics to evolutionary theory: Detecting the direct effects of natural selection on particular polymorphic loci. *Genetics* 79: 101–13.

Clausen, J. 1951. *Stages in the Evolution of Plant Species.* Ithaca, New York: Cornell University Press.

Clausen, J. and W. M. Hiesey. 1958. Experimental studies on the nature of species. IV. Genetic structure of ecological races. *Carnegie Inst. Washington Publ.* 615: 1–312.

Clausen, R. E. and T. H. Goodspeed. 1925. Interspecific hybridization in *Nicotiana.* II. A tetraploid *glutinosa*–*Tabacum* hybrid, an experimental verification of Winge's hypothesis. *Genetics* 10: 279–84.

Clausen, J., D. D. Keck, and W. M. Hiesey. 1940. Experimental studies on the nature of species. I. The effect of varied environments on western North American plants. *Carnegie Inst. Washington Publ.* 520: 1–452.

Clausen, J., D. D. Keck, and W. M. Hiesey. 1945. Experimental studies on the nature of species. II. Plant evolution through amphiploidy and autoploidy, with examples from the Madiinae. *Carnegie Inst. Washington Publ.* 564: 1–174.

Clausen, J., D. D. Keck, and W. M. Hiesey. 1948. Experimental studies on the nature of species. III. Environmental responses of climatic races of *Achillea.* *Carnegie Inst. Washington Publ.* 581: 1–129.

Clegg, M. T. and R. W. Allard. 1972. Patterns of genetic differentiation in the slender wild oat species *Avena barbata.* *Proc. Nat. Acad. Sci. USA* 69: 1820–24.

Clegg, M. T. and R. W. Allard. 1973. Viability versus fecundity selection in the slender wild oat, *Avena barbata* L. *Science* 181: 667–68.

Clegg, M. T., C. R. Horch, and G. L. Church. 1976. Extreme genetic similarity among Northeastern species of wild rye. *Genetics* 83: 515.

Clegg, M. T., A. L. Kahler, and R. W. Allard, 1978. Estimation of life cycle components of selection in an experimental plant population. *Genetics* 89: 765–92.

Cleland, R. E. 1936. Some aspects of the cytogenetics of *Oenothera.* *Bot. Rev.* 2: 316–48.

Clements, E. S. 1905. The relation of leaf structure to physical factors. *Trans. Amer. Microsc. Soc.* 26: 19–102.

Clements, F. E. 1916. Plant succession. *Carnegie Inst. Washington Publ.* 242: 1–512.

Clements, F. E., J. E. Wavers, and H. C. Hanson. 1929. Plant Competition. *Carnegie Inst. Washington Publ.* 398: 1–340.

Cody, M. L. 1971. Ecological aspects of reproduction. In D. S. Farmer and J. R. King (eds.), *Avian Biology.* New York: Academic Press.

Cohen, D. 1966. Optimizing reproduction in a randomly varying environment. *J. Theor. Biol.* 12: 119–29.

Cohen, D. 1967. Optimizing reproduction in a randomly varying environment when a correlation may exist between the conditions at the time a choice has to be made and the subsequent outcome. *J. Theor. Biol.* 16: 1–14.

Cohen, D. 1968. A general model of optimal reproduction in a randomly varying environment. *J. Ecol.* 56: 219–28.

Cohen, D. 1970. The expected efficiency of water utilization in plants under different competition and selection regimes. *Isr. J. Bot.* 19: 50–54.

Cohen, D. 1971. Maximizing final yield when growth is limited by time or by limiting resources. *J. Theor. Biol.* 33: 299–307.

Cole, C. V., G. S. Innis, and J. W. B. Stewart. 1977. Simulation of phosphorus cycling in semiarid grasslands. *Ecology* 77: 1–15.

Cole, L. C. 1954. The population consequences of life history phenomena. *Quart. Rev. Biol.* 29: 103–37.

Coleman, D. C. 1976. A review of root production processes and their influence on soil biota in terrestrial ecosystems. In J. M. Anderson and A. Macfadyen (eds.), *The Role of Terrestrial and Aquatic Organisms in Decomposition Processes.* Oxford: Blackwell.

Coleman, P. 1977. Spatial and temporal variation in population structure of *Lolium multiflorum* Law. (Italian ryegrass). Ph.D. dissertation, University of California, Davis.

Colwell, R. K. 1973. Competition and coexistence in a simple tropical community. *Amer. Nat.* 107: 737–60.

Colwell, R. K. 1974. Predictability, constancy, and contingency of periodic phenomena. *Ecology* 55: 1148–53.

Connell, J. H. 1970. On the role of natural enemies in preventing competitive exclusion in some marine animals and in rain forest trees. In P. J. den Boer and G. R. Gradwell (eds.), *Dynamics of Populations.* Wageningen: Centre for Agricultural Publication and Documentation.

Cook, R. E. 1976. Photoperiod and the determination of potential seed number in *Chenopodium rubrum* L.   *Ann. Bot.* 40: 1085–99.

Cook, R. E. 1978. The biology of seeds in the soil. In O. T. Solbrig (ed.), *Demography and Plant Population.* Oxford: Blackwell Scientific.

Cook, S. A. and M. P. Johnson. 1968. Adaptation to heterogeneous environments. I. Variation in heterophylly in *Ranunculus flammula* L. *Evolution* 22: 496–516.

Cooper-Driver, G. A. and T. Swain. 1976. Cyanogenic polymorphism in bracken in relation to herbivore predation. *Nature* (London) 260: 604.

Cornell, H. 1976. Search strategies and the adaptive significance of switching in some general predators. *Amer. Nat.* 110: 317–20.

Covich, A. 1974. Ecological economics of foraging among coevolving animals and plants. *Ann. Mo. Bot. Gard.* 61: 794–805.

Covich, A. 1976. Analyzing shapes of foraging areas: some ecological and economic theories. *Ann. Rev. Ecol. Syst.* 7: 235–57.

Cowan, I. R. 1965. Transport of water in the soil–plant–atmosphere system. *J. Appl. Ecol.* 2: 221–39.

Cox, T. L. 1975. Seasonal respiration rates of yellow-poplar roots by diameter classes. *For. Sci.* 21: 185–90.

Crafts, A. S. 1975. *Modern Weed Control.* Berkeley: University of California Press.

Craig, I. L., B. E. Murray, and T. Rajhathy. 1974. *Avena canariensis:* Morphological and electrophoretic polymorphism and relationship to the *A. magna–A. murphy* complex and *A. sterilis.* *Canad. J. Genet. Cytol.* 16: 677–89.

Crawford, R. 1966. The control of anaerobic respiration as a determining factor in the distribution of the genus *Senecio.* *J. Ecol.* 54: 403–13.

Crawford, R. 1972. Some metabolic aspects of ecology. *Trans. Edin. Bot. Soc.* 41: 309–22.

Crocker, W. 1916. Mechanics of dormancy in seeds. *Amer. J. Bot.* 3: 99–120.

Cromartie, W. J. 1975. The effect of stand size and vegetational background on the colonization of cruciferous plants by herbivorous insects. *J. Appl. Ecol.* 12: 517–33.

Croze, H. 1970. Searching image in carrion crows: Hunting strategy in a predator and some antipredator devices in a camouflaged prey. *Tierpsychol.* 5: 1–86.

Cruden, R. W. 1976. Intraspecific variation in pollen–ovule ratios and nectar secretion—preliminary evidence of ecotypic differentiation. *Ann. Mo. Bot. Gard.* 63: 277–89.

Cruden, R. W. 1977. Pollen–ovule ratios: A conservative indicator of breeding systems in the flowering plants. *Evolution* 31: 32–46.

Cumber, R. A. 1949. The biology of bumble-bees, with special reference to the production of the worker caste. *Trans. Roy. Entomol. Soc. London* 100: 1–45.

Cunningham, G. L. and B. R. Strain. 1969. An ecological significance of seasonal leaf variability in a desert shrub. *Ecology* 50: 400–08.

Dadd, N. W. 1973. Insect nutrition: Current developments and metabolic implications. *Ann. Rev. Entomol.* 18: 381–420.

Darlington, C. D. 1929. Ring formation in *Oenothera* and other genera. *J. Genet.* 20: 345–63.

Darlington, C. D. 1931. The cytological theory of inheritance in *Oenothera*. *J. Genet.* 24: 405–74.

Darlington, C. D. 1937. Recent advances in cytology. 2d ed. Philadelphia: Blakiston.

Darlington, C. D. 1939, 1958. *The Evolution of Genetic Systems.* 1st ed., London: Cambridge Univ. Press; 2d ed., New York: Basic Books.

Darwin, C. 1876, 1888. *The Effects of Cross and Self-fertilization in the Vegetable Kingdom.* London: Murray. 1st ed., 1876; 2nd ed., 1888.

Daumer, K. 1956. Reizmetrische Untersuchung des Farbensehens der Bienen. *Z. Vergl. Physiol.* 41: 413–78.

Daumer, K. 1963. Kontrastempfindlichkeit der Bienen für "Weiss" verschiedenen UV-Gehalts. *Z. Vergl. Physiol.* 46: 336–50.

Davidson, J. L. and J. R. Phillip. 1958. Light and pasture growth. In *Climatology and Microclimatology.* Paris: UNESCO.

Davidson, R. L. 1969. Effects of soil nutrients and moisture on root/shoot ratios in *Lolium perenne* L. and *Trifolium repens* L. *Ann. Bot.* 33: 571–77.

Davies, M. S. and R. W. Snaydon. 1976. Rapid population differentiation in a mosaic environment. III. Coefficients of selection. *Heredity* 36: 59–66.

Day, P. R., P. S. Carlson, O. L. Gamborg, E. G. Jaworski. A. Maretzki, O. E. Nelson, I. M. Sussex, and J. G. Torrey. 1977. Somatic cell genetic manipulation in plants. *BioScience* 27: 116–18.

Day, T., P. Hiller, and B. Clarke. 1975. The relative quantities and catalytic activities of enzymes produced by alleles at the alcohol dehydrogenase locus in *D. melanogaster*. *Biochem. Genet.* 11: 155–71.

Deevey, E. S. 1947. Life tables for natural populations of animals. *Quart. Rev. Biol.* 22: 283–314.

Demetrius, L. 1975. Reproductive strategies and natural selection. *Amer. Nat.* 109: 243–49.

Dengler, N. G. and L. B. MacKay. 1975. The leaf anatomy of beech, *Fagus grandifolia*. *Canad. J. Bot.* 53: 2202–11.

Denmead, O. T. 1964. Evaporation sources and apparent diffusivities in a forest canopy. *J. Appl. Meteorol.* 3: 383–89.

Denmead, O. T. 1973. Relative significance of soil and plant evaporation in estimating evapotranspiration. In R. O. Slatyer (ed.), *Plant Response to Climatic Factors: Proceedings of the Uppsala Symposium*. Paris: UNESCO.

Dennis, D. and T. Coultate. 1966. Phosphofructokinase: A regulatory enzyme in plants. *Biochem. Biophys. Res. Commun.* 25: 187–91.

Detling, J. K. and L. G. Klikoff. 1973. Physiological response to moisture stress as a factor in halophyte distribution. *Amer. Midl. Nat.* 90: 307–18.

de Wit, C. T. 1960. On competition. *Versl. Landbouwk. Onderz. Rijkslandb. Proefstn.* 66 (8): 1–82.

de Wit, C. T. 1965. *Photosynthesis of Leaf Canopies*. Agric. Res. Rep. 663, Inst. Biol. Chem. Res. Field Crops Herb., Wageningen.

de Wit, C. T. and J. Goudriaann. 1974. *Simulation of Ecological Processes*. Wageningen: Centre for Agricultural Publishing and Documentation. Simulation Monograph Series.

Dickinson, H. and J. Antonovics. 1973. Theoretical considerations of sympatric divergence. *Amer. Nat.* 107: 256–74.

Digby, L. 1912. The cytology of *Primula kewensis* and of other related *Primula* hybrids. *Ann. Bot.* 26: 357–88.

Dina, S. and L. G. Klikoff. 1973. Carbon dioxide exchange by several streamside and scrub oak community species of Red Butte Canyon, Utah. *Amer. Midl. Nat.* 89: 70–80.

Dittmer, H. J. 1938. A quantitative study of the subterranean members of three field grasses. *Amer. J. Bot.* 25: 654–57.

Dobzhansky, T. 1950. Evolution in the tropics. *Amer. Sci.* 38: 209–21.

Dobzhansky, T. 1968. On some fundamental concepts of Darwinian biology. *Evol. Bio.* 2: 1–34.

Dobzhansky, T. and F. Ayala. 1973. Temporal frequency changes of enzyme and chromosomal polymorphisms in natural populations of *Drosophila*. *Proc. Nat. Acad. Sci. USA* 70: 680–83.

Dobzhansky, T., F. J. Ayala, G. L. Stebbins, and J. W. Valentine. 1977. *Evolution*. San Francisco: W. H. Freeman.

Dodson, C. H. 1970. The role of chemical attractants in orchid pollination. In K. L. Chambers (ed.), *Biochemical Coevolution*. Corvalis, Ore.: Oregon State University Press.

Dodson, C. H. 1975. Coevolution of orchids and bees. In L. E. Grant and P. H. Raven (eds.), *Coevolution of Animals and Plants*. Austin: University of Texas Press.

Dodson, C. H., R. L. Dressler, H. G. Hills, R. M. Adams, and N. H. Williams. 1969. Biologically active compounds in orchid fragrances. *Science* 164: 1243–49.

Donald, C. M. 1961. *Competition for Light in Crops and Pastures*. Society for Experimental Biology, Symposium XV. New York: Academic Press.

Dorf, E. 1969. Paleobotanical evidence of mesozoic and cenozoic climatic changes. *Proc. N. Amer. Paleontol. Conv.*, pp. 323–46.

Downton, W. J. S. 1975. The occurrence of C_4 photosynthesis among plants. *Photosynthetica* 9: 96–105.

Duncan, W. C., R. S. Loomis, W. A. Williams, and R. Hanau. 1967. A model for simulating photosynthesis in plant communities. *Hilgardia* 38: 181–205.

Dunn, E. L. 1975. Environmental stresses and inherent limitations affecting CO_2 exchange in evergreen sclerophylls in Mediterranean climates. In D. M. Gates and R. B. Schmerl (eds.), *Perspectives of Biophysical Ecology*. New York: Springer-Verlag.

Dustin, A. P. 1934. Action de la colchicine sur le sarcome greffé, type Crocker, de la souris. *Bull. Acad. Roy. Med. Belgique* 14: 487–88.

Dustin, A. P., L. J. Havas, and F. Lits. 1937. Action de la colchicine sur les divisions cellulaires chez les végétaux. *C. R. Assoc. Anatomistes* (Marseilles): 1–5.

East, E. M. 1916. Studies in size inheritance in *Nicotiana*. *Genetics* 1: 164–76.

East, E. M. 1935a. Genetic reactions in *Nicotiana*. II. Phenotypic reaction patterns. *Genetics* 20: 414–42.

East, E. M. 1935b. Genetic reactions in *Nicotiana*. III. Dominance. *Genetics* 20: 443–51.

Edwards, J. H. and S. A. Barber. 1976. Phosphorus uptake rate of soybean roots as influenced by plant age, root trimming, and solution P concentration. *Agron. J.* 68: 973–75.

Edwards, N. T. and W. F. Harris. 1977. Carbon cycling in a mixed deciduous forest floor. *Ecology* 58: 431–37.

Effer, W. and S. Ranson. 1967. Some effects of oxygen concentration on levels of respiratory intermediates in buckwheat seedlings. *Plant Physiol.* 42: 1053–58.

Ehleringer, J. 1976. Leaf absorptance and photosynthesis as affected by pubescence in the genus *Encelia*. *Carnegie Inst. Washington Yearb.* 75: 413–18.

Ehleringer, J. 1978. Implications of quantum yield differences on the distributions of C_3 and C_4 grasses. *Oecologia* 31: 255–67.

Ehleringer, J. and O. Björkman. 1977. Quantum yields for CO_2 uptake in C_3 and C_4 plants. *Plant Physiol.* 59: 86–90.

Ehleringer, J., and P. C. Miller. 1975a. A simulation model of plant water relations and production in the alpine tundra, Colorado. *Oecologia* 19: 177–93.

Ehleringer, J., and P. C. Miller. 1975b. Water relations of selected plant species in the alpine tundra, Colorado. *Ecology* 56: 370–80.

Ehleringer, J., O. Björkman, and H. A. Mooney. 1976. Leaf pubescence: Effects on absorptance and photosynthesis in a desert shrub. *Science* 192: 376–77.

Ehrendorfer, F. 1959. Differentiation–hybridization cycles and polyploidy in *Achillea*. *Cold Spring Harbor Symp. Quant. Biol.* 24: 141–52.

Ehrendorfer, F. 1965. Dispersal mechanisms, genetic systems, and colonizing ability in some flowering plant families. In H. G. Baker and G. L. Stebbins (eds.), *The Genetics of Colonizing Species*. New York and London: Academic Press.

Ehrendorfer, F. 1973. Adaptive significance of major taxonomic characters and morphological trends in angiosperms. In V. H. Heywood (ed.), *Taxonomy and Ecology*. London and New York: Academic Press.

Ehrendorfer, F. 1976. Concluding remarks. *Plant Syst. Evol.* 125: 189–94.

Ehrlich, P. R. 1964. Some axioms of taxonomy. *Syst. Zool.* 13: 109–23.

Ehrlich, P. R. and R. W. Holm. 1962. Patterns and populations. *Science* 137: 652–57.

Ehrlich, P. R. and P. H. Raven. 1965. Butterflies and plants: A study in coevolution. *Evolution* 18: 586–608.

Ehrlich, P. R. and P. H. Raven. 1967. Butterflies and plants. *Sci. Amer.* 216 (6): 104–13.

Ehrlich, P. R. and P. H. Raven. 1969. Differentiation of populations. *Science* 165: 1228–32.

Eigsti, O. J. 1938. A cytological study of colchicine effects in the induction of polyploids in plants. *Proc. Nat. Acad. Sci. USA* 24: 56–63.

Elfving, D. C., M. R. Kaufmann, and A. E. Hall. 1972. Interpreting leaf water potential measurements with a model of the soil–plant atmosphere continuum. *Physiol. Plant.* 27: 161–68.

Emden, H. F. van, ed. 1973. Insect/plant relationships. *Symp. Roy. Entomol. Soc. London* 6: i–viii, 1–213.

Emlen, J. M. 1968. Optimal choice in animals. *Amer. Nat.* 102: 385–90.

Emlen, J. M. 1970. Age-specificity and ecological theory. *Ecology* 57: 588–601.

Emlen, J. M. 1973. *Ecology: An Evolutionary Approach.* Reading, Mass.: Addison-Wesley.

Enama, M. 1976. Genetic variation affecting metabolic phenotypes: An approach to analyzing photosynthetic carbon reduction in a C_4 plant. *Carnegie Inst. Washington Yearb.* 75: 407–9.

Endler, J. A. 1973. Gene flow and population differentiation. *Science* 179: 243–50.

Engler, A. 1913. Einfluss der Provenienz des Samens auf die Eigenschaften der forstlichen Holzgewächse. *Mitt. Schweiz. Zentralanst. forst. Versuchswesen* 10: 191–386.

Engler, A. and K. Prantl. 1931. *Die natürlichen Pflanzenfamilien,* vol. 19a, 2d ed. Leipzig: Wilhelm Engelmann.

Epling, C. and T. Dobzhansky. 1942. Genetics of natural populations. VI. Microgeographic races in *Linanthus parryae.* *Genetics* 27: 317–22.

Epling, C., H. Lewis, and F. Ball. 1960. The breeding group and seed storage: A study in population dynamics. *Evolution* 14: 238–55.

Erickson, R. O. and F. J. Michelini. 1957. The plastochron index. *Amer. J. Bot.* 44: 297–305.

Estabrook, G. A. and A. E. Dunham. 1976. Optimal diet as a function of absolute abundance, relative abundance, and relative value of available prey. *Amer. Nat.* 110: 401–13.

Estes, J. R. and R. W. Thorp. 1975. Pollination ecology of *Pyrrhopappus carolinanus*.   *Amer. J. Bot.* 62: 148–59.

Estes, J. R., R. W. Thorp, and D. L. Briggs. Foraging patterns of the solitary bee *Lasioglossum titusi* on *Agoseris heterophylla*. In prep.

Eurola, S. 1972. Germination of seeds collected in Spitsbergen. *Ann. Bot. Fenn.* 9: 149–59.

Evans, G. C. 1972. *The Quantitative Analysis of Plant Growth*. Berkeley: University of California Press.

Evenari, M. 1961. A survey of the work done in seed physiology by the Department of Botany, Hebrew University, Jerusalem (Israel). *Proc. Int. Seed Test. Assoc.* 26: 597–658.

Evenari, M., L. Shannan, and N. Tadmor. 1971. *The Negev: The Challenge of a Desert*. Cambridge, Mass.: Harvard University Press.

Faegri, K. and L. van der Pijl. 1971. *The Principles of Pollination Ecology*, 2d ed. Oxford: Pergamon Press.

Favarger, C. 1964. Die zytotaxonomische Erforschung der Alpenflora. *Ber. Deutsch. Bot. Ges.* 77: 73–83.

Feldman, M. W. and F. B. Christiansen. 1975. The effect of population subdivision on two loci without selection. *Genet. Res.* 24: 151–62.

Fellows, R. J. and J. S. Boyer. 1976. Structure and activity of chloroplasts of sunflower leaves having various water potentials. *Planta* 132: 229–39.

Felsenstein, J. 1976. The theoretical population genetics of variable selection and migration. *Ann. Rev. Genet.* 10: 253–80.

Feret, P. P. 1974. Genetic difference among three small stands of *Pinus pungens*.   *Theor. Appl. Genet.* 44: 173–77.

Fernald, M. L. 1950. *Gray's Manual of Botany*, 8th ed. Boston: American Book Co.

Fernandez, O. A. and M. M. Caldwell. 1975. Phenology and dynamics of root growth of three cool semi-desert shrubs under field conditions. *J. Ecol.* 63: 703–14.

Finlay, K. W. and G. N. Wilkinson. 1963. The analysis of adaptation in a plant breeding programme. *Aust. J. Agr. Res.* 14: 742–54.

Finnerty, V. and G. B. Johnson. 1978. Post translational modification as a potential explanation of high levels of enzyme polymorphism: Xanthine dehydrogenase and aldehyde oxidase in *Drosophila melanogaster*. *Genetics* (in press).

Fisher, R. A. 1930. *The Genetical Theory of Natural Selection*. Oxford: Clarendon Press.

Fogg, G. E. 1972. *Photosynthesis*. New York: American Elsevier.

Foin, T. C. and S. K. Jain. 1977. Ecosystems analysis and population biology: Lessons for the development of community ecology. *BioScience* 27: 532–38.

Forcier, L. K. 1975. Reproductive strategies and the co-occurrence of climax tree species. *Science* 189: 808–10.

Ford, E. B. 1975. *Ecological Genetics*. New York: Wiley.

Foster, R. 1973. Seasonality of fruit production and seed fall in a tropical forest ecosystem in Panama. Ph.D. dissertation, Duke University, Durham, N.C.

Fowler, D. P. 1964. Effects of inbreeding in red pine (*Pinus resinosa*). *Silvae Genet*. 13: 170–77.

Fowler, D. P. 1965. Effects of inbreeding in red pine: Pollination studies. *Silvae Genet*. 14: 12–22.

Fowler D. P. and D. T. Lester. 1970. Genetics of red pine, *USDA Forest Service Res. Paper* WO-8.

Fowler, D. P. and R. W. Morris. 1977. Genetic diversity in red pine: Evidence for low heterozygosity. *Canad. J. For. Res.* 7: 343–47.

Fox. J. F. 1977. Alternation and coexistence of tree species. *Amer. Nat*. 111: 69–89.

Frankel, R. and E. Galun. 1977. *Pollination Mechanisms, Reproduction and Plant Breeding*. Berlin, Heidelberg, and New York: Springer-Verlag.

Frankie, G. W. 1976. Pollination of widely dispersed trees by animals in Central America, with an emphasis on bee pollination systems. In J. Burley and B. T. Styles (eds.), *Tropical Trees: Variation, Breeding and Conservation*. pp. 151–159; *Linnaean Soc. Symp. Ser*. 2.

Frankie, G. W. and H. G. Baker. 1974. The importance of pollinator behavior in the reproductive biology of tropical trees. *Ann. Inst. Univ. Nac. Auton. México*, 45, Ser. Bot. 1: 10.

Frankie, G. W., P. A. Opler, and K. S. Bawa. 1976. Foraging behavior of solitary bees: Implications for outcrossing of a neotropical forest tree species. *J. Ecol.* 64: 1049–57.

Free, J. B. 1968. Dandelion as a competitor to fruit trees for bee visits. *J. Appl. Ecol.* 5: 169–78.

Free, J. B. 1970. *Insect Pollination of Crops.* New York: Academic Press.

Freeman, C. E. 1973. Germination responses of a Texas population of ocotillo (*Fouquieria splendens* Engelm.) to constant temperature, water stress, pH and salinity. *Amer. Midl. Nat.* 89: 252–56.

Friedman, J. and W. T. Elberse. 1976. Competition between two desert varieties of *Medicago laciniata* (L.) Mill. under controlled conditions. *Oecologia* 22: 321–39.

Friedman, J. and G. Orshan. 1975. The distribution, emergence and survival of seedlings of *Artemisia herba-alba* Asso in the Negev Desert of Israel in relation to distance from the adult plants. *J. Ecol.* 63: 627–32.

Frisch, K. von. 1914. Der Farbensinn und Formensinn der Bienen. *Zool. Jahrb. Abt. Allg. Zool. Physiol.* 35: 1–182.

Frisch, K. von. 1919. Über den Geruchsinn der Bienen und seine blütenbiologische Bedeutung. *Zool. Jahrb. Abt. Allg. Zool. Physiol.* 37: 1–238.

Gaastra, P. 1959. Photosynthesis of crop plants as influenced by light, carbon dioxide, temperature and stomatal diffusion resistance. *Meded. Landbouwk Hogesch. Wageningen* 59: 1–68.

Gadgil, M. 1971. Dispersal: Population consequences and evolution. *Ecology* 52: 253–61.

Gadgil, M. and W. H. Bossert. 1970. Life historical consequences of natural selection. *Amer. Nat.* 104: 1–24.

Gadgil, M. and O. T. Solbrig. 1972. The concept of r and K selection: Evidence from wild flowers and some theoretical considerations. *Amer. Nat.* 106: 14–31.

Gaines, M. S., K. J. Vogt, J. L. Hamrick, and J. Caldwell. 1974. Reproductive strategies and growth patterns in sunflowers (*Helianthus*). *Amer. Nat.* 108: 889–94.

Gajewski, W. 1946. Cytogenetic investigations on *Anemone*. I. *Anemone janczewskii*, a new amphi-diploid species of hybrid origin. *Acta Soc. Bot. Polon.* 17: 129–94.

Ganders, F. R. 1976. Pollen flow in distylous populations of *Amsinckia* (Boraginaceae). *Canad. J. Bot.* 54: 2530–35.

Gardner, W. R. 1960. Dynamic aspects of water availability to plants. *Soil Sci.* 89: 63–73.

Gartside, D. W. and T. McNeilly. 1974. The potential for evolution of heavy metal tolerance in plants. II. Copper tolerance in normal populations of different plant species. *Heredity* 32: 335–48.

Garwood, N. C. 1977. Seasonal constraints on seedling appearance in a tropical moist forest in Panama. *Proceedings of the IV International Symposium of Tropical Ecology.*

Gates, D. M. 1962. *Energy Exchange in the Biosphere.* New York: Harper and Row.

Gates, D. M. 1965. Energy, plants and ecology. *Ecology* 46: 1–13.

Gates, D. M. 1968. Transpiration and leaf temperature. *Ann. Rev. Plant Physiol.* 19: 211–38.

Gates, D. M. and L. E. Papian. 1971. *An Atlas of the Leaf Energy Budgets.* New York: Plenum Press.

Gates, D. M. and R. B. Schmerl, eds. 1975. *Perspectives of Biophysical Ecology.* New York: Springer-Verlag.

Gates, D. M., R. Alderfer, and S. E. Taylor. 1968. Leaf temperature of desert plants. *Science* 159: 994–95.

Gauhl, E. 1968. Differential photosynthetic performance of *Solanum dulcamara* ecotypes from shaded and exposed habitats. *Carnegie Inst. Washington Yearb.* 67: 482–87.

Gaul, H. 1954. Asynapsis und ihre Bedeutung für die Genomanalyse. *Zeitschr. Abst. Vererbungsl.* 86: 69–100.

Gentry, A. H. 1976. Bignoneaceae of southern Central America: Distribution and ecological specificity. *Biotropica* 8: 117–31.

Gibson, R. W. 1971. Glandular hairs providing resistance to aphids in certain wild potato species. *Ann. Appl. Biol.* 68: 113–19.

Giesel, J. T. 1976. Reproductive strategies as adaptations to life in temporally heterogeneous environments. *Ann. Rev. Ecol. Syst.* 7: 57–80.

Gifford, R. M. 1974. A comparison of potential photosynthesis, productivity and yield of plant species with differing photosynthetic metabolism. *Aust. J. Plant Physiol.* 1: 107–17.

Gilbert, L. E. 1971. Butterfly–plant coevolution: Has *Passiflora adenopoda* won the selection race with heliconiine butterflies? *Science* 172: 585–86.

Gilbert, L. E. 1975. Ecological consequences of a coevolved mutualism between butterflies and plants. In L. B. Gilbert and P. H. Raven (eds.), *Coevolution of Animals and Plants.* Austin, Texas: University of Texas Press.

Gilbert, L. E. and P. H. Raven, (eds.) 1975. *Coevolution of Animals and Plants.* Austin, Texas: University of Texas Press.

Gilbert, L. E. and M. C. Singer. 1975. Butterfly ecology. *Ann. Rev. Ecol. Syst.* 6: 365–97.

Giles, K. L., M. F. Beardsell, and D. Cohen. 1974. Cellular and ultrastructural changes in mesophyll and bundle sheath cells of maize in response to water stress. *Plant Physiol.* 54: 208–12.

Giles, K. L., D. Cohen, and M. F. Beardsell. 1976. Effects of water stress on the ultrastructure of leaf cells of *Sorghum bicolor. Plant Physiol.* 57: 11–14.

Gill, F. B. and L. L. Wolf. 1975a. Foraging strategies and energetics of east African sunbirds at mistletoe flowers. *Amer. Nat.* 109: 491–510.

Gill, F. B. and L. L. Wolf. 1975b. Economics of feeding territoriality in the golden-winged sunbird. *Ecology* 56: 331–45.

Gillespie, J. H. 1974. Polymorphism in patchy environments. *Amer. Nat.* 108: 145–51.

Gillespie, J. H. 1976. The role of migration in the genetic structure of populations in temporally and spatially varying environments. II. Island models. *Theor. Popul. Biol.* 10: 112–238.

Givnish, T. J. 1976. Leaf form in relation to environment: A theoretical study. Ph.D. dissertation, Princeton University.

Givnish, T. J. 1978. On the adaptive significance of compound leaves, with particular reference to tropical trees. In P. B. Tomlinson and M. H. Zimmermann (eds.), *Tropical Trees as Living Systems.* Cambridge: Cambridge University Press.

Givnish, T. J. 1978. Ecological aspects of plant morphology: Leaf form in relation to environment. *Quart. Rev. Biol.* (in press).

Givnish, T. J. and G. J. Vermeij. 1976. Sizes and shapes of liane leaves. *Amer. Nat.* 110: 743–76.

Glass, A. D. M. 1976. The allelopathic potential of phenolic acids associated with the rhizosphere of *Pteridium aquilinum. Canad. J. Bot.* 54: 2440–44.

Goldberg, L. D. and P. R. Atsatt. 1975. Frequency of reflection and absorption of ultraviolet light in flowering plants. *Amer. Midl. Nat.* 93: 35–43.

Good, R. 1974. *The Geography of the Flowering Plants,* 4th ed. London: Longmans.

Goodall, D. W. 1970. Statistical plant ecology. *Ann. Rev. Ecol. Syst.* 1: 99–124.

Goodman, D. 1974. Natural selection and a cost ceiling on reproductive effort. *Amer. Nat.* 108: 247–68.

Goodman, P. J., E. M. Braybrook, C. J. Marchant and J. M. Lambert. 1969. *Spartina* × *townsendii* H. & J. Groves *sensu lato*. In Biological Flora of the British Isles, *J. Ecol.* 59: 298–313.

Goodspeed, T. H. 1954. *The Genus Nicotiana*. Waltham, Mass: Chronica Botanica.

Goodwin, T. W. 1973. Carotenoids. In L. P. Miller (ed.), *Phytochemistry*, vol. 1. New York: Van Nostrand Reinhold Co.

Goodwin, T. W. 1976. Distribution of carotenoids. In T. W. Goodwin, (ed.), *Chemistry and Biochemistry of Plant Pigments*, vol. 1, 2d ed. New York: Academic Press.

Gottlieb, L. D. 1971. Evolutionary relationships in the outcrossing diploid annual species of *Stephanomeria* (Compositae). *Evolution* 25: 312–29.

Gottlieb, L. D. 1973a. Genetic differentiation, sympatric speciation, and the origin of a diploid species of *Stephanomeria*. *Amer. J. Bot.* 60: 545–53.

Gottlieb, L. D. 1973b. Enzyme differentiation and phylogeny in *Clarkia franciscana, C. rubicunda* and *C. amoena*. *Evolution* 27: 205–14.

Gottlieb, L. D. 1974a. Genetic stability in a peripheral isolate of *Stephanomeria exigua* ssp. *coronaria* that fluctuates in population size. *Genetics* 76: 551–56.

Gottlieb, L. D. 1974b. Genetic conformation of the origin of *Clarkia lingulata*. *Evolution* 28: 244–50.

Gottlieb, L. D. 1975. Allelic diversity in the outcrossing annual plant *Stephanomeria exigua* ssp. *carotifera* (Compositae). *Evolution* 29: 213–25.

Gottlieb, L. D. 1976a. Biochemical consequences of speciation in plants. In F. J. Ayala (ed.), *Molecular Evolution*. Sunderland, Mass.: Sinauer Assoc.

Gottlieb, L. D. 1976b. Review of "Genetics of Flowering Plants." *Evolution* 30: 192–94.

Gottlieb, L. D. 1977a. Phenotypic variation in *Stephanomeria exigua* ssp. *coronaria* (Compositae) and its recent derivative species "Malheurensis." *Amer. J. Bot.* 64: 873–80.

Gottlieb, L. D. 1977b. Genotypic similarity of large and small individuals in a natural population of the annual plant *Stephanomeria exigua* ssp. *coronaria* (Compositae). *J. Ecol.* 65: 127–34.

Gottlieb, L. D. 1977c. Electrophoretic evidence and plant systematics. *Ann. Mo. Bot. Gard.* (in press).

Gottlieb, L. D. 1978a. *Stephanomeria malheurensis* (Compositae), a new species from Oregon. *Madroño* 25: 44–46.

Gottlieb, L. D. 1978b. Allocation, growth rates, and gas exchange in seedlings of *Stephanomeria exigua* ssp. *coronaria* and its recent derivative S. *malheurensis.* *Amer. J. Bot.* 65 (in press).

Gottlieb, L. D. and G. Pilz. 1976. Genetic similarity between *Gaura longiflora* and its obligately outcrossing derivative *G. demareei.* *Syst. Bot.* 1: 181–87.

Gottsberger, G. 1971. Color changes of petals in *Malvaviscus arboreus.* *Acta Bot. Neerl.* 20: 381–88.

Goudriaan, J., and P. E. Waggoner. 1972. Simulating both aerial microclimate and soil temperature from observations above the foliar canopy. *Neth. J. Agr. Sci.* 20: 104–24.

Grant, K. A. and V. Grant. 1968. *Hummingbirds and Their Flowers.* New York: Columbia University Press.

Grant, M. C. and J. B. Mitton. 1976. Genetic variation associated with morphological variation in Englemann Spruce at tree line. *Genetics* 83: 528.

Grant, P. R. 1972. Convergent and divergent character displacement. *Biol. J. Linn. Soc.* 4: 39–68.

Grant, P. R. 1975. The classical case of character displacement. *Evol. Biol.* 8: 237–337.

Grant, V. 1949. Pollination systems as isolating mechanisms in flowering plants. *Evolution* 3: 82–97.

Grant, V. 1958. The regulation of recombination in plants. *Cold Spring Harbor Symp. Quant. Biol.* 23: 337–63.

Grant, V. 1963. *The Origin of Adaptations.* New York: Columbia University Press.

Grant, V. 1964. *The Architecture of the Germplasm.* New York: John Wiley.

Grant, V. 1971. *Plant Speciation.* New York: Columbia University Press.

Grant, V. 1975. *Genetics of Flowering Plants.* New York: Columbia University Press.

Gray, J. R., D. E. Fairbrothers, and J. A. Quinn. 1973. Biochemical and anatomical population variation in the *Danthonia sericea* complex. *Bot. Gaz.* 134: 166–73.

Greenwood, D. 1967. Studies on the transport of oxygen through the stems and roots of vegetable seedlings. *New Phytol.* 66: 337–47.

Greenwood, D. 1969. Effects of oxygen distribution in the soil on plant growth. In W. Whittington (ed.), *Root Growth*. London: Butterworths.

Gregor, J. W. 1938. Experimental taxonomy. II. Initial population differentiation in *Plantago maritima* L. of Britain. *New Phytol.* 37: 15–49.

Gregor, J. W. 1939. Experimental taxonomy. IV. Population differentiation in North American and European sea plantains allied to *Plantago maritima* L. *New Phytol.* 38: 293–322.

Gregor, J. W., V. McM. Davey, and J. M. S.Lang. 1936. Experimental taxonomy. Experimental garden techique in relation to the recognition of small taxonomic units. *New Phytol.* 35: 323–50.

Greig-Smith, P. 1961. Data on pattern within plant communities. II. *Ammophila arenaria* (L.) Link. *J. Ecol.* 49: 703–48.

Greig-Smith, P. 1964. *Quantitative Plant Ecology*, 2d ed. London: Butterworth.

Grime, J. P. and R. Hunt. 1975. Relative growth-rate: Its range and adaptive significance in a local flora. *J. Ecol.* 63: 393–422.

Grime, J. P. and B. C. Jarvis. 1974. Shade avoidance and shade tolerance in flowering plants. II. Effects of light on the germination of species of contrasted ecology. In R. E. Bainbridge, C. Evans, and O. Rackham (eds.), *Light as an Ecological Factor*, vol. II; *Brit. Ecol. Soc. Symp.* 16: 525–32.

Grover, N. S. and O. R. Byrne. 1975. Genetic control of acid phosphate isozymes in *Arabidopsis thaliana* (L.) Heynh. *Biochem. Genet.* 13: 527–31.

Grubb, P. J. 1976. A theoretical background to the conservation of ecologically distinct groups of annuals and biennials in the chalk grassland ecosystem. *Biol. Cons.* 10: 53–76.

Grubb, P. J. 1977. The maintenance of species-richness in plant communities: The importance of the regeneration niche. *Biol. Rev.* 52: 107–45.

Grubb, P. J., J. R. Lloyd, T. D. Pennington, and T. C. Whitmore. 1963. A comparison of montane and lowland rain forest in Ecuador. I. Forest structure, physiognomy, and floristics. *J. Ecol.* 51: 567–602.

Hackett, C. 1969. Quantitative aspects of cereal root systems. In W. J. Whittington, (ed.), *Root Growth*. London: Butterworths.

Hagerup, O. 1927. *Empetrum hermaphroditicum* (Lge.) Hagerup, a new tetraploid, bisexual species. *Dansk Bot. Arkiv.* 5: 1–17.

Hagerup, O. 1932. Über Polyploidie in Beziehung zu Klima, Ökologie und Phylogenie. *Hereditas* 16: 19–40.

Hainsworth, F. R. and L. L. Wolf. 1976. Nectar characteristics and food selection by hummingbirds. *Oecologia* 25: 101–13.

Hairston, N. G., D. W. Tinkle, and H. M. Wilbur. 1970. Natural selection and the parameters of population growth. *J. Wildlife Manag.* 34: 681–90.

Haldane, J. B. S. 1930. Theoretical genetics of autopolyploids. *J. Genet.* 32: 359–372.

Hall, A. E. and O. Björkman. 1975. Model of leaf photosynthesis and respiration. In D. M. Gates and R. B. Schmerl (eds.), *Perspectives in Biophysical Ecology*. Berlin: Springer-Verlag.

Hall, A. E. and M. R. Kaufmann. 1975. Stomatal response to environment with *Sesamum indicum* L. *Plant Physiol.* 55: 455–59.

Hall, A. E., E. D. Schulze, and O. L. Lange. 1976. Current perspectives of steady-state stomatal responses to environment. In O. L. Lange, L. Kappen, and E. D. Schulze (eds.), *Water and Plant Life*. Berlin: Springer-Verlag.

Halliwell, B. 1974. Superoxide dismutase, catalase, and glutathione peroxidase: Solutions to the problems of living with oxygen. *New Phytol.* 73: 1075–82.

Hamilton, E. L. and P. B. Rowe. 1949. *Rainfall Interception by Chaparral in California*. Calif. Dept. of Nat. Resources, Div. of Forestry.

Hamilton, W. D. and R. M. May. 1977. Dispersal in stable habitats. *Nature* (London) 269: 578–81.

Hamrick, J. L. 1976. Variation and selection in western montane species. II. Variation within and between populations of white fir on an elevational transect. *Theor. Appl. Genet.* 47: 27–34.

Hamrick, J. L. and R. W. Allard. 1972. Microgeographical variation in allozyme frequencies in *Avena barbata*. *Proc. Nat. Acad. Sci. USA* 69: 2100–104.

Hamrick, J. L. and R. W. Allard. 1975. Correlations between quantitative characters and enzyme genotypes in *Avena barbata*. *Evolution* 29: 438–42.

Hamrick, J. L. and L. R. Holden. 1979. The influence of microhabitat heterogeneity on gene frequency distribution and gametic phase disequilibrium in *Avena barbata*. *Evolution* (in press).

Hansen, G. K. 1974. Resistance to water transport in soil and young wheat plants. *Acta Agr. Scand.* 24: 37–48.

Hansen, G. K. and C. R. Jensen. 1977. Growth and maintenance respiration in whole plants, tops, and roots of *Lolium multiflorum*. *Physiol. Plant.* 39: 155–64.

Harberd, D. 1961. Observations on population structure and longevity of *Festuca rubra* L. *New Phytol.* 60: 184–206.

Harborne, J. B. 1967. *Comparative Biochemistry of the Flavonoids.* New York: Academic Press.

Harborne, J. B., ed. 1972. *Phytochemical Ecology.* London and New York: Academic Press.

Harding, J. and C. B. Mankinen. 1972. Genetics of *Lupinus*. IV. Colonization and genetic variability in *Lupinus succulentus*. *Theor. Appl. Genet.* 42: 267–71.

Hare, F. K. 1972. The observed annual water balance over North America south of 60°N. and inferred convective heat exchange. *Publ. Climatol. Lab. Climatol.* 25(3): 7–15.

Harlan, H. V. and M. L. Martini. 1938. The effect of natural selection on a mixture of barley varieties. *J. Agr. Res.* 57: 189–99.

Harlan, J. R. and J. M. J. de Wet. 1975. On O. Winge and a prayer: The origins of polyploidy. *Bot. Rev.* 41: 361–90.

Harland, S. C. 1939. Genetical studies on the genus *Gossypium* and their relationship to evolutionary and taxonomic problems. *Proc. Seventh Int. Congr. Genetics:* 138–44.

Harper, J. L. 1957. The ecological significance of dormancy and its importance in weed control. *Proc. 4th Int. Congr. Crop Protection*, Hamburg; Braunschweig Hamburg 1957, vol. I, pp. 415–20.

Harper, J. L. 1965. Establishment, aggression and cohabitation in weedy species. In H. G. Baker and G. L. Stebbins (eds.), *The Genetics of Colonizing Species.* New York: Academic Press.

Harper, J. L. 1967. A Darwinian approach to plant ecology. *J. Ecol.* 55: 247–70.

Harper, J. L. 1977. *Population Biology of Plants.* London: Academic Press.

Harper, J. L. and R. A. Benton. 1966. The behavior of seeds in soil. II. The germination of seeds on the surface of a water supplying substrate. *J. Ecol.* 54: 151–66.

Harper, J. L. and A. P. Chancellor. 1959. The comparative biology of closely related species living in the same area. IV. *Rumex*: Interference between individuals in populations of one and two species. *J. Ecol.* 47: 679–95.

Harper, J. L. and J. N. Clatworthy. 1963. The comparative biology of closely related species. VI. Analysis of the growth of *Trifolium repens* and *T. fragiferum* in pure and mixed populations. *J. Expt. Bot.* 14: 172–90.

Harper, J. L. and I. H. McNaughton. 1962. The comparative biology of closely related species living in the same area. VII. Interference between individuals in pure and mixed populations of *Papaver* species. *New Phytol.* 61: 175–88.

Harper, J. L. and J. Ogden. 1970. The reproductive strategy of higher plants. I. The concept of strategy with special reference to *Senecio vulgaris* L. *Ecology* 58: 681–98.

Harper, J. L. and J. White. 1971. The dynamics of plant populations. In P. J. den Boer and G. R. Gradwell (eds.), *Dynamics of Population Numbers.* Wageningen: Centre for Agricultural Publication and Documentation.

Harper, J. L. and J. White. 1974. The demography of plants. *Ann. Rev. Ecol. Syst.* 5: 419–63.

Harper, J. L., J. N. Clatworthy, I. H. McNaughton, and G. R. Sagar. 1961. The evolution of closely related species living in the same area. *Evolution* 15: 209–27.

Harper, J. L., J. T. Williams, and G. R. Sagar. 1965. The behavior of seeds in soil. I. The heterogeneity of soil surfaces and its role in determining the establishment of plants from seed. *J. Ecol.* 53: 273–86.

Harper, J. L., P. H. Lovell, and K. G. Moore. 1970. The shapes and sizes of seeds. *Ann. Rev. Ecol. Syst.* 1: 327–56.

Harris, H. 1966. Enzyme polymorphism in man. *Proc. Roy. Soc. (London) (B)* 164: 298–310.

Harrison, A. T. 1971. Temperature related effects on photosynthesis in *Heteromeles arbutifolia* M. Roem. Ph.D. dissertation, Stanford University, Palo Alto, California.

Harrison, A. T., E. Small, and H. Mooney. 1971. Drought relationships and distribution of two Mediterranean-climate California plant communities. *Ecology* 52: 869–75.

Hart, R. 1977. Why are biennials so few? *Amer. Nat.* 111: 792–99.

Hartshorn, G. 1975. A matrix model of tree population dynamics. In F. B. Golley and E. Medina (eds.), *Tropical Ecological Systems.* Berlin: Springer-Verlag.

Hartshorn, G. 1978. Tree falls and tropical rain forest dynamics. In P. B. Tomlinson and M. H. Zimmermann (eds.), *Tropical Trees as Living Systems.* Cambridge: Cambridge University Press.

Hasfurther, V. R. and R. D. Burman. 1974. Soil temperature modeling using air temperature as a driving mechanism. *Trans. Am. Soc. Agr. Eng.* 17: 78–81.

Hawthorn, W. R. and P. B. Cavers. 1976. Populations dynamics of the perennial herbs *Plantago major* L. and *P. rugelii* Decne. *J. Ecol.* 64: 511–27.

Hayward, H. E., W. M. Blair, and P. E. Skaling. 1942. Device for measuring entry of water into roots. *Bot. Gaz.* 104: 152–60.

Head, G. C. 1973. Shedding of roots. In T. T. Kozlowski (ed.), *Shedding of Plant Parts.* New York: Academic Press.

Hedberg, O. 1957. Afroalpine vascular plants: A taxonomic version. *Symb. Bot. Upsal.* 15: 1–411.

Hedberg, O., O. Martensson, and S. Rudberg. 1952. Botanical investigations in the Paltsa region of northernmost Sweden. *Bot. Not. Suppl.* 3(2): 1–208.

Hedin, P. A., A. C. Thompson, and R. C. Gueldner. 1976. Cotton plant and insect constituents that control boll weevil behavior and development. In J. W. Wallace and R. L. Mansell (eds.), *Biochemical Interaction between Plants and Insects.* New York and London: Plenum Press.

Hedrick, P. W., M. E. Ginevan, and E. P. Ewing. 1976. Genetic polymorphism in heterogeneous environments. *Ann. Rev. Ecol. Syst.* 7: 1–32.

Heed, W. B., W. T. Starmer, M. Miranda, M. W. Miller, and H. J. Phaff. 1976. An analysis of the yeast flora associated with cactiphilic *Drosophila* and their host plants in the Sonoran Desert and its relation to temperate and tropical associations. *Ecology* 57: 151–60.

Heinrich, B. 1975. Energetics of pollination. *Ann. Rev. Ecol. Syst.* 6: 139–70.

Heinrich, B. 1976a. The foraging specializations of individual bumblebees. *Ecol. Monogr.* 46: 105–28.

Heinrich, B. 1976b. Resource partitioning among some unsocial insects: Bumblebees. *Ecology* 57: 874–89.

Heinrich, B. and P. H. Raven. 1972. Energetics and pollination ecology. *Science* 176: 597–602.

Heiser, C. B. 1949. Natural hybridization with reference to introgression. *Bot. Rev.* 15: 645–87.

Heiser, C. B. 1973. Introgression re-examined. *Bot. Rev.* 39: 347–66.

Heithaus, E. R. 1974. The role of plant–pollinator interactions in determining community structure. *Ann. Mo. Bot. Gard.* 61: 675–91.

Heithaus, E. R., P. A. Opler, and H. G. Baker. 1974. Bat activity and pollination of *Bauhinia pauletia*: Plant–pollinator coevolution. *Ecology* 55: 412–519.

Heithaus, E. R., T. H. Fleming, and P. A. Opler. 1975. Foraging patterns and resource utilization on seven species of bats in a seasonal tropical forest. *Ecology* 56: 841–54.

Henry, J. D. and J. M. A. Swan. 1974. Reconstructing forest history from live and dead plant material—an approach to the study of forest succession in southwest New Hampshire. *Ecology* 55: 772–83.

Herrera, R., T. Merida, N. Stark, and C. F. Jordan. 1978. Direct phosphorus transfer from leaf litter to roots. *Naturwissenschaften* 65: 208–9.

Heslop-Harrison, J. 1964. Forty years of genecology. *Adv. Ecol. Res.* 2: 159–247.

Hett, J. M. 1971. A dynamic analysis of age in sugar maple seedlings. *Ecology* 52: 1071–74.

Hett, J. M. and O. L. Loucks. 1971. Sugar maple (*Acer saccharum* Marsh.) seedling mortality. *J. Ecol.* 59: 507–20.

Heydecker, W., ed. 1972. *Seed Ecology.* University Park, Pa., and London: Pennsylvania State University Press.

Heywood, V. H., ed. 1973. *Taxonomy and Ecology.* London and New York: Academic Press.

Hickey, D. A. and T. McNeilly. 1975. Competition between metal tolerant and normal plant populations. A field experiment on normal soil. *Evolution* 29: 458–64.

Hickman, J. C. 1970. Seasonal course of xylem sap tension. *Ecology* 51: 1052–56.

Hickman, J. C. 1975. Environmental unpredictability and plastic energy allocation strategies in the annual *Polygonum cascadense* (Polygonaceae). *J. Ecol.* 63: 689–701.

Hickman, J. C. 1976. Non-forest vegetation of the central Western Cascade Mountains of Oregon. *Northwest Sci.* 50: 145–55.

Hickman, J. C. 1977. Energy allocation and niche differentiation of four co-existing annual species of *Polygonum* in western North America. *J. Ecol* 65: 317–26.

Hiebert, R. D. 1977. The population biology of bristlecone pine (*Pinus longaeva*) in the eastern Great Basin. Ph.D. dissertation, University of Kansas.

Hiesey, W. M. and M. A. Nobs. 1970. Genetic and transplant studies on contrasting species and ecological races of the *Achillea millefolium* complex. *Bot. Gaz.* 131: 245–59.

Hills, N. G., N. H. Williams, and C. H. Dodson. 1972. Floral fragrances and isolating mechanisms in the genus *Catasetum* (Orchidaceae). *Biotropica* 4: 61–76.

Hirschfield, M. F. and D. W. Tinkle. 1975. Natural selection and the evolution of reproductive effort. *Proc. Nat. Acad. Sci. USA* 72: 2227–31.

Hitchcock, A. S. and A. Chase. 1971. *Manual of the Grasses of the United States*, 2d ed. New York: Dover Press.

Hochachka, P. W. and G. N. Somero. 1973. *Strategies of Biochemical Adaptation*. Philadelphia: W. B. Saunders.

Hocking, B. 1968. Insect–flower associations in the high Arctic with special reference to nectar. *Oikos* 19: 359–88.

Holdsworth, M. 1972. Phytochrome and seed germination. *New Phytol.* 71: 105.

Holler, L. C. and W. G. Abrahamson. 1977. Seed and vegetative reproduction in relation to density in *Fragaria virginiana* (Rosaceae). *Amer. J. Bot.* 64: 1003–7.

Hollingshead, L. and E. B. Babcock. 1930. Chromosomes and phylogeny in *Crepis*. *Univ. Calif. Publ. Agr. Sci.* 6: 1–53.

Holmgren, P., P. G. Jarvis, and M. S. Jarvis. 1965. Resistances to carbon dioxide and water vapor transfer in leaves of different plant species. *Physiol. Plant.* 18: 557–73.

Holt, B. R. 1972. Effect of arrival time on recruitment, mortality, and reproduction in successional plant populations. *Ecology* 53: 668–73.

Honert, T. H. van den. 1948. Water transport in plants as a catenary process. *Disc. Faraday Soc.* 3: 146–53.

Horn, H. S. 1971. *The Adaptive Geometry of Trees*. Monographs on Population Biology, 3. Princeton: Princeton University Press.

Horn, H. S. 1975. Forest succession. *Sci. Amer.* 232(5): 90–98.

Horn, H. S. and R. H. Mac Arthur. 1972. Competition among fugitive species in a harlequin environment. *Ecology* 53: 749–52.

Horovitz, A. 1976. Edaphic factors and flower colour distribution in the Anemoneae (Ranunculaceae). *Plant Syst. Evol.* 126: 239–42.

Horovitz, A. and Y. Cohen. 1972. Ultraviolet reflectance characteristics in flowers of crucifers. *Amer. J. Bot.* 59: 706–13.

Howard, R. A. 1969. The ecology of an elfin forest in Puerto Rico. 8. Studies of stem growth and form and of leaf structure *J. Arnold Arb.* 50: 225–26.

Howland, H. C. 1962. Structural, hydraulic, and "economic" aspects of leaf venation and shape. In E. E. Bernard, and M. R. Kare (eds.),

Biological Prototypes and Synthetic Systems, vol. I. Ithaca, N.Y.: Cornell University Press.

Hsiao, T. C. 1973. Plant responses to water stress. *Ann. Rev. Plant Physiol.* 24: 519–70.

Hu, S. 1967. The evolution and distribution of the species of Aquifoliaceae in the Pacific area. I. *Jap. J. Bot.* 42: 13–27.

Hubby, J. L. and R. C. Lewontin. 1966. A molecular approach to the study of genic heterozygosity in natural populations. I. The number of alleles at different loci in *Drosophila pseudoobscura. Genetics* 54: 577–94.

Huck, M. G., B. Klepper, and H. M. Taylor. 1970. Diurnal variations in root diameter. *Plant Physiol.* 45: 529–30.

Huether, C. A., Jr. 1966. The extent of variability for a canalized character (corolla lobe number) in natural populations of *Linanthus* (Benth.). Ph.D. dissertation, University of California, Davis.

Huether, C. A., Jr. 1968. Exposure of natural genetic variability underlying the pentamerous corolla constance in *Linanthus androsaceus* ssp. *androsaceus. Genetics* 60: 123–46.

Huether, C. A., Jr. 1969. Constancy of the pentamerous corolla phenotype in natural populations of *Linanthus. Evolution* 23: 572–88.

Huffaker, C. B. 1971. The phenomenon of predation and its roles in nature. In P. J. den Boer and G. R. Gradwell (eds.), *Dynamics of Population Numbers.* Wageningen: Centre for Agricultural Publication and Documentation.

Huffaker, R. C., T. Radin, G. E. Kleinkopf, and E. L. Cox. 1970. Effects of mild water stress on enzymes of nitrate assimilation and of the carboxylative phase of photosynthesis in barley. *Crop Sci.* 10: 471–76.

Hunt, R. 1975. Further observations on root/shoot equilibria in perennial ryegrass (*Lolium perenne* L.). *Ann. Bot.* 39: 745–55.

Hurlbert, S. H. 1970. Flower number, flowering time, and reproductive isolation among ten species of *Solidago* (Compositae). *Bull. Torrey Bot. Club* 97: 189–95.

Hutchinson, G. E. 1959. Homage to Santa Rosalia or why are there so many kinds of animals? *Amer. Nat.* 93: 145–59.

Idso, S. B. and C. T. de Wit. 1970. Light relations in plant canopies. *Appl. Opt.* 9: 177–84.

Ingram, V. 1963. *The Hemoglobins in Genetics and Evolution.* New York: Columbia University Press.

Inouye, D. W. 1976. Resource partitioning and community structures: A study of bumblebees in the Colorado Rocky Mountains. Ph.D. dissertation, University of North Carolina, Chapel Hill.

Jackson, R. C. 1976. Evolution and systematic significance of polyploidy. Ann. Rev. Ecol. Syst. 7: 209–34.

Jain, S. K. 1969. Comparative ecogenetics of two Avena species occurring in central California. Evol. Biol. 3: 73–118.

Jain, S. K. 1976a. Patterns of survival and microevolution in plant populations. In S. Karlin and E. Nevo (eds.), Population Genetics and Ecology. New York: Academic Press.

Jain, S. K. 1976b. Evolution of inbreeding plants. Ann. Rev. Ecol. Syst. 7: 469–95.

Jain, S. K. 1976c. Evolutionary studies in the meadowfoam genus Limnanthes: An overview. In S. K. Jain, (ed.), Vernal Pools: Their Ecology and Conservation. Davis, Calif.: Institute of Ecology Publ. 9, pp. 50–57.

Jain, S. K. and A. D. Bradshaw. 1966. Evolutionary divergence among adjacent plant populations. I. Evidence and its theoretical analysis. Heredity 21: 407–41.

Jain, S. K. and D. R. Marshall. 1967. Populations studies in predominantly self-pollinating species. X. Variation in natural populations of Avena fatua and A. barbata. Amer. Nat. 101: 19–33.

Jain, S. K. and K. N. Rai. 1974. Population biology of Avena. IV. Polymorphism in small populations of Avena fatua. Theor. Appl. Genet. 44: 7–11.

Jalloq, M. C. 1975. The invasion of molehills by weeds as a possible factor in the degeneration of reseeded pasture. I. The buried viable seed population of molehills from four reseeded pastures in West Wales. J. Appl. Ecol. 12: 643–57.

Jameson, D. A. 1963. Responses of individual plants to harvesting. Bot. Rev. 29: 532–94.

Janssen, J. G. M. 1973. Effects of light, temperature and seed age on the germination of the winter annuals Veronica arvensis L. and Myosotis ramosissima Rochel ex. Schult. Oecologia 12: 141–46.

Janzen, D. H. 1967. Synchronization of sexual reproduction of trees within the dry season in Central America. Evolution 21: 620–37.

Janzen, D. H. 1969. Seed-eaters versus seed size, number, toxicity and dispersal. Evolution 23: 1–27.

Janzen, D. H. 1970. Herbivores and the number of tree species in tropical forests. Amer. Nat. 104: 501–28.

Janzen, D. H. 1971a. Seed predation by animals. *Ann. Rev. Ecol. Syst.* 2: 465–92.

Janzen, D. H. 1971b. Euglossine bees as long-distance pollinators of tropical plants. *Science* 171: 203–5.

Janzen, D. H. 1972a. Association of a rainforest palm and seedeating beetles in Puerto Rico. *Ecology* 53: 258–61.

Janzen, D. H. 1972b. Escape in space by *Sterculia apetala* seeds from the bug *Dysdercus fasciatus* in a Costa Rican deciduous forest. *Ecology* 53: 350–61.

Janzen, D. H. 1972c. Interfield and interplant spacing in tropical insect control. *Proc. Ann. Tall Timbers Conf. Ecol., Animal Control by Habitat Management,* pp. 1–6.

Janzen, D. H. 1974. Tropical black water rivers animals, and insect fruiting by the Dipterocarpaceae. *Biotropica* 6: 69–103.

Janzen, D. H. 1976. Why bamboo wait so long to flower. *Ann. Rev. Ecol. Syst.* 7: 347–91.

Janzen, D. H. 1977a. A note on optimal mate selection by plants. *Amer. Nat.* 111: 365–71.

Janzen, D. H. 1977b. What are dandelions and aphids? *Amer. Nat.* 111: 586–89.

Jarvis, P. G. and M. S. Jarvis. 1963a. The water relations of tree seedlings. II. Transpiration in relation to soil water potential. *Physiol. Plant.* 16: 236–53.

Jarvis, P. G. and M. S. Jarvis. 1963b. The water relations of tree seedlings. IV. Some aspects of the tissue water relations and drought resistance. *Physiol. Plant.* 16: 501–16.

Jaynes, R. A. 1968. Breaking seed dormancy of *Kalmia hirsuta* with high temperatures. *Ecology* 49: 1196–98.

Jeffrey, C. 1966. Notes on Compositae: The Cichorieae in East Tropical Africa. *Kew Bull.* 18: 427–36.

Jennings, P. R. and R. C. Aquino. 1968. Studies on competition in rice. III. The mechanism of competition among phenotypes. *Evolution* 22: 529–42.

Jensen, R. D., S. A. Taylor, and H. H. Wiebe. 1961. Negative transport and resistance to water flow through plants. *Plant Physiol.* 36: 633–38.

Jermy, T., ed. 1976. *The Host-plant in Relation to Insect Behavior and Reproduction.* New York and London: Plenum Press.

Johnson, B. L. and M. M. Thein. 1970. Assessment of evolutionary affinities in *Gossypium* by protein electrophoresis. *Amer. J. Bot.* 57: 1081–92.

Johnson, D. A. and M. M. Caldwell. 1975. Gas exchange of four arctic and alpine tundra plant species in relation to atmospheric and soil moisture stress. *Oecologia* 21: 93–108.

Johnson, D. A. and L. L. Tieszen. 1976. Aboveground biomass allocation, leaf growth, and photosynthesis patterns in tundra plant forms in arctic Alaska. *Oecologia* 24: 159–73.

Johnson, F. and H. Schaffer. 1973. Isozyme variability in species of the genus *Drosophila*. VII. Genotype–environment relationships in population of *D. melanogaster* from the eastern U.S. *Biochem. Genet.* 10: 149–63.

Johnson, F., H. Shaffer, J. Gillaspy, and E. Rockwood. 1969. Isozyme genotype–environment relationships in natural populations of the harvester ant, *Pogonomyrmex barbatus*, from Texas. *Biochem. Genet.* 3: 429–50.

Johnson, G. B. 1973. Enzyme polymorphism and biosystematics: The hypothesis of selective neutrality. *Ann. Rev. Ecol. Syst.* 4: 93–116.

Johnson, G. B. 1974. Enzyme polymorphism and metabolism. *Science* 184: 28–37.

Johnson, G. B. 1975a. The use of internal standards in electrophoretic surveys of enzyme polymorphism. *Biochem. Genet.* 13: 833–47.

Johnson, G. B. 1975b. Enzyme polymorphism and adaptation. *Stadler Symposium* 7, pp. 1–29. Columbia: University of Missouri Press.

Johnson, G. B. 1976a. Polymorphism and predictability at the α-GPdH locus in *Colias* butterflies: Gradients in allele frequencies within a single population. *Biochem. Genet.* 14: 403–26.

Johnson, G. B. 1976b. Enzyme polymorphism and adaptation in alpine butterflies. *Ann. Mo. Bot. Gard.* 63: 248–61.

Johnson, G. B. 1976c. Hidden alleles at the alpha-glycerophosphate dehydrogenase locus in *Colias* butterflies. *Genetics* 83: 149–67.

Johnson, G. B. 1976d. Genetic polymorphism and enzyme function. In F. J. Ayala (ed.), *Molecular Evolution*. Sunderland, Mass.: Sinauer Assoc.

Johnson, G. B. 1977a. Characterization of electrophoretically cryptic variation in the alpine butterfly *Colias meadii*. *Biochem. Genet.* 15: 665–93.

Johnson, G. B. 1977b. Assessing electrophoretic similarity: The problem of hidden heterogeneity. *Ann. Rev. Ecol. Syst.* 8: 309–28.

Johnson, G. B. 1978. Improving the resolution of polyacrylamide gel electrophoresis by varying the degree of crosslinking. *Biochem. Genet.* (in press).

Johnson, G. B and D. Hartl. 1978. Post translational modification in natural populations of *Drosophila melanogaster*. *Genetics*. (in press).

Johnson, H. B. 1975. Gas exchange strategies in desert plants. In D. M. Gates and R. B. Schmerl, *Perspectives of Biophysical Ecology*. New York: Springer-Verlag.

Johnson, L. K. and S. P. Hubbell. 1975. Contrasting foraging strategies and coexistence of two bee species in a single resource. *Ecology* 56: 1398–1406.

Johnson, R. R., N. M. Frey, and D. N. Moss. 1974. Effect of water stress on photosynthesis and transpiration of flag leaves and spikes of barley and wheat. *Crop Sci.* 14: 728–31.

Jones, D. A. 1966. On the polymorphism of cyanogenesis in *Lotus corniculatus*. I. Selection by animals. *Canad. J. Genet. Cytol.* 8: 556–67.

Jones, D. A. 1973. Co-evolution and cyanogenesis. In V. H. Heywood (ed.), *Taxonomy and Ecology*. London and New York: Academic Press.

Jones, H. G. 1973a. Limiting factors in photosynthesis. *New Phytol.* 72: 1089–94.

Jones, H. G. 1973b. Moderate term water stress and associated changes in some photosynthetic parameters in cotton. *New Phytol.* 72: 1095–1105.

Jones, H. G. 1977. Transpiration in barley lines with differing stomatal frequencies. *J. Exp. Bot.* 28: 162–68.

Jones, K. and P. E. Brandham, eds. 1976. *Current Chromosome Research*. Amsterdam, New York, Oxford: North-Holland Publ.

Jorgenson, C. A. 1928. The experimental formation of heteroploid plants in the genus *Solanum*. *J. Genet.* 19: 133–271.

Juhren, M., F. W. Went, and E. Phillips. 1956. Ecology of desert plants. IV. Combined field and laboratory work on germination of annuals in the Joshua Tree National Monument, California. *Ecology* 37: 318–30.

Kahler, A. L. and R. W. Allard. 1970. Genetics of isozyme variants in barley. I. Esterases. *Crop Sci.* 10: 444–48.

Kahler, A. L. and R. W. Allard. 1978. Worldwide patterns of variation of esterase allozymes in barley (Hordeum vulgare L. and H.
spontaneum) (in press).

Kannenberg, L. W. and R. W. Allard. 1967. Population studies in
predominantly self-pollinated species. VIII. Genetic variability in
the Festuca microstachys complex. Evolution 21: 227–40.

Kappert, H. 1954. Experimentelle Untersuchungen über die Variabilität eines Totalapomikten. Ber. Deutsch. Bot. Ges. 67: 325–34.

Karpechenko, G. D. 1927. Polyploid hybrids of Raphanus sativus L.
× Brassica oleracea L. Bull. Appl. Bot. Genet., Pl. Breed. 17:
305–410.

Katz, P. O. 1974. A long-term approach to foraging optimization.
Amer. Nat. 108: 758–82.

Kaufmann, M. R. 1975. Leaf water stress in Englemann spruce. Plant
Physiol. 56: 841–44.

Kaufmann, M. R. and A. E. Hall. 1974. Plant water balance: Its
relationship to atmospheric and edaphic conditions. Agr. Meteorol. 14: 85–98.

Kausch, W. and H. Ehrig. 1959. Beziehungen zwischen Transpiration und Wurzelwerk. Planta 53: 434–48.

Kawano, S. 1970. Species problems viewed from productive and
reproductive biology. I. Ecological life histories of some representative members associated with temperate deciduous forests in
Japan. J. College Liberal Arts Toyama Univ. Japan 3: 181–213.

Kawano, S. 1975. The productive and reproductive biology of flowering plants. II. The concept of life history strategy in plants. J.
College Liberal Arts Toyama Univ. Japan 8: 51–86.

Kawano, S. and Y. Nagai. 1975. The productive and reproductive
biology of flowering plants. I. Life history strategies of three
Allium species in Japan. Bot Mag. Tokyo 88: 281–318.

Kay, Q. O. N. 1976. Preferential pollination of yellow-flowered
morphs of Raphanus raphanistrum by Pieris and Eristalis species. Nature (London) 261: 230–32.

Keast, A. 1970. Adaptive evolution and shifts in niche occupation in
island birds. Biotropica 2: 61–75.

Keck, R. W. and J. S. Boyer. 1974. Chloroplast response to low leaf
water potenials. III. Differing inhibition of electron transport and
photophosphorylation. Plant Physiol. 53: 474–79.

Keeler, K. H. 1975. Electrophoretic variation in vernal pool Veronica
peregrina. Ph.D. dissertation, University of California, Berkeley.

Keller, E. C., R. H. T. Mattoni, and M. S. B. Seiger. 1966. Preferential return of artificially displaced butterflies. *Ann. Behav.* 14: 197–200.

Kelly, G. and J. Turner. 1968a. Inhibition of pea-seed phosphofructokinase by phosphoenolpyruvate. *Biochem. Biophys. Res. Commun.* 30: 195–99.

Kelly, G. and J. Turner. 1968b. The regulation of pea-seed phosphofructokinase by phosphoenolpyruvate. *Biochem. J.* 115: 481–87.

Kelly, G., E. Latzo, and M. Gibbs. 1976. Regulatory aspects of photosynthetic carbon metabolism. *Ann. Rev. Plant Phys.* 27: 181–205.

Kemperman, J. A. and B. V. Barnes. 1976. Clone size of American aspens. *Canad. J. Bot.* 54: 2603–7.

Kendall, W. A. and R. T. Sherwood. 1975. Palatability of leaves of tall fescue and reed canarygrass and some of their alkaloids to meadow voles. *Agron. J.* 67: 667–71.

Kershaw, K. A. 1958. An investigation of the structure of a grassland community. I. Pattern of *Agrostis tenuis*. *J. Ecol.* 46: 571–92.

Kershaw, K. A. 1973. *Quantitative and Dynamic Plant Ecology*. 2d ed. London: Edward Arnold.

Keulen, H. van, N. G. Seligman, and J. Goudriaan. 1975. Availability of anions in the growth medium of roots of an actively growing plant. *Neth. J. Agr. Sci.* 23: 131–38.

Kevan, P. G. 1972. Floral colors in the high Arctic with reference to insect flower relations and pollination. *Canad. J. Bot.* 50: 2289–2316.

Kihara, H. 1940. Verwandtschaft der *Aegilops*-Arten im Lichte der Genomanalyse. Ein Überlick. *Der Züchter* 12: 49–62.

Kihara, H. and I. Nishiyama. 1930. Genomanalyse bei *Triticum* und *Aegilops*. I. Genomaffinitäte in tri- tetra- und pentaploiden Weizenbastarde. *Cytologia* 1: 263–84.

Kihara, H. and T. Ono. 1926. Chromosomenzahlen und systematische Gruppierung der *Rumex*-Arten. *Zeitschr. Zellf. Mikr. Anat.* 4: 475–81.

Kimura, M. and P. F. Crow. 1963. The measurement of effective population number. *Evolution* 17: 279–88.

Kimura, M. and T. Maruyama. 1971. Pattern of neutral polymorphism in a geographically structured population. *Genet. Res.* 18: 125–31.

Kimura, M. and T. Ohta. 1971. *Theoretical Aspects of Population Genetics*. Princeton, N.J.: Princeton University Press.

Kimura, M. and G. Weiss. 1964. The stepping-stone model of population structure and the decrease of genetic correlation with distance. Genetics 49: 561–76.

Kinerson, R. S., C. W. Ralston, and C. G. Wells. 1977. Carbon cycling in a loblolly pine plantation. Oecologia 29: 1–10.

King, C. E. and W. W. Anderson. 1971. Age-specific selection. II. The interaction between r and K during population growth. Amer. Nat. 105: 137–56.

King, C. E., E. E. Gallaher, D. A. Levin. 1975. Equilibrium diversity in plant–pollinator systems. J. Theor. Biol. 53: 263–75.

King, M. and A. Wilson. 1975. Evolution at two levels in humans and chimpanzees. Science 188: 107–16.

King, T. J. 1975. Inhibition of seed germination under leaf canopies in Arenaria serpyllifolia, Vernonica arvensis and Cerastium holosteoides. New. Phytol. 75: 87–90.

King, T. J. 1977a. The plant ecology of ant-hills in calcareous grasslands. I. Patterns of species in relation to ant-hills in southern England. J. Ecol. 65: 235–56.

King, T. J. 1977b. The plant ecology of ant-hills in calcareous grasslands. II. Succession on the mounds. J. Ecol. 65: 257–78.

King, T. J. 1977c. The plant ecology of ant-hills in calcareous grasslands. III. Factors affecting the population sizes of selected species. J. Ecol. 65: 279–316.

Kivillaan, A. and R. S. Bandurski. 1973. The ninety-year period for Dr. Beal's seed viability experiment. Amer. J. Bot. 60: 140–45.

Knuth, P. 1906–9. Handbook of Flower Pollination. 3 vols. Oxford: Oxford Univ. Press.

Kobr, M. and H. Beevers. 1970. Gluconeogenesis in the castor bean endosperm. I. Changes in glycolytic intermediates. Plant Physiol. 47: 48–52.

Koehn, R., J. Perez, and R. Merritt. 1971. Esterase enzyme function and genetical structure of populations of the freshwater fish N. stramingus. Amer. Nat. 105: 51.

Koehn, R., F. Turano, and J. Mitton. 1973. Population genetics of marine pelecypods. II. Genetic zonation of Modiolus demissus. Evolution 27: 100–105.

Koeppe, D. E., R. J. Miller, and D. T. Bell. 1973. Drought-affected mitochondrial processes as related to tissue and whole-plant responses. Agron. J. 65: 566–69.

Kojima, K., P. Smouse, S. Yang, P. Nair, and D. Brncic. 1972. Isozyme frequency patterns in *Drosophila pavani* associated with geographical and seasonal variables. *Genetics* 72: 721–31.

Koller, D. 1956. Germination regulation mechanisms in some desert seeds. III. *Calligonum comosum* L'Her. *Ecology* 37: 430–33.

Koller, D. 1972. Environmental control of seed germination. In T. T. Kozlowski (ed.), *Seed Biology*, vol. II. New York: Academic Press.

Koller, D. and M. Negbi. 1959. The regulation of germination in *Oryzopsis miliacea*. *Ecology* 40: 20–36.

Kormondy, E. J. 1969. *Concepts of Ecology*. Englewood Cliffs, N.J.: Prentice-Hall.

Kozlowski, T. T., ed. 1972. *Seed Biology*, vols. 1–3, New York: Academic Press.

Kozlowski, T. T. 1976. Water supply and leaf shedding. In T. T. Kozlowski (ed.), *Water Deficits and Plant Growth:* vol. 4: *Soil Water Measurement, Plant Responses, and Breeding for Drought Resistance*. New York: Academic Press.

Kramer, P. J. 1969. *Plant and Soil Relationships: A Modern Synthesis*. New York: McGraw–Hill.

Kramer, P. J. and H. C. Bullock. 1966. Seasonal variations in the proportions of suberized and unsuberized roots of trees in relation to the absorption of water. *Amer. J. Bot.* 53: 200–204.

Krause, D. 1975. Xeromorphic structure and soil moisture in the chaparral. M.S. thesis, San Diego State University.

Krause, D. and J. Kummerow. 1977. Xeromorphic structure and soil moisture in the chaparral. *Oecol. Plant.* 12: 133–48.

Krebs, J. R. 1973. Behavioral aspects of predation. In P. P. G. Bateson and P. H. Klopfer (eds.), *Perspectives in Ethology*. New York: Plenum Press.

Krugman, S. L., W. I. Stein, and D. M. Schmitt. 1974. Seed biology. In *Seeds of Woody Plants in the United States*. U.S.D.A. Handbook No. 450.

Kucera, C. L., R. C. Dahlman, and M. R. Koelling. 1967. Total net productivity and turnover on an energy basis for tallgrass prairie. *Ecology* 48: 536–41.

Kugler, H. 1963. UV-Musterung auf Blüten and ihr Zustandekommen. *Planta* 59: 296–329.

Kugler, H. 1966. UV Male auf Blüten. *Ber. Deutsch. Bot. Ges.* 79: 57–70.

Kuhn, T. S. 1970. *The Structure of Scientific Revolutions*. 2d ed. Chicago: University of Chicago Press.

Kullenberg, B. and G. Bergstrom. 1973. The pollination of *Ophrys* orchids. In *Chemistry in Botanical Classification*, Nobel Symposium 25. New York: Academic Press.

Kullenberg, B. and G. Bergstrom. 1976. The pollination of *Ophrys* orchids. *Bot. Not.* 129: 11–19.

Kummerow, J. and K. Fishbeck, eds. 1977. *Chile–California Mediterranean Scrub Atlas*. Part II. Stroudsburg, Pa.: Dowden, Hutchinson and Ross.

Ladiges, P. Y. 1976. Variation in drought resistance in adjacent edaphic populations of *Eucalyptus viminalis* Labill. *Aust. J. Ecol.* 1: 67–76.

Laessle, A. M. 1965. Spacing and competition in natural stands of sand pine. *Ecology* 46: 65–72.

Lakovaara, S. and A. Saura. 1971. Genetic variation in natural populations of *Drosophila obscura*. *Genetics* 69: 377–84.

Lambert, J. R. and F. W. T. Penning de Vries. 1973. Dynamics of water in the soil–plant atmosphere system: A model named Troika. In A. Hadas, D. Schwartzendruber, P. E. Rijtema, M. Fuchs, and B. Yaron (eds.), *Ecological Studies, Analysis and Synthesis*, vol. 4. Berlin: Springer-Verlag.

Langlet, O. 1936. Studien über die physiologische Variabilität der Kiefer und deren Zusammenhang mit dem Klima. Beiträge zur Kenntnis der Ökotypen von *Pinus sylvestris* L. Meddel. Statens Skogforsoksanstalt 29: 219–470.

Langridge, J. 1977.Genetic engineering in plants. *Search* 8: 13–15.

Larcher, W. 1975. *Physiological Plant Ecology*. Berlin: Springer-Verlag.

Law, R., A. D. Bradshaw, and P. D. Putvain. 1977. Life-history variation in *Poa annua*. *Evolution* 31: 233–46.

Lawlor, D. W. 1972. Growth and water use of *Lolium perenne*. I Water transport. *J. Appl. Ecol.* 9: 79–98.

Lawlor, D. W. 1976. Water stress induced changes in photosynthesis, photorespiration, respiration and CO_2 compensation concentration of wheat. *Photosynthetica* 10: 378–87.

Lawlor, D. W. and H. Fock. 1975. Photosynthesis and photorespiratory CO_2 evolution of water-stressed sunflower leaves. *Planta* 126: 247–58.

Lawrence, L., R. Bartschot, E. Zavarin, and J. R. Griffin. 1975. Natural hybridization of *Cupressus sargentii* and *macnabiana* and the composition of the derived essential oils. *Biochem. Syst. Ecol.* 2: 113–19.

Lawrence, W. T. 1975. A radiation model for chaparral canopies. M.S. thesis, San Diego State University.

Leach, G. J. 1968. The growth of the lucerne plant after cutting, the effects of cutting at different stages of maturity and at different intensities. *Aust. J. Agr. Res.* 19: 517–30.

Leach, G. J. 1969. Shoot numbers, shoot size and yield of regrowth in three lucerne cultivars. *Aust. J. Agr. Res.* 20: 425–34.

Leach, G. J. 1970. Growth of the lucerne plant after defoliation. In *Proceedings of the XI International Grassland Congress,* 1970, pp. 562–66.

Ledig, F. T. 1969. A growth model for tree seedlings based on the rate of photosynthesis and distribution of photosynthate. *Photosynthetica* 3: 263–75.

Ledig, F. T., A. P. Drew, and J. G. Clark. 1976. Maintenance and constructive respiration, photosynthesis, and net assimilation rate in seedlings of pitch pine, *Pinus rigida* Mill. *Ann. Bot.* 40: 289–300.

Lee, D. W. and R. L. Postle. 1975. Isozyme variation in *Colobanthus quitensis* (Kunth) Bartl.: Methods and preliminary analysis. *Brit. Antarctic Surv. Bull.* 41: 133–317.

Leigh, E. G. 1972. The golden section and spiral leaf arrangement. In E. S. Deevey, Jr., (ed.), *Growth by Intussusception.* Hamden, Conn.: Archon Books.

Leigh, E. G. 1975. Structure and climate in tropical rain forests. *Ann. Rev. Ecol. Syst.* 6: 67–86.

Leleji, O. I. 1973. Apparent preference by bees for different flower colours in cowpeas (*Vigna sinensis* (L.) Savi ex Hassk.). *Euphytica* 22: 150–53.

Lemon, E. R. 1967. The impact of the atmospheric environment of the integument of plants. *Int. J. Biometeorol.* 3: 57–69.

Lerner, I. M. 1954. *Genetic Homeostasis.* New York: John Wiley.

Leverich. W. S., and D. A. Levin. 1979. Age-specific survivorship and reproduction in *Phlox drummondii. Amer. Nat.* (in press).

Levin, D. A. 1969. The effect of color and outline on interspecific pollen flow in *Phlox. Evolution* 23: 444–55.

Levin, D. A. 1970. Reinforcement of reproductive isolation: Plants versus animals. *Amer. Nat.* 104: 571–81.

Levin, D. A. 1972a. The adaptedness of corolla-color variants in experimental and natural populations of *Phlox drummondii*. *Amer. Nat.* 106: 57–70.

Levin, D. A. 1972b. Low frequency disadvantage in the exploitation of pollinators by corolla variants in *Phlox*. *Amer. Nat.* 106: 453–60.

Levin, D. A. 1973. The role of trichomes in plant defense. *Quart. Rev. Biol.* 48: 3–15.

Levin, D. A. 1975a. Genic heterozygosity and protein polymorphism among local populations of *Oenothera biennis*. *Genetics* 79: 477–91.

Levin, D. A. 1975b. Somatic cell hybridization: Application in plant systematics. *Taxon* 24: 261–70.

Levin, D. A. 1975c. Genetic correlates of translocation heterozygosity in plants. *Bioscience* 25: 724–28.

Levin, D. A. 1975d. Interspecific hybridization, heterozygosity and gene exchange in *Phlox*. *Evolution* 29: 37–51.

Levin, D. A. 1976a. The chemical defenses of plants to pathogens and herbivores. *Ann. Rev. Ecol. Syst.* 7: 121–59.

Levin, D. A. 1976b. Consequences of long-term artificial selection, inbreeding and isolation in *Phlox*. II. The organization of allozymic variability. *Evolution* 30: 463–72.

Levin, D. A. and W. W. Anderson. 1970. Competition for pollinators between simultaneously flowering species. *Amer. Nat.* 104: 455–67.

Levin, D. A. and W. L. Crepet. 1973. Genetic variation in *Lycopodium lucidulum*: A phylogenetic relic. *Evolution* 27: 622–32.

Levin, D. A. and H. W. Kerster. 1969. The dependence of bee-mediated pollen and gene dispersal upon plant density. *Evolution* 23: 560–71.

Levin, D. A. and H. W. Kerster. 1971. Neighborhood structure in plants under diverse reproductive methods. *Amer. Nat.* 105: 345–54.

Levin, D. A. and H. W. Kerster. 1974. Gene flow in seed plants. *Evol. Biol.* 7: 139–220.

Levin, D. A. and H. W. Kerster. 1975. The effect of gene dispersal on the dynamics and statics of gene substitution in plants. *Heredity* 35: 317–36.

532   LITERATURE CITED

Levin, D. A., H. W. Kerster, and M. Niedzlek. 1971. Pollinator flight directionality and its effect on pollen flow. *Evolution* 25: 113–18.
Levin, D. A., G. P. Howland, and E. Steiner. 1972. Protein polymorphism and genic heterozygosity in a population of the permanent translocation heterozygote, *Oenothera biennis*. *Proc. Nat. Acad. Sci. USA* 69: 1475–77.
Levin, I., B. Bravdo, and R. Assaf. 1973. Relation between apple root distribution and soil water extraction in different irrigation regimes. In A. Hadas, D. Swartzendruber, P. E. Rijtema, M. Fuchs, and B. Yaron (eds.), *Physical Aspects of Soil Water and Salts in Ecosystems*. Berlin: Springer-Verlag.
Levin, S. A. 1974. Dispersion and population interactions. *Amer. Nat.* 168: 207–28.
Levin, S. A. 1976a. Population dynamic models in heterogeneous environments. *Ann. Rev. Ecol. Syst.* 7: 287–310.
Levin, S. A. 1976b. Spatial patterning and the structure of ecological communities. In S. A. Levin (ed.), *Some Mathematical Questions in Biology*, vol. 7. Providence, R.I.: American Mathematical Society.
Levin, S. A. and R. T. Paine. 1974. Disturbance, patch formation, and community structure. *Proc. Nat. Acad. Sci. USA* 71: 2744–47.
Levins, R. 1964. The theory of fitness in a heterogeneous environment. IV. The adaptive significance of gene flow. *Evolution* 18: 635–38.
Levins, R. 1966. The strategy of model building in population biology. *Amer. Sci.* 54: 421–31.
Levins, R. 1968. *Evolution in Changing Environments*. Princeton, N.J.: Princeton University Press.
Levins, R. 1969. Dormancy as an adaptive strategy. *Symp. Soc. Exp. Biol.* 21: 1–10.
Levins, R. 1970. Extinction. In M. Gerstenhaba (ed.), *Some Mathematical Problems in Biology*. Providence, R.I.: American Mathematical Society.
Levins, R. and R. Mac Arthur. 1969. An hypothesis to explain the incidence of monophagy. *Ecology* 50: 910–11.
Levinton, J. 1973. Genic variation in gradient of environmental variability. *Science* 180: 75–76.
Levitt, J. 1972. *Responses of Plants to Environmental Stresses*. New York: Academic Press.

Levitzky, G. A. 1931a. The morphology of chromosomes. *Bull. Appl. Bot. Genet. Plant Breed.* 27: 19–174.

Levitzky, G. A. 1931b. The karyotype in systematics. *Bull. Appl. Bot. Genet. Plant Breed.* 27: 220–40.

Levy, M. and D. A. Levin. 1975. Genetic heterozygosity and variation in permanent translocation heterozygotes of the *Oenothera biennis* complex. *Genetics* 79: 493–512.

Lewin, D. C. 1974. The vegetation of the ravines of the Southern Finger Lakes, New York, region. *Amer. Midl. Nat.* 91: 315–42.

Lewis, H. 1962. Catastrophic selection as a factor in evolution. *Evolution* 16: 257–71.

Lewis, H. 1966. Speciation in flowering plants. *Science* 152: 167–72.

Lewis, R. F. and W. J. Crotty. 1977. The primary root epidermis of *Panicum virgatum* L. II. Fine structural evidence suggestive of a plant–bacterium–virus symbiosis. *Amer. J. Bot.* 64: 190–98.

Lewontin, R. C. 1961. Evolution and the theory of games. *J. Theor. Biol.* 1: 382–403.

Lewontin, R. C. 1965. Selection for colonizing ability. In H. G. Baker and G. L. Stebbins (eds.), *The Genetics of Colonizing Species.* New York: Academic Press.

Lewontin, R. C. 1974. *The Genetic Basis of Evolutionary Change.* New York and London: Columbia University Press.

Libby, W. F., R. F. Stettler and F. W. Seitz. 1969. Forest genetics and forest tree breeding. *Ann. Rev. Genet.* 3: 469–94.

Lilley, S. R. and J. R. Wall. 1972. Determination of geographic variation in allelic frequencies for leucine aminopeptidase and alcohol dehydrogenase isozymes in *Cucurbita foetidissima. Va. J. Sci.* 23: 112.

Lindauer, M. 1972. Foraging and homing flight of the honey-bee: Some general problems of orientation. *Symp. Roy. Entomol. Soc.* 7: 199–216.

Lindley, M. 1977. Plants and animals interact. *Nature (London)* 266: 776–77.

Linhart, Y. B. 1974. Intra-population differentiation in annual plants. I. *Veronica peregrina* L. raised under non-competitive conditions. *Evolution* 28: 232–43.

Linhart, Y. B. 1976. Density-dependent seed germination strategies in colonizing versus non-colonizing plant species. *J. Ecol.* 64: 375–80.

Linsley, E. G. 1958. The ecology of solitary bees. *Hilgardia* 27: 453–599.

Linsley, E. G., C. G. Rick, and S. G. Stephens. 1966. Observations on the floral relationships of the Galápagos carpenter bee. *Pan Pacific Entomol.* 1: 1–18.

Lister, G. R., V. Slankis, G. Krotkov, and C. D. Nelson. 1967. Physiology of *Pinus strobus* L. seedlings grown under high or low moisture conditions. *Ann. Bot.* 31: 121–32.

Little, T. M. 1945. Gene segregation in autotetraploids. *Bot. Rev.* 11: 60–85.

Livingston, R. B. and M. L. Allessio. 1968. Buried viable seed in successional field and forest stands, Harvard Forest, Massachusetts. *Bull. Torrey Bot. Club* 95: 58–69.

Lloyd, D. G. 1965. Evolution of self-compatibility and racial differentiation in *Leavenworthia* (Cruciferae). *Contr. Gray Herb.* (Harvard Univ.) 195: 3–195.

Lloyd, R. M. and R. S. Mitchell. 1973. *A Flora of the White Mountains, California and Nevada.* Berkeley: University of California Press.

Lockhart, J. A. 1976. Plant growth, assimilation, and development: A conceptual framework. *Bioscience* 26: 332–38.

Lommen, P. W., S. K. Smith, C. S. Yocum, and D. M. Gates. 1975. Photosynthetic model. In Gates, D. M. and R. B. Schmerl, *Perspectives of Biophysical Ecology.* New York: Springer-Verlag.

Longman, K. A. 1969. The dormancy and survival of plants in the humid tropics. In *Dormancy and Survival; Soc. Exp. Biol. Symp.* 23: 471–88.

Lopushinsky, W. and G. O. Klock. 1974. Transpiration in conifer seedlings in relation to soil water potential. *For. Sci.* 20: 181–86.

Lotsy, J. P. 1916. *Evolution by Means of Hybridization.* The Hague: M. Nijhoff.

Loveless, A. R. and G. F. Asprey. 1957. The dry evergreen formations of Jamaica. I. The limestone hills of the South Coast. *J. Ecol.* 45: 799–822.

Lowry, O. and J. Passonneau. 1964. A comparison of the kinetic properties of phosphofructokinase from bacterial, plant, and animal sources. *Arch. Exp. Pathol. Pharmakol.* 248: 185–94.

Ludwig, J. W. and J. L. Harper. 1958. The influence of the environment on seed and seedling mortality. VIII. The influence of soil color. *J. Ecol.* 46: 381–91.

Lutz, A. M. 1907. A preliminary note on the chromosomes of Oenothera lamarckiana and one of its mutants, O. gigas. Science 26: 151–52.

Lutz, F. E. 1924. The colors of flowers and the vision of insects, with special reference to ultraviolet. Ann. New York Acad. Sci. 29: 233–83.

Mac Arthur, R. H. 1972a. Coexistence of species. In J. A. Behnke (ed.), Challenging Biological Problems. New York: Oxford University Press.

Mac Arthur, R. H. 1972b. Geographical Ecology. New York: Harper and Row.

Mac Arthur, R. H. and J. H. Connell. 1967. The Biology of Populations. New York: John Wiley.

Mac Arthur, R. H. and R. Levins. 1967. The limiting similarity, convergence and divergence of coexisting species. Amer. Nat. 101: 377–85.

Mac Arthur, R. H. and E. O. Wilson, 1967. The Theory of Island Biogeography. Princeton, N. J.: Princeton University Press.

McClintock, B. 1931. Cytological observations of deficiencies involving known genes, and an inversion in Zea mays. Res. Bull. Mo. Agr. Exp. Sta., no. 163.

McClure, S. 1973. Allozyme variability in natural populations of the yellow monkeyflower, Mimulus guttatus, located in the North Yuba River Drainage. Ph.D. dissertation, University of California, Berkeley.

McCree, K. J. 1974. Equations for the rate of dark respiration of white clover and grain sorghum, as functions of dry weight, photosynthetic rate, and temperature. Crop Sci. 14: 509–14.

McCree, K. J. and J. H. Troughton. 1966. Prediction of growth rate at different light levels from measured photosynthesis and respiration rates. Plant Physiol. 41: 559–66.

McDonough, W. T. 1970. Germination of 21 species collected from a high-elevation rangeland in Utah. Amer. Midl. Nat. 84: 551–53.

McDougall, W. B. 1921. Thick-walled root hairs of Gleditsia and related genera. Amer. J. Bot. 8: 171–75.

Macior, L. W. 1971. Coevolution of plants and animals: Systematic insights from plant–insect interactions. Taxon 20: 17–28.

Macior, L. W. 1973. The pollination ecology of Pedicularis on Mount Rainer. Amer. J. Bot. 60: 363–71.

Macior, L. W. 1974. Behavioral aspects of coadaptations between flowers and insect pollinators. *Ann. Mo. Bot. Gard.* 61: 760–69.

Macior, L. W. 1975. The pollination ecology of *Pedicularis* (Scrophulariaceae) in the Yukon Territory. *Amer. J. Bot.* 62: 1065–72.

Mack, R. N. 1976. Survivorship of *Cerastium atrovirens* at Aberffraw, Anglesey. *J. Ecol.* 64: 309–12.

Mack, R. N. and J. L. Harper. 1977. Interference in dune annuals: Spatial pattern and neighborhood effects. *J. Ecol.* 65: 345–364.

McKey, D. 1974. Adaptive patterns in alkaloid physiology. *Amer. Nat.* 108: 305–20.

McManmon, M. and R. Crawford. 1970. A metabolic theory of flooding tolerance: The significance of enzyme distribution and behavior. *New Phytol.* 70: 299–306.

McNaughton, S. J. 1968. Autotoxic feedback in relation to germination and seedling growth in *Typha latifolia.* *Ecology* 49: 367–69.

McNaughton, S. J. 1975. r- and K-selection in *Typha.* *Amer. Nat.* 109: 251–61.

McPherson, J. K. and C. H. Muller. 1969. Allelopathic effects of *Adenostoma fasciculatum* "chamise" in the California chaparral. *Ecol. Monogr.* 39: 177–98.

Magasanik, B., M. Prival, J. Brenchley, B. Tyler, A. Deleo, S. Streicher, R. Bender, and C. Paris. 1974. Glutamine synthetase as a regulator of enzyme synthesis. *Curr. Top. Cell. Reg.* 8: 119–34.

Malecot, G. 1969. *The Mathematics of Heredity.* San Francisco: W. H. Freeman.

Mankinen, C. B., J. Harding, and M. Elliott. 1975. Genetics of Lupinus. VIII. Variations in the occurrence of alkaloids in natural populations of *Lupinus nanus.* *Taxon* 24: 415–29.

Manly, B. F. J. 1973. A linear model for frequency-dependent selection by predators. *Res. Popul. Ecol.* 14: 137–50.

Manton, I. and W. A. Sledge. 1954. Observations on the cytology and taxonomy of the Pteridophyte flora of Ceylon. *Phil. Trans. Roy. Soc. London* 238: 125–80.

Margalef, R. 1958. Information theory in ecology. *Genet. Syst.* 3: 36–71.

Marks, P. L. 1974. The role of pin cherry (*Prunus pennsylvanica* L.) in the maintenance of stability in northern hardwood ecosystems. *Ecol. Monogr.* 44: 73–88.

Marshall, D. R. and R. W. Allard. 1970. Isozyme polymorphisms in natural populations of *Avena fatua* and *A. barbata. Heredity* 25: 373–82.

Marshall, D. R. and S. K. Jain. 1968. Phenotypic plasticity of *Avena fatua* and *A. barbata. Amer. Nat. 102: 457–67.*

Marshall, D. R. and S. K. Jain. 1969a. Genetic polymorphism in natural populations of *Avena fatua* and *A. barbata. Nature (London)* 221: 276–78.

Marshall, D. R. and S. K. Jain. 1969b. Interference in pure and mixed populations of *Avena fatua* and *A. barbata. J. Ecol.* 57: 251–70.

Martin, E. C. and S. E. McGregor. 1973. Changing trends in insect pollination of commercial crops. *Amer. Rev. Entomol.* 18: 207–26.

Martin, J. T. and B. E. Juniper. 1970. *The Cuticles of Plants.* Edinburgh: E. Arnold Publ.

Martins, P. S. and S. K. Jain. 1979. Role of genetic variation in the colonizing ability of rose clover (*Trifolium hirtum* All.). *Amer. Nat.* (in press).

Maruyama, T. 1971. The rate of decrease of heterozygosity in a population occupying a circular or linear habitat. *Genetics* 67: 437–54.

Maruyama, T. 1972. The rate of decrease of genetic variability in a two-dimensional continuous population of finite size. *Genetics* 70: 639–51.

Mather, K. 1943. Polygenic inheritance and natural selection. *Biol. Rev.* 18: 32–64.

Mather, K. 1953. The genetical structure of populations. *Symp. Soc. Exp. Biol.* 7: 66–95.

Mather, K. 1955. Polymorphism as an outcome of disruptive selection. *Evolution* 9: 52–61.

Mattson, W. J. and N. D. Addy. 1975. Phytophagous insects as regulators of forest primary production. *Science* 190: 515–21.

Maximov, N. A. 1929. *The Plant in Relation to Water* (trans. by R. H. Yapp). New York: Macmillan Co.

May, R. M., 1976. Estimating r: A pedagogical note. *Amer. Nat.* 110: 496–99.

May, R. M., J. A. Endler, and R. E. McMurtrie. 1975. Gene frequency clines in the presence of selection opposed by gene flow. *Amer. Nat.* 109: 659–76.

Mayer, A. M. and A. Poljakoff-Mayber. 1975. *The Germination of Seeds.* 2d ed. London: Pergamon Press.

Maynard Smith, J. 1971. What use is sex? *J. Theor. Biol.* 30: 319–35.

Maynard Smith, J. 1974. The theory of games and the evolution of animal conflict. *J. Theor. Biol.* 47: 209–21.

Maynard Smith, J. 1976. Evolution and the theory of games. *Amer. Sci.* 64: 41–45.

Maynard Smith, J. 1977. Why the genome does not congeal. *Nature (London)* 268: 693–96.

Maynard Smith, J. and G. R. Price. 1973. The logic of animal conflicts. *Nature (London)* 246: 15–18.

Mayr, E. 1942. *Systematics and the Origin of Species.* New York: Columbia University Press.

Mayr, E. 1963. *Animal Species and Evolution.* Cambridge, Mass.: Harvard University Press, Belknap Press.

Mazokhin-Porshnyakov, G. A. 1962. Colorimetric index of trichromatic bees. *Biofizika* 7: 211–17.

Medina, E. 1970. Relationships between nitrogen level, photosynthetic capacity and carboxydismutase activity in *Atriplex patula* leaves. *Carnegie Inst. Washington Yearb.* 69: 655–62.

Medina, E. 1971. Effect of nitrogen supply and light intensity during growth on the photosynthetic capacity and carboxydismutase activity of leaves of *Atriplex patula* ssp. *hastata. Carnegie Inst. Washington Yearb.* 70: 551–59.

Mejnartowicz, L. and F. Bergmann. 1975. Genetic studies on European larch (*Larix decidua* Mill.) employing isoenzyme polymorphisms. *Genet. Polon.* 16: 29–36.

Merritt, R. 1972. Geographic distribution and enzymatic properties of lactate dehydrogenase allozymes in the fathead minnow. *Amer. Nat.* 106: 173–94.

Merz, R. W. and S. G. Boyce. 1956. Age of oak "seedlings." *J. For.* 54: 774–75.

Miles, J. 1972. Experimental establishment of seedlings on a southern English heath. *J. Ecol.* 60: 225–34.

Miles, J. 1973. Early mortality and survival of self-sown seedlings in Glenfeshire, Invernessshire. *J. Ecol.* 61: 93–98.

Miller, P. C. 1972. Bioclimate, leaf temperature, and primary production in red mangrove canopies in south Florida. *Ecology* 53: 22–45.

Miller, P. C. 1974. Resource utilization and production processes. Paper presented at the Third Annual Symposium on Analysis of Ecological Systems. Ohio State University.

Miller, P. C. 1975a. Simulation of water relations and net photosynthesis in mangroves in south Florida. In *Proceedings of First International Symposium on Mangrove Biology and Management,* Honolulu, Hawaii.

Miller, P. C. 1975b. A comparison of short-term effects of thermal addition on photosynthesis and plant-water stress in three ecosystems. In *Environmental Effects of Cooling Systems at Nuclear Power Plants.* International Atomic Energy Agency, Vienna, Austria. (Proceedings of a Symposium on the Physical and Biological Effects on the Environment of Cooling Systems and Thermal Discharges at Nuclear Power Stations, Oslo, 26–30 August 1974.)

Miller, P. C. 1979. Quantitative plant ecology. In D. Horn (ed.), *Symposium on Analysis of Ecological Systems. Columbus: Ohio State University Press.*

Miller, P. C., and H. A. Mooney. 1974. The origin and structure of American arid-zone ecosystems. The producers: Interactions between environment, form, and function. In *Proceedings of First International Congress of Ecology,* The Hague, Netherlands.

Miller, P. C., J. R. Ehleringer, B. Hynum, and W. A. Stoner. 1974. Digital simulation of potential reforestation problems in the Rung Sat Delta, Viet Nam. In H. T. Odum, M. Sell, M. Brown, J. Zucchetto, C. Swallows, J. Browder, T. Ahlstrom, and L. Peterson (eds.), *The Effects of Herbicides in South Vietnam. Part B: Working papers: Models of herbicide, mangroves, and war in Vietnam.* Washington, D.C.: National Academy of Sciences, National Research Council.

Miller, P. C., W. A. Stoner, and L. L. Tieszen. 1976. A model of stand photosynthesis for the wet meadow tundra at Barrow, Alaska. *Ecology* 57: 411–30.

Miller, P. C., D. E. Bradbury, E. Hajek, V. La Marche, and N. J. W. Thrower. 1977. Past and present environment. In H. A. Mooney (ed.), *Convergent Evolution in California and Chile.* Stroudsburg, Pa.: Dowden, Hutchinson, and Ross.

Miller, R. D. 1977. Genetic variability in the slender wild oat *Avena barbata* in California. Ph.D. dissertation, University of California, Davis.

Mitchell, R. S. 1976. Submergence experiments on nine species of semi-aquatic *Polygonum. Amer. J. Bot.* 63: 1158–65.

Mitra, R., R. Gross, and J. Varner. 1978. An intact tissue assay for enzymes that labilize CH bonds (in press).

Mitton, J. B., Y. B. Linhart, J. L. Hamrick, and J. S. Beckman. 1977.

Observations on the genetic structure and mating system of Ponderosa pine in the Colorado Front Range. *Theor. Appl. Genet.* 51: 5–13.

Moldenke, A. R. 1975. Niche specialization and species diversity along a California transect. *Oecologia* 21: 219–42.

Molz, F. J. and C. M. Peterson. 1976. Water transport from roots to soil. *Agron. J.* 68: 901–4.

Monk, C. D. 1966. An ecological significance of evergreenness. *Ecology* 47: 504–5.

Monsi, M. 1968. Mathematical models of plant communities. In F. Eckardt (ed.), *Functioning of Terrestrial Ecosystems at the Primary Production Level*. Paris: UNESCO.

Monsi, M. and T. Saeki. 1953. Über den Lichtfactor in den Pflanzengesellschaften und seine Bedeutung für die Stoffproduktion. *Jap. J. Bot.* 14: 22–52.

Monteith, J. L. 1965. Light distribution and photosynthesis in field crops. *Ann. Bot.*, n.s. 29: 17–37.

Monteith, J. L. 1973. *Principles of Environmental Physics*. New York: American Elsevier.

Mooney, H. A. 1963. Physiological ecology of coastal, subalpine, and alpine populations of *Polygonum bistortioides*. *Ecology* 44: 813–16.

Mooney, H. A. 1969. Dark respiration of related evergreen and deciduous Mediterranean plants during induced drought. *Bull. Torrey Bot. Club* 96: 550–55.

Mooney, H. A. 1972. Carbon balance of plants. *Ann. Rev. Ecol. Syst.* 3: 315–46.

Mooney, H. A. 1975. Plant physiological ecology—a synthetic view. In F. J. Vernberg (ed.), *Physiological Adaptation to the Environment*. New York: Intext Publ.

Mooney, H. A. 1976. Some contributions of physiological ecology to plant population biology. *Syst. Bot.* 1: 269–83.

Mooney, H. A., ed. 1977. *Convergent Evolution in California and Chile*. Stroudsburg, Pa.: Dowden, Hutchinson, and Ross.

Mooney, H. A. and W. D. Billings. 1961. Comparative physiological ecology of arctic and alpine populations of *Oxyria digyna*. *Ecol. Monogr.* 31: 1–29.

Mooney, H. A. and E. L. Dunn. 1970a. Photosynthetic systems of Mediterranean-climate shrubs and trees of California and Chile. *Amer. Nat.* 104: 447–53.

Mooney, H. A. and E. L. Dunn, 1970b. Convergent evolution of Mediterranean climate evergreen sclerophyll shrubs. *Evolution* 24: 292–303.

Mooney, H. A. and A. Gigon. 1973. Some requirements for a physiological model to predict the carbon gain of plants under natural conditions. *Eastern Deciduous Forest Biome* IBP-73-6: 147–59.

Mooney, H. A. and F. Shropshire. 1967. Population variability in temperature related photosynthetic acclimation. *Oecologia* 2: 1–13.

Mooney, H. A. and M. West. 1964. Photosynthetic acclimation of plants of diverse origin. *Amer. J. Bot.* 51: 825–27.

Mooney, H. A., D. J. Parsons, and J. Kummerow. 1974. Plant development in Mediterranean climates. In H. Lieth (ed.), *Phenology and Seasonality Modeling*. New York: Springer Verlag.

Mooney, H. A., J. Troughton, and J. Berry. 1974. Arid climates and photosynthetic systems. *Carnegie Inst. Washington Yearb.* 73: 793–805.

Mooney, H. A., J. Ehleringer, and J. Berry. 1976. High photosynthetic capacity of a winter annual in Death Valley. *Science* 194: 322–23.

Mooney, H. A., O. Björkman, J. Ehleringer, and J. Berry. 1976. Photosynthetic capacity of *in situ* Death Valley plants. *Carnegie Inst. Washington Yearb.* 75: 410–13.

Mooney, H. A, O. Björkman, and J. Collatz. 1977. Photosynthetic acclimation to temperature and water stress in the desert shrub, *Larrea divaricata*. *Carnegie Inst. Washington Yearb.* 76: 328–35.

Mooney, H. A., J. Ehleringer, and O. Björkman. 1977. The energy balance of leaves of the evergreen desert shrub *Atriplex hymenelytra*. *Oecologia* 29: 301–10.

Mooney, H. A., J. Troughton, and J. Berry. 1977. Carbon isotope ratio measurements of succulent plants in Northern Africa. *Oecologia* 30: 295–305.

Mooney, H. A., P. J. Ferrar, and R. O. Slatyer. 1978. Photosynthetic capacity and carbon allocation patterns in diverse growth forms of *Eucalyptus*. *Oecologia* 36: 103–11.

Moore, P. J. 1976. How far does pollen travel? *Nature (London)* 260: 388–89.

Mooring, J. 1958. A cytogenetic study of *Clarkia unguiculata*. I. Translocations. *Amer. J. Bot.* 45: 233–42.

Moraes, C. F. 1972. Ecogenetic studies on variation and population structure in soft chess, *Bromus mollis* L. Ph.D. Dissertation, University of California, Davis.

Morata, G. and P. A. Lawrence. 1977. Homeotic genes, compartments and cell determination in *Drosophila. Nature (London)* 265: 211–16.

Morley, F. H. W. and J. Katznelson. 1965. Colonization in Australia by *Trifolium subterraneum* L. In H. G. Baker and G. L. Stebbins (eds.), *The Genetics of Colonizing Species.* New York and London: Academic Press.

Morley, F. H. W., C. I. Davern, V. E. Rogers, and J. W. Peak. 1962. Natural selection among strains of *Trifolium subterraneum.* In G. W. Leeper (ed.), *The Evolution of Living Organisms.* Melbourne, Aust.: Melbourne University Press.

Morrison, R. G. and G. A. Yarranton. 1973. Diversity, richness, and evenness during a primary sand dune succession at Grand Bend, Ontario. *Canad. J. Bot.* 51: 2401–11.

Morton, J. K. 1955. The incidence of polyploidy in a tropical flora. *Recent Adv. Bot.* (University of Toronto Press) 9: 900–903.

Mosquin, T. 1966. Toward a more useful taxonomy for chromosomal races. *Brittonia* 18: 203–14.

Mosquin, T. 1971. Competition for pollinators as a stimulus for the evolution of flowering time. *Oikos* 22: 398–402.

Mountford, M. D. 1971. Population survival in a variable environment. *J. Theor. Biol.* 32: 75–79.

Muller, C. H. 1974. Allelopathy in the environmental complex. In B. R. Strain and W. D. Billings (eds.), *Vegetation and Environment.* The Hague: Dr. W. Junk.

Muller, C. H. and C. H. Chou. 1972. Phytotoxins: An ecological phase of phytochemistry. In J. B. Harbourne (ed.), *Phytochemical Ecology.* London and New York: Academic Press.

Muller, C. H., R. B. Hanawalt, and J. K. McPherson. 1968. Allelopathic control of herb growth in the fire cycle of California chaparral. *Bull. Torrey Bot. Club* 95: 225–31.

Mulligan, G. A. 1972. Autogamy, allogamy and pollination in some lava weeds. *Canad. J. Bot.* 50: 1767–71.

Mulligan, G. A. and P. G. Kevan. 1973. Color brightness and other floral characteristics attracting insects to the blossoms of some Canadian weeds. *Canad. J. Bot.* 51: 1939–52.

Mulroy, T. W. and P. W. Rundel. 1977. Annual plants: Adaptations to desert environments. *BioScience* 27: 109–14.

Müntzing, A. 1930a. Outlines to a genetic monograph of *Galeopsis*. *Hereditas* 13: 185–341.

Müntzing, A. 1930b. Über Chromosomenvermehrung in *Galeopsis*-Kreuzungen und ihre Phylogenetische Bedeutung. *Hereditas* 14: 153–72.

Müntzing, A. 1934. Chromosome fragmentation in a *Crepis* hybrid. *Hereditas* 19: 284–302.

Müntzing, A. 1939. Studies on the properties and the ways of production of rye–wheat amphidiploids. *Hereditas* 25: 387–430.

Munz, P. A. 1974. *A Flora of Southern California*. Los Angeles: University of California Press.

Murdoch, W. W. 1966. Population stability and life history phenomena. *Amer. Nat.* 100: 45–51.

Murdoch, W. W. 1973. The functional response of predators. *J. Appl. Ecol.* 10: 335–42.

Murphy, C. E. and K. R. Knoerr. 1970. *Modeling the Energy Balance Processes of Natural Ecosystems*. Oak Ridge, Tenn.: U.S. IBP (Int. Biol. Program) Anal. Ecosyst. Program, Eastern Deciduous Forest Biome Subproject, Res. Rep. Final 1969–1970, Oak Ridge National Laboratory.

Murphy, C. E. and K. R. Knoerr. 1972. Modeling the energy balance processes of natural ecosystems. Oak Ridge Tenn.: U.S. IBP (Int. Biol. Program) Anal. Ecosyst. Program, Eastern Deciduous Forest Biome Subproject, Res. Rep. EDFB-IBP 72-10, Oak Ridge National Laboratory.

Murphy, G. I. 1968. Pattern in life history and the environment. *Amer. Nat.* 102: 390–404.

Musselman, R. C., D. T. Lester, and M. S. Adams. 1975. Localized ecotypes of *Thuja occidentalis* L. in Wisconsin. *Ecology* 56: 647–55.

Nagylaki, T. 1976a. Clines with variable migration. *Genetics* 83: 867–86.

Nagylaki, T. 1976b. The decay of genetic variability in geographically structured populations. *Theor. Popul. Biol.* 10: 70–82.

Nagylaki, T. 1976c. Dispersion–selection balance in localized plant populations. *Heredity* 37: 59–67.

Natr, L. 1975. Influence of mineral nutrition on photysynthesis and the use of assimilates. In J. P. Cooper (ed.), *Photosynthesis and Productivity*. Cambridge: Cambridge University Press.

Navashin, M. S. 1932. The dislocation hypothesis of evolution of chromosome numbers. Zeitschr. Abst. Vererbungsl. 63: 224–31.

Naylor, R. E. L. 1972. Aspects of the population dynamics of the weed Alopecurus myosurides Huds. in winter cereal crops. J. Appl. Ecol. 9: 127–39.

Nebel, B. R. and Ruttle, M. L. 1938. The cytological and genetical significance of colchicine. J. Hered. 29: 3–9.

Negisi, K. 1966. Photosynthesis, respiration and growth in one-year-old seedlings of Pinus densiflora, Cryptomeria japonica and Chamaecyparis obtusa. Bull. Tokyo Univ. For., 62.

Nei, M. 1972. Genetic distance between populations. Amer. Nat. 106: 283–92.

Nei, M. 1975. Molecular Population Genetics and Evolution. Oxford, Amsterdam: North-Holland Publ. Co.

Nei, M. and W. Li. 1973. Linkage disequilibrium in subdivided populations. Genetics 75: 213–19.

Nevo, E. 1976. Adaptive strategies of genetic systems in constant and varying environments. In S. Karlin and E. Nevo (eds.), Population Genetics and Ecology. New York: Academic Press.

Newman, E. I. 1963. Factors controlling the germination date of winter annuals. J. Ecol. 51: 625–38.

Newman, E. I. 1969. Resistance to water flow in soil and plant. I. Soil resistance in relation to amounts of root: Theoretical estimates. J. Appl. Ecol. 6: 1–12.

Newman, E. I. 1974. Root and soil water relations. In E. W. Carson (ed.), The Plant Root and its Environment. Charlottesville, Va.: University Press of Virginia.

Newman, E. I. and A. D. Rovira. 1975. Allelopathy among some British grassland species. J. Ecol. 63: 727–37.

Newton, W. C. F. and C. Pellew. 1929. Primula kewensis and its derivatives. J. Genet. 20: 405–67.

Ng, E. and P. C. Miller. 1975. A model of the effect of tundra vegetation on soil temperatures. In G. Weller and S. A. Bowling (eds.), Climate of the Arctic. 24th Alaska Science Conference, Fairbanks, Alaska, August 15–17, 1973. Fairbanks, Alaska: Geophysical Institute, University of Alaska.

Ng, E. and P. C. Miller. 1977. Validation of a model of the effect of tundra vegetation on soil temperatures. Arc. Alpine Res. 9: 89–104.

Nichols, J. D., W. Conley, B. Batt, and A. R. Tipton. 1976. Temporally dynamic reproductive strategies and the concept of r- and K-selection. *Amer. Nat.* 110: 995–1005.

Nilsson, H. 1947. Totale Inventierung der Mikrotypen eines Minimal areals von *Taraxacum officinale*. *Hereditas* 33: 119–142.

Nilsson-Ehle, H. 1909. Kreuzungsuntersuchungen an Hafer und Weizen. *Lunds Univ. Årsskr.*, ser. 2, 5 (2): 1–122.

Nir, I. and A. Poljakoff-Mayber. 1967. Effect of water stress on the photochemical activity of chloroplasts. *Nature (London)* 213: 418–19.

Nobel, P. S. 1977. Water relations of flowering of *Agave deserti*. *Bot. Gaz.* 138: 1–6.

Nobel, P., L. Zaragosa, and W. Smith. 1975. Relation between meso-phyll surface area, photosynthetic rate, and illumination level during development of leaves of *Plectranthus parviflorus* Henckel. *Plant Physiol.* 55: 1067–70.

Ogden, John. 1974. The reproductive strategy of higher plants. II. The reproductive strategy of *Tussilago farfara* L. *J. Ecol.* 62: 291–324.

Olmstead, C. E. 1944. Growth and development in range grasses. IV. Photoperiodic responses in twelve geographic strains of side-oats grama. *Bot. Gaz.* 106: 46–74.

Olson, D. F., Jr., and W. J. Gabriel. 1974. *Acer* L. In *Seeds of Woody Plants in the United States*. U.S.D.A. Handbook 450.

Opler, P. A. and K. S. Bawa. 1978. Sex ratios of tropical dioecious trees: Selective pressures and ecological fitness. *Evolution* (in press).

Opler, P. A., H. G. Baker and G. W. Frankie. 1975. Reproductive biology of some Costa Rican *Cordia* species (Boraginaceae). *Biotropica* 7: 234–47.

Oppenheimer, H. R. 1949. The water turnover of the Valonea oak. *Palest. J. Bot. Rehovot Ser.* 7: 177–79.

Oppenheimer, H. R. 1960. Adaptation to drought xerophytism. In *Plant-Water Relationships in Arid and Semi-Arid Conditions* (UNESCO). Lucerne: C. J. Bucher.

Orians, G. H. and O. T. Solbrig. 1977. A cost–income model of leaves and roots with special reference to arid and semi-arid areas. *Amer. Natur.* 111: 677–90.

Ornduff, R. 1969. Reproductive biology in relation to systematics. *Taxon* 18: 121–33.

Ornduff, R. 1970. Incompatibility and the pollen economy of *Jepsonia parryi*. *Amer. J. Bot.* 57: 1036–41.

Oster, G. and B. Heinrich. 1976. Why do bumblebees major? A mathematical model. *Ecol. Monogr.* 46: 129–33.

O'Toole, J. C. 1975. Photosynthetic response to water stress in *Phaseolus vulgaris* L. Ph.D. dissertation, Cornell University, Ithaca, N.Y.

Palmblad, I. G. 1968. Competition in experimental populations of weeds with emphasis on the regulation of population size. *Ecology* 49: 26–34.

Parenti, R. L. and E. L. Rice. 1969. Inhibitional effects of *Digitaria sanguinalis* and possible role in old field succession. *Bull. Torrey Bot. Club* 96: 70–78.

Parker, G. A. and R. A. Stuart. 1976. Animal behavior as a strategy optimizer: Evolution of threshold assessment strategies and optimal emigration thresholds. *Amer. Nat.* 110: 1055–76.

Parkhurst, D. and O. Loucks. 1972. Optimal leaf size in relation to environment. *J. Ecol.* 60: 505–37.

Parrish, J. A. D. and F. A. Bazzaz. 1976a. Underground niche separation in successional plants. *Ecology* 57: 1281–88.

Parrish, J. A. D. and F. A. Bazzaz. 1976b. Niche differences in use of pollinators between plant species of an early successional community. *Bull. Ecol. Soc. Amer.* 57: 35.

Parsons, D. J. 1973. A comparative study of vegetation structure in the Mediterranean scrub communities of California and Chile. Ph.D. dissertation, Stanford University, Palo Alto, California.

Payne, W. W. 1976. Biochemistry and species problems in *Ambrosia* (Asteraceae-Ambrosieae). *Plant Syst. Evol.* 125: 169–78.

Payne, W. W., T. A. Geissman, A. J. Lucas, and T. Saitoh. 1973. Chemosystematics and taxonomy of *Ambrosia chamissonis*. *Biochem. Syst.* 1: 21–33.

Pearcy, R. W., O. Björkman, A. T. Harrison, and H. A. Mooney. 1971. Photosynthetic performance of two desert species with C_4 photosynthesis in Death Valley, California. *Carnegie Inst. Washington Yearb.* 70: 540–50.

Pemadasa, M. A. and P. H. Lovell. 1975. Factors controlling germination of some dune annuals. *J. Ecol.* 63: 41–60.

Penning de Vries, F. W. T. 1972a. A model for simulating transpiration of leaves with special attention to stomatal functioning. *J. Appl. Ecol.* 9: 57–77.

Penning de Vries, F. W. T. 1972b. Respiration and growth. In A. R. Rees, K. E. Cockshull, D. W. Hand, and R. G. Hurd (eds.), *Crop Processes in Controlled Environments*. London: Academic Press.

Penning de Vries, F. W. T. 1973. Substrate utilization and respiration in relation to growth and maintenance in higher plants. Ph.D. dissertation, Agricultural University, Wageningen.

Penning de Vries, F. W. T. 1974. Substrate utilization and respiration in relation to growth and maintenance in higher plants. *Neth. J. Agr. Sci.* 22: 40–44.

Penning de Vries, F. W. T. 1975a. The cost of maintenance processes in plant cells. *Ann. Bot.* 39: 77–92.

Penning de Vries, F. W. T. 1975b. *Use of Assimilates in Higher Plants in Photosynthesis and Productivity in Different Environments*. IBP, vol. 3. Cambridge: Cambridge University Press.

Penning de Vries, F. W. T., A. H. M. Brunsting, and H. H. van Laar. 1974. Products, requirements and efficiency of biosynthesis: A quantitative approach. *J. Theor. Biol.* 45: 339–77.

Penning de Vries, F. W. T., C. E. Murphy, Jr., C. G. Wells, and J. R. Jörgensen. 1974. Simulation of nitrogen distribution in time and space in even-aged loblolly pine plantations and its effect on productivity. In IBP Contribution no. 156, *Proc. Symp. Mineral Cycling in Southern Ecosystems*, Augusta, Georgia, May 1–3.

Percival, M. S. 1961. Types of nectar in angiosperms. *New Phytol.* 60: 235–81.

Percival, M. S. 1965. *Floral Biology*. London: Pergamon Press.

Percival, M. S. 1974. Floral ecology of coastal scrub in southeast Jamaica. *Biotropica* 6: 104–29.

Peters, R. A. and S. Dunn. 1971. Life history studies as related to weed control in the Northeast. 6. Large and small crabgrass. *Conn. Agr. Exp. Sta. Bull.* 415.

Philip, J. R. 1966. Plant water relations: Some physical aspects. *Ann. Rev. Plant Physiol.* 17: 245–68.

Pianka, E. R. 1970. On r- and K-selection. *Amer. Nat.* 104: 592–97.

Pianka, E. R. 1972. r and K selection or b and d selection? *Amer. Nat.* 106: 581–88.

Pianka, E. R. 1974. *Evolutionary Ecology*. New York: Harper and Row.

Pianka, E. R. and W. S. Parker. 1975. Age-specific reproductive tactics. *Amer. Nat.* 109: 453–64.

Pickett, S. T. A. and F. A. Bazzaz. 1976. Divergence of two cooccurring successional annuals on a soil moisture gradient. *Ecology* 57: 169–76.

Pielou, E. C. 1959. The use of point-to-plant distances in the study of the pattern of plant populations. *J. Ecol.* 47: 607–13.

Pielou, E. C. 1960. A single mechanism to account for regular, random and aggregated populations. *J. Ecol.* 48: 575–84.

Pielou, E. C. 1966. Species-diversity and pattern diversity in the study of succession. *J. Theor. Biol.* 10: 370–83.

Pillemer, E. A. and W. M. Tingey. 1976. Hooked trichomes: A physical plant barrier to a major agricultural pest. *Science* 193: 482–84.

Platt, W. J. 1975. The colonization and formation of equilibrium plant species associations on badger disturbances in a tallgrass prairie. *Ecol. Monogr.* 45: 285–305.

Platt, W. J. 1976. The natural history of a fugitive prairie plant [*Mirabilis hirsuta* (Pursh) MacM.]. *Oecologia* 22: 399–409.

Platt, W. J. and I. M. Weis. 1977. Resource partitioning and competition within a guild of fugitive prairie plants. *Amer. Nat.* 111: 479–513.

Plaut, Z. and B. Bravdo. 1973. Response of carbon dioxide fixation to water stress. *Plant Physiol.* 52: 28–32.

Popay, A. I. and E. H. Roberts. 1970. Factors involved in the dormancy and germination of *Capsella bursa-pastoris* (L.) Medik. and *Senecio vulgaris* L. *J. Ecol.* 58: 103–22.

Post, C. T., Jr. and T. H. Goldsmith. 1969. *Ann. Entomol. Soc. Amer.* 62: 1497–98.

Powell, J. R. 1975. Protein variation in natural populations of animals. *Evol. Biol.* 8: 79–119.

Prat, H. 1932. L'épiderme des Graminées: Étude anatomique et systematique. *Ann. Sci. Nat.*, ser. 10, Bot., 14: 117–234.

Proctor, M. and P. Yeo. 1972. *The Pollination of Flowers.* New York: Taplinger Publ. Co.

Provine, W. B. 1971. *The Origins of Theoretical Population Genetics.* Chicago and London: University of Chicago Press.

Prout, T. 1969. The estimation of fitness from population data. *Genetics* 63: 949–67.

Prout, T. 1973. Appendix to population genetics of marine pelecypods. III. Epistasis between functionally related isozymes in *Mytilus edulis* J. B. Mitton and R. C. Koehn. *Genetics* 73: 487–96.

Pulliam, H. R. 1974. On the theory of optimal diets. *Amer. Nat.* 108: 59–74.

Pulliam, H. R. 1975. Diet optimization with nutrient constraints. *Amer. Nat.* 109: 765–68.

Putwain, P. D. and J. L. Harper. 1970. Studies in the dynamics of plant populations. III. The influence of associated species on populations of *Rumex acetosella* L. in grassland. *J. Ecol.* 58: 251–64.

Putwain, P. D., D. Machin, and J. L. Harper. 1968. Studies in the dynamics of plant populations. II. Components and regulation of a natural population of *Rumex acetosella* L. *J. Ecol.* 56: 421–31.

Pyke, G. H., H. R. Pulliam, and E. L. Charnov. 1977. Optimal foraging: A selective review of theory and tests. *Quart. Rev. Biol.* 52: 137–52.

Quarrie, S. A. and H. G. Jones. 1977. Effects of abscisic acid and water stress on development and morphology of wheat. *J. Exp. Bot.* 28: 192–203.

Quarterman, E. 1970. Germination of seeds of certain tropical species. In H. T. Odum (ed.), *A Tropical Rain Forest: A Study of Irradiation and Ecology at El Verde, Puerto Rico.* New York: U.S. Atomic Energy Commission.

Rabinowitch, E. I. 1951. *Photosynthesis.* New York: Interscience.

Rabotnov, T. A. 1974. Differences between fluctuations and successions. In R. Knapp (ed.), *Handbook of Vegetation Science, part VIII: Vegetation Dynamics.* The Hague: Dr. W. Junk.

Rapport, D. J. 1971. An optimization model of food selection. *Amer. Nat.* 105: 575–87.

Raschke, K. 1960. Heat transfer between the plant and the environment. *Ann. Rev. Plant Physiol.* 11: 111–26.

Raschke, K. 1975. Stomatal action. *Ann. Rev. Plant Physiol.* 26: 309–40.

Ratcliffe, D. 1961. Adaptation to habitat in a group of annual plants. *J. Ecol.* 49: 187–203.

Rathcke, B. J. and R. W. Poole. 1975. Coevolutionary race continues: Butterfly larval adaptation to plant trichomes. *Science* 187: 175–76.

Raunkiaer, C. 1934. *The Life Forms of Plants and Statistical Plant Geography: Being the Collected Papers of C. Raunkiaer.* Oxford: Clarendon Press.

Raven, P. H. 1972. Why are bird-visited flowers predominantly red? *Evolution* 26: 674.

Raven, P. H. 1973. The evolution of Mediterranean floras. In F.

diCastri and H. A. Mooney (eds.), *Mediterranean Type Ecosystems: Origin and Structure.* New York: Springer-Verlag.

Raven, P. H. 1976. Systematics and plant population biology. *Syst. Bot.* 1: 284–316.

Raven, P. H. 1977. The systematics and evolution of higher plants. In C. E. Goulden (ed.), *Changing Scenes in the Natural Sciences, 1776–1976.* Academy of Natural Sciences, Special Publication no. 12. Philadelphia.

Rawlins, S. L. 1963. Resistance to water flow in the transpiration stream. In I. Zelitch (ed.), *Stomata and Water Relations in Plants. Conn. Agr. Exp. Sta. Bull.* 664.

Raynal, D. J. and F. A. Bazzaz. 1975. Interference of winter annuals with *Ambrosia artemisiifolia* in early successional fields. *Ecology* 56: 35–49.

Reader, R. J. 1975. Competitive relationships of some bog ericads for major insect pollinators. *Canad. J. Bot.* 53: 1300–1305.

Reddingius, J. and P. J. Den Boer. 1970. Simulation experiments illustrating stabilization of animal number by spreading of risk. *Oecologia* 5: 240–48.

Reeder, J. R. 1946. Additional evidence of affinities between *Eragrostis* and certain Chlorideae. (Abstract.) *Amer. J. Bot.* 33: 843.

Reeder, J. R. 1957. The embryo in grass systematics. *Amer. J. Bot.* 44: 756–68.

Rees, N. and R. N. Jones. 1972. The origin of wide species variation in nuclear DNA content. *Int. Rev. Cytol.* 32: 53–92.

Remington, C. L. 1968. Suture-zones of hybrid interaction between recently joined biotas. *Evol. Biol.* 2: 321–428.

Renner, O. 1917. Versuche über die gametische Konstitution der Oenotheren. *Zeit. Abst. Verebungsl.* 18: 121–294.

Renner, O. 1921. Heterogamie im weiblichen Geschlecht und Embryosackentwicklung ber den Oenotheren. *Zeitschr. Bot.* 13: 609–21.

Reuther, W. 1973. Climate and citrus behavior. In W. Reuther (ed.), *The Citrus Industry;* vol. 3: *Production Technology.* Univ. of Calif. Div. of Agricultural Sci.

Reynolds, E. R. C. 1975. Tree rootlets and their distribution. In J. G. Torrey and D. T. Clarkson (eds.), *The Development and Function of Roots.* New York: Academic Press.

Ribbands, C. R. 1955. The scent perception of the honeybee. *Proc. Roy. Soc., Ser. B* 143: 367–79.

Richardson, J. V. and J. H. Borden. 1972. Hostfinding behavior of *Coelonoides brunneri* (Hymenoptera: Braconidae). *Canad. Entomol.* 104: 1235–50.

Rick, C. M. and J. F. Fobes. 1975. Allozyme variation in the cultivated tomato and closely related species. *Bull. Torrey Bot. Club* 102: 376–84.

Rick, C. M., E. Kesicki, J. F. Fobes, and M. Holle. 1976. Genetic and biosystematic studies on two new sibling species of *Lycopersicon* from interandean Peru. *Theor. Appl. Genet.* 47: 55–68.

Ricklefs, R. E. 1977. Environmental heterogeneity and plant species diversity: A hypothesis. *Amer. Nat.* 111: 376–81.

Riley, R. 1960. The diploidization of polyploid wheat. *Heredity* 15: 407–29.

Roberts, E. H. 1972. Dormancy: A factor affecting seed survival in the soil. In E. H. Roberts (ed.), *Viability of Seeds.* London: Chapman and Hall.

Roberts, S. W. and P. C. Miller. 1977. Interception of solar radiation as affected by canopy organization in two Mediterranean shrubs. *Oecologia* 12: 273–90.

Rockwood, W. 1969. Enzyme variation in natural population of *Drosophila mimica*. *Stud. Genet. V; Univ. Tex. Publ.* 6918: 111–32.

Rockwood-Sluss, E., J. Johnston, and W. Heed. 1973. Allozyme genotype–environment relationships. I. Variation in natural populations of *Drosophila pachea*. *Genetics* 73: 135–46.

Rodriguez, E. and D. A. Levin. 1976. Biochemical parallelisms of repellents and attractants in higher plants and arthropods. In J. W. Wallace and R. L. Mansell (eds.), *Biochemical Interactions between Plants and Insects*. New York and London: Plenum Press.

Roeske, C. N., J. N. Seiber, L. P. Brower, and C. M. Moffitt. 1976. Milkweed cardenolides and their comparative processing by monarch butterflies (*Danaus plexuppus* L.). In J. W. Wallace and R. L. Mansell (eds.), *Biochemical Interaction between Plants and Insects*. New York and London: Plenum Press.

Roos, F. H. and J. A. Quinn. 1977. Phenology and reproductive allocation in *Andropogon scoparius* (Gramineae) populations in communities of different successional stages. *Amer. J. Bot.* 64: 535–40.

Roose, M. L. and L. D. Gottlieb. 1976. Genetic and biochemical consequences of polyploidy in *Tragopogon*. *Evolution* 30: 818–30.

Root, R. B. 1973. Organization of a plant–arthropod association in simple and diverse habitats: The fauna of collards. *Ecol. Monogr.* 43: 95–124.

Rosen, R. 1967. *Optimality Principles in Biology.* New York: Butterworths.

Rosenberg, O. 1909. Cytologische und Morphologische Studien an *Drosera longifolia* × *rotundifolia*. *K. Svensk. Vet. Akad. Handl.* 43 (11): 1–64.

Ross, M. A. and J. L. Harper. 1972. Occupation of biological space during seedling establishment. *J. Ecol.* 61: 77–88.

Roughgarden, J. 1971. Density-dependent natural selection. *Ecology* 52: 453–68.

Roughgarden, J. 1972. Evolution of niche width. *Amer. Nat.* 106: 683–718.

Roughgarden, J. 1974. Population dynamics in a spatially varying environment: How population size "tracks" spatial variation in carrying capacity. *Amer. Nat.* 108: 649–64.

Roughgarden, J. 1976. Resource partitioning among competing species: A coevolutionary approach. *Theor. Popul. Biol.* 9: 388–424.

Roughgarden, J. 1977. Basic ideas in ecology. *Science* 196: 51.

Rouse, W. R. and R. B. Stewart. 1972. A model for determining evaporation from high latitude upland sites. *J. Appl. Meteorol.* 11: 1063–70.

Royama, T. 1970. Factors governing the hunting behavior and selection of food by the Great Tit (*Parus major* L.). *J. Amer. Ecol.* 39: 619–68.

Rudin, D. 1974. Gene and genotype frequencies in Swedish Scots pine populations studied by the isozyme technique. *Hereditas* 78: 325.

Rudin, D., G. Eriksson, I. Ekberg, and M. Rasmuson. 1974. Studies of allele frequencies and inbreeding in Scots pine populations by the aid of the isozyme technique. *Silvae Genet.* 23: 10–13.

Rudolf, P. O. and W. B. Leak. 1974. *Fagus* L. In *Seeds of Woody Plants in the United States.* U.S.D.A. Handbook no. 450, pp. 401–5.

Russell, R. S. 1970. Root systems and plant nutrition: Some new approaches. *Endeavour* 39: 60–66.

Sagar, G. R. and J. L. Harper. 1960. Factors affecting the germination and early establishment of plantains (*Plantago lanceolata, P. media* and *P. major*). In J. L. Harper (ed.), *The Biology of Weeds.* 1st *Symp. Brit. Ecol. Soc.*

Sagar, G. R. and J. L. Harper. 1961. Controlled interference with natural populations of *Plantago lanceolata*, *P. major* and *P. media*. *Weed Res.* 1: 163–76.

Sakai, K.-I. 1965. Contributions to the problem of species colonization from the viewpoint of competition and migration. In H. G. Baker and G. L. Stebbins (eds.), *The Genetics of Colonizing Species*. New York and London: Academic Press.

Sale, P. F. 1974. Overlap in resource use and interspecific competition. *Oecologia* 17: 245–56.

Salisbury, E. J. 1930. Mortality amongst plants and its bearing on natural selection. *Nature (London)* 126: 95–96.

Salisbury, E. J. 1942. *The Reproductive Capacity of Plants*. London: Bell and Sons.

Salisbury, E. J. 1974. Seed size and mass in relation to environment. *Proc. Roy. Soc., Ser. B* 186: 83–88.

Sanchez-Diaz, M. F. and P. J. Kramer. 1971. Behavior of corn and sorghum under water stress and during recovery. *Plant Physiol.* 48: 613–16.

Sanders, T. B. and J. L. Hamrick. Allozyme variation in *Elymus canadensis* from the tall grass prairie region. 1. Geographic variation. Submitted to *Amer. Midl. Nat.*

Sarukhán, J. 1974. Studies on plant demography: *Ranuculus repens* L., *R. bulbosus* L. and *R. acris* L. II. Reproductive strategies and seed population dynamics. *J. Ecol.* 62: 151–77.

Sarukhán, J. 1976. On selective pressures and energy allocation in populations of *Ranunculus repens* L., *R. bulbosus* L., and *R. acris* L. *Ann. Mo. Bot. Gard.* 63: 290–308.

Sarukhán, J. and M. Gadgil. 1974. Studies on plant demography: *Ranunculus repens* L., *R. bulbosus* L. and *R. acris* L. III. A mathematical model incorporating multiple modes of reproduction. *J. Ecol.* 62: 921–36.

Sarukhán, J. and J. L. Harper. 1973. Studies on plant demography: *Ranunculus repens* L., *R. bulbosus* L. and *R. acris* L. I. Population flux and survivorship. *J. Ecol.* 61: 675–716.

Sayers, R. L. and R. T. Ward. 1966. Germination responses in alpine species. *Bot. Gaz.* 127: 11–16.

Scala, J., C. Patrick, and G. Macbeth. 1968. FDPases of the castor bean endosperm and leaf: Properties and partial purification. *Arch. Biochem. Biophys.* 127: 576–84.

Schaal, B. A. 1974. Isolation by distance in *Liatris cylindracea*. *Nature (London)* 252: 703.

Schaal, B. A. 1975. Population structure and local differentiation in *Liatris cylindracea*. *Amer. Nat.* 109: 511–28.

Schaal, B. A. and D. A. Levin. 1976. The demographic genetics of *Liatris cylindracea* Michx. (Compositae). *Amer. Nat.* 110: 191–206.

Schafer, D. E. and D. O. Chilcote. 1969. Factors influencing persistence and depletion in buried seed populations. I. A model for analysis of parameters of buried seed persistence and depletion. *Crop Sci.* 9: 417–19.

Schaffer, W. M. 1974a. Selection for optimal life histories: The effects of age structure. *Ecology* 55: 291–303.

Schaffer, W. M. 1974b. Optimal reproductive effort in fluctuating environments. *Amer. Nat.* 108: 783–90.

Schaffer, W. M. and M. D. Gadgil. 1975. Selection for optimal life histories in plants. In M. L. Cody and J. M. Diamond (eds.), *Ecology and Evolution of Communities*. Cambridge and London: Harvard University Press, Belknap Press.

Schaffer, W. M. and E. G. Leigh. 1976. The prospective role of mathematical theory in plant ecology. *Syst. Bot.* 1: 209–32.

Schaffer, W. M. and M. L. Rosenzweig. 1977. Selection for life histories. II. Multiple equilibria and the evolution of alternative reproductive strategies. *Ecology* 58: 60–72.

Schemske, D. W. 1976. Pollinator specificity in *Lantana camara* and *L. trifolia* (Verbenaceae). *Biotropica* 8: 260–64.

Schimper, A. F. S. 1898. *Pflanzengeographie auf physiologischer Grundlage*. Jena: G. Fischer.

Schlesinger, W. H. and B. F. Chabot. 1977. The use of water and minerals by evergreen and deciduous shrubs in Okefenokee Swamp. *Bot. Gaz.* 138: 490–97.

Schoener, T. W. 1969. Optimal size and specialization in constant and fluctuating environments: An energy–time approach. *Brookhaven Symp. Biol.* 22: 103–14.

Schoener, T. W. 1971. Theory of feeding strategies. *Ann. Rev. Ecol. Syst.* 2: 369–404.

Schoener, T. W. 1974a. The compression hypothesis and temporal resource partitioning. *Proc. Nat. Acad. Sci. USA* 71: 4169–72.

Schoener, T. W. 1974b. Resource partitioning in ecological communities. *Science* 185: 27–39.

Schonherr, J. 1976. Water permeability of cuticular membranes. In

O. L. Lange, L. Kappen, and E. D. Schulze (eds.), *Water and Plant Life*. Berlin: Springer-Verlag.

Schopf, T. and J. Gooch. 1971. A natural experiment using deep-sea invertebrates to test the hypothesis that genetic homozygosity is proportional to environmental stability. (Abstract.) *Biol. Bull.* 141: 401.

Schreiner, E. J. 1974. *Populus* L. In *Seeds of Woody Plants in the United States*. U.S.D.A. Handbook no. 450, pp. 645–55.

Schulze, E. D., O. L. Lange, and W. Koch. 1972. Ökophysiologische Untersuchungen an Wild- und Kulturpflanzen der Negev-Wuste. II. Die Wirkung der Aubenfaktoren auf CO_2-Gaswechsel und Transpiration am Ende der Trockenzeit. *Oecologia* 8: 334–55.

Schwartz, D. 1971. Genetic control of alcohol dehydrogenase: A competition model for regulation of gene action. *Genetics* 67: 411–25.

Schwemer, J. and R. Paulsen. 1973. Three visual pigments in *Deilephila elpenor* (Lepidoptera, Sphingidae). *J. Comp. Physiol.* 86: 215–29.

Scogin, R. 1969. Isoenzyme polymorphism in natural populations of the genus *Baptisia* (Leguminosae). *Phytochemistry* 8: 1733–37.

Scogin, R. 1973. Leucine aminopeptidase polymorphism in the genus *Lupinus* (Leguminosae). *Bot. Gaz.* 134: 73–76.

Seabrook, J. A. E. and L. A. Dionne. 1976. Studies on the genus *Apios*. I. Chromosome number and distribution of *Apios americana* and *A. priceana*. *Canad. J. Bot.* 54: 2567–72.

Sears, E. R. 1952. Homeologous chromosomes in *Triticum aestivum*. (Abstract.) *Genetics* 37: 624.

Sears, E. R. 1976. Genetic control of chromosome pairing in wheat. *Ann. Rev. Genet.* 10: 31–51.

Selander, R. K. 1976. Genic variation in natural populations. In F. J. Ayala (ed.), *Molecular Evolution*. Sunderland, Mass.: Sinauer Publ.

Selander, R. K. and D. W. Kaufman. 1973a. Self-fertilization and genetic population structure in a colonizing land snail. *Proc. Nat. Acad. Sci. USA* 70: 1186–90.

Selander, R. K. and D. W. Kaufman. 1973b. Genic variability and strategies of adaptation in animals. *Proc. Nat. Acad. Sci. USA* 70: 1875–77.

Selander, R. K., M. Smith, S. Yang, W. Johnson, and J. Gentry. 1971. Biochemical polymorphism and systematics in the genus *Pero-*

myscus. I. Variation in the old field mouse. Texas Univ. Publ. 7103: 49–73.

Senadhira, D. 1976. Genetic variation in corn and its relatives. Ph.D. dissertation, University of California, Davis.

Shaffer, B., J. Rytka, and G. R. Fink. 1969. Nonsense mutations affecting the his 4 enzyme complex of yeast. Proc. Nat. Acad. Sci. USA 63: 1198–1205.

Sharitz, R. R. and J. F. McCormick. 1973. Population dynamics of two competing annual plant species. Ecology 54: 723–40.

Shaver, G. R. and W. D. Billings. 1975. Root production and root turnover in a wet tundra ecosystem, Barrow, Alaska. Ecology 56: 401–9.

Sheldon, J. C. 1974. The behavior of seeds in soil. III. The influence of seed morphology and the behavior of seedlings on the establishment of plants from surface lying seeds. J. Ecol. 62: 47–66.

Sherman, M. 1946. Karyotype evolution: A cytogenetic study of seven species and six interspecific hybrids of Crepis. Univ. Calif. Publ. Bot. 18: 369–408.

Shields, L. M. 1950. Leaf xeromorphy as related to physiological and structural influences. Bot. Rev. 16: 399–447.

Shiroya, T., G. R. Lister, V. Slankis, G. Krotkov, and C. D. Nelson. 1966. Seasonal changes in respiration, photosynthesis, and translocation of the ^{14}C labelled products of photosynthesis in young Pinus strobus L. plants. Ann. Bot. 30: 81–91.

Sims, P. L. and J. S. Singh. 1971. Herbage dynamics and net primary production in certain ungrazed and grazed grasslands in North America. In N. R. French (ed.), Preliminary Analysis of Structure and Function in Grasslands. Range Sci. Dept., Sci. Ser. No. 10.

Singh, R. S. and S. K. Jain. 1971. Population biology of Avena. II. Isoenzyme polymorphisms in populations of the Mediterranean region and central California. Theor. Appl. Genet. 41: 79–84.

Singh, R., R. C. Lewontin, and A. A. Felton. 1976. Genetic heterogeneity within electrophoretic "alleles" of xanthine dehydrogenase in Drosophila pseudoobscura. Genetics 84: 609–29.

Sinnott, E. W. and I. W. Bailey. 1915. Investigations on the phylogeny of the angiosperms. V. Foliar evidence as to the ancestry and early climatic environment of the angiosperms. Amer. J. Bot. 2: 1–22.

Slatkin, M. 1973. Gene flow and selection in a cline. Genetics 75: 733–56.

Slatkin, M. 1974. Competition and regional coexistence. *Ecology* 55: 128–34.

Slatkin, M. 1975. Gene flow and selection in a two-locus system. *Genetics* 81: 787–802.

Slatkin, M. and T. Maruyama. 1975. Genetic drift in a cline. *Genetics* 81: 209–22.

Slavik, B. 1975. Water stress, photosynthesis and the use of photosynthates. In J. P. Cooper (ed.), *Photosynthesis and Productivity in Different Environments*. Cambridge: Cambridge University Press.

Slayter, R. O. 1967. *Plant–Water Relationships*. New York: Academic Press.

Slayter, R. O. 1973. The effect of internal water status on plant growth, development and yield. In R. O. Slayter (ed.), *Plant Response to Climatic Factors*, Proc. Uppsala Symp. Paris: UNESCO.

Slayter, R. O. 1978. Altitudinal variation in the photosynthetic characteristics of snow gum, *Eucalyptus pauciflora* Sieb. ex Spreng. VII. Relationship between gradients of field temperature and photosynthetic temperature optima in the Snowy Mountains area. *Austr. J. Bot.* 26: 111–21.

Slobodchikoff, C. N., ed. 1976. *Concepts of Species*. Stroudsburg, Pa.: Dowden, Hutchinson, and Ross.

Small, E. 1972a. Ecological significance of four critical elements in plants of raised *Sphagnum* bogs. *Ecology* 53: 498–503.

Small, E. 1972b. Photosynthetic rates in relation to nitrogen recycling as an adaption to nutrient deficiency in peat bog plants. *Canad. J. Bot.* 50: 2227–33.

Smith, B. N. and W. V. Brown. 1973. The Kranz syndrome in the Gramineae as indicated by carbon isotopic ratios. *Amer. J. Bot.* 60: 505–13.

Smith, B. N. and S. Epstein. 1971. Two categories of $^{13}C/^{12}C$ ratios for higher plants. *Plant Physiol.* 47: 380–84.

Smith, D. M. 1951. The influence of seedbed conditions on the regeneration of eastern white pine. *Conn. Agr. Sta. Bull.* 545.

Smith, H. 1973. Light quality and germination: Ecological implications. In W. Heydecker (ed.) *Seed Ecology*. University Park, Pa.: Penn State University Press.

Smith, J. M., See Maynard Smith, J.

Smith-White, S. 1959. Cytological evolution in the Australian flora. *Cold Spring Harbor Symp. Quant. Biol.* 24: 273–89.

Smouse, P. and K. Kojima. 1972. Maximum likelihood analysis of population differences in allelic frequencies. Genetics 72: 709–19.

Snaydon, R. W. 1970. Rapid population differentiation in a mosaic environment. I. The response of Anthoxanthum odoratum population to soils. Evolution 24: 257–69.

Snaydon, R. W. and M. S. Davies. 1972. Rapid population differentiation in a mosaic environment. II. Morphological variation in Anthoxanthum odoratum. Evolution 26: 390–405.

Snaydon, R. W. and M. S. Davies. 1976. Rapid population differentiation in a mosaic environment. IV. Populations of Anthoxanthum odoratum at sharp boundries. Heredity 37: 9–25.

Snow, B. K. and D. W. Snow. 1972. Feeding niches of hummingbirds in a Trinidad valley. J. Anim. Ecol. 41: 471–85.

Snow, R. 1960. Chromosomal differentiation in Clarkia dudleyana. Amer. J. Bot. 47: 302–9.

Snow, R. 1963. Cytogenetic studies in Clarkia, section Primigenia: I. A cytological survey of Clarkia amoena. Amer. J. Bot. 50: 337–48.

Snyder, A. W. and W. H. Miller. 1972. Fly color vision. Vision Res. 12: 1389–96.

Sokal, R. R. and T. J. Crovello. 1970. The biological species concept: A critical evaluation. Amer. Nat. 104: 127–53.

Solbrig, O. T. 1971. The population biology of dandelions. Amer. Sci. 59: 686–94.

Solbrig, O. T. 1976a. On the relative advantages of cross- and self-fertilization. Ann. Mo. Bot. Gard. 63: 262–76.

Solbrig, O. T. 1976b. Plant population biology: An overview. Syst. Bot. 1: 202–8.

Solbrig, O. T. and R. C. Rollins. 1977. The evolution of autogamy in species of the mustard genus Leavenworthia. Evolution 31: 265–81.

Solbrig, O. T. and B. B. Simpson. 1974. Components of regulation of a population of dandelions in Michigan. J. Ecol. 62: 473–86.

Solbrig, O. T. and B. B. Simpson. 1977. A garden experiment on competition between biotypes of the common dandelion (Taraxacum officinale). J. Ecol. 65: 427–30.

Somero, G. and M. Soule. 1974. Genetic variation in marine fish as a test of the niche-variation hypothesis. Nature (London) 249: 670–72.

Soo Hoo, C. F. and G. Fraenkel. 1966a. The selection of food plants in a polypagous insect, *Prodenia eridania* (Cramer). *J. Insect Physiol.* 12: 693–709.

Soo Hoo, C. F. and G. Fraenkel. 1966b. The consumption, digestion and utilization of food plants by a polyphagous insect, *Prodenia eridania* (Cramer). *J. Insect Physiol.* 12: 711–30.

Soule, M. 1976. Allozyme variation: Its determinants in space and time. In F. J. Ayala (ed.), *Molecular Evolution*. Sunderland, Mass.: Sinauer Publ.

Southwood, T. R. E., R. M. May, M. P. Hassell, and G. R. Conway. 1974. Ecological strategies and population parameters. *Amer. Nat.* 108: 791–804.

Sparrow, A. H., H. J. Price, and A. G. Underbrink. 1972. A survey of DNA content per cell and per chromosome of prokaryotic and eukaryotic organisms: Some evolutionary considerations. In *Basic Mechanisms of Morphogenesis; Brookhaven Symp. Biol.* 23: 451–94.

Specht, R. L. 1963. Dark Island heath (Ninety-Mile Plain, South Australia). VII. The effect of fertilizers on composition and growth, 1950–60. *Aust. J. Bot.* 11: 67–94.

Specht, R. L. 1969a. A comparison of the sclerophyllous vegetation characteristic of Mediterranean type climates in France, California, and southern Australia. *Aust. J. Bot.* 17: 277–92.

Specht, R. L. 1969b. A comparison of the sclerophyllous vegetation characteristic of Mediterranean type climates in France, California, and southern Australia. II. Dry matter, energy, and nutrient accumulation. *Aust. J. Bot.* 17: 293–308.

Standley, P. C. 1937. *Flora of Costa Rica*. Field Mus. Nat. Hist., Bot. Ser., Publ. 391. V. 18.

Staniforth, R. J. and P. B. Cavers. 1977. The importance of cottontail rabbits in the dispersal of *Polygonum* spp. *J. Appl. Ecol.* 14: 261–67.

Stanley, R. G. and H. F. Linskens. 1974. *Pollen: Biology, Biochemistry, Management*. New York: Springer-Verlag.

Stearns, S. C. 1976. Life-history tactics: A review of the ideas. *Quart. Rev. Biol.* 51: 3–47.

Stebbins, G. L. 1949. The evolutionary significance of natural and artificial polyploids in the family Gramineae. Proc. 8th Int. Congr. Genet., *Hereditas*, Suppl. Vol. 461–85.

Stebbins, G. L. 1950. *Variation and Evolution in Plants*. New York: Columbia University Press.

Stebbins, G. L. 1953. A new classification of the tribe Cichorieae, family Compositae. Madroño 12: 33–64.

Stebbins, G. L. 1957. Self-fertilization and population variability in the higher plants. Amer. Nat. 91: 337–54.

Stebbins, G. L. 1958. Longevity, habit and release of genetic variability in the higher plants. Cold Spring Harbor Symp. Quant. Biol. 23: 365–78.

Stebbins, G. L. 1959. The role of hybridization in evolution. Proc. Amer. Phil. Soc. 103: 231–51.

Stebbins, G. L. 1960. The comparative evolution of genetic systems. In S. Tax (ed.), Evolution after Darwin; vol. 2: The Evolution of Life. Chicago: University of Chicago Press.

Stebbins, G. L. 1965. Colonizing species of the native California Flora. In H. G. Baker and G. L. Stebbins (eds.), The Genetics of Colonizing Species. New York and London: Academic Press.

Stebbins, G. L. 1966a. Chromosome variation and evolution. Science 152: 1463–69.

Stebbins, G. L. 1966b. Processes of Organic Evolution. Englewood Cliffs, N.J.: Prentice-Hall.

Stebbins, G. L. 1970. Adaptive radiation in angiosperms. I. Pollination mechanisms. Ann. Rev. Ecol. Syst. 1: 307–26.

Stebbins, G. L. 1971a. Chromosomal Evolution in Higher Plants. Reading, Mass.: Addison-Wesley.

Stebbins, G. L. 1971b. Adaptive radiation of reproductive characteristics in angiosperms. II. Seeds and seedlings. Ann. Rev. Ecol. Syst. 2: 237–60.

Stebbins, G. L. 1974. Flowering Plants: Evolution above the Species Level. Cambridge, Mass.: Harvard University Press, Belknap Press.

Stebbins, G. L. 1976a. Seed and seedling ecology in annual legumes. I. A comparison of seed size and seedling development in some annual species. Oecol. Plant. 11: 321–31.

Stebbins, G. L. 1976b. Chromosomes, DNA and plant evolution. Evol. Biol 9: 1–34.

Stebbins, G. L. and B. Crampton. 1961. A suggested revision of the grass genera of temperate North America. Recent Adv. Bot. (University of Toronto Press): 133–45.

Stebbins, G. L., B. L. Harvey, E. L. Cox, J. N. Rutger, G. Jelencovic, and E. Yagil. 1963. Identification of the ancestry of an amphidiploid Viola with the aid of paper chromatography. Amer. J. Bot. 50: 830–38.

Steenberg, W. F. and C. H. Lowe. 1969. Critical factors during the first years of life of the Saguaro (Cereus giganteus) at Saguaro National Monument, Arizona. Ecology 50: 825–34.

Stern, K. and L. Roche. 1974. Genetics of Forest Ecosystems. New York: Springer-Verlag.

Stern, W. L. 1971. Adaptive Aspects of Insular Evolution. Pullman, Wash.: Washington State University Press.

Stewart, D. W. and E. R. Lemon. 1972. The energy budget at the earth's surface: A simulation of net photosynthesis of field corn. ECOM Atmospheric Sciences Lab., Fort Huachuca, Ariz. Int. Rep. 69–3.

Stiles, F. G. 1975. Ecology, flowering phenology and hummingbird pollination of some Costa Rican Heliconia. Ecology 56: 285–301.

Stiles, F. G. 1976. Taste preferences, color preferences, and flower choice in hummingbirds. Condor 78: 10–26.

Stone, D. E. 1959. A unique balanced breeding system in the vernal pool Mouse-tails. Evolution 13: 151–74.

Stone, E. C. and Juhren, G. 1951. The effect of fire on the germination of Rhus ovata Wats. Amer. J. Bot. 38: 368.

Stoner, W. A. and P. C. Miller. 1975. Water relations of plant species in wet coastal tundra at Barrow, Alaska. Arct. Alp. Res. 7: 109–24.

Stoner, W. A., P. C. Miller, S. P. Richards, and S. A. Barkley. 1978. Internal nutrient recycling as related to plant life form: A simulation approach. In Proc. Environmental Chemistry and Cycling Processes Symposium, 28–30 April 1976, Athens, Georgia.

Stowe, L. G. and J. A. Teeri. 1978. The geographic distribution of C_4 species of the Dicotyledonae in relation to climate. Amer. Nat. 112: 609–23.

Strain, B. R. and V. C. Chase. 1966. Effect of past and prevailing temperatures on the carbon dioxide exchange capacities of some woody desert perennials. Ecology 47: 1043–45.

Strandberg, J. 1973. Spatial distribution of cabbage black rot and the estimation of diseased plant populations. Phytopathology 63: 998–1002.

Street, H. E. 1969. Factors influencing the initiation and activity of meristems in roots. In W. J. Whittington (ed.), Root Growth. London: Butterworths.

Sukatschew, W. 1928. Einige experimentelle Untersuchungen über den Kampf ums Dasein zwischen Biotypen derselben Art. Zeitschr. Abst. Vererbungsl. 47: 54–74.

Summerfield, R. J. 1972. Factors affecting the germination and seedling establishment of *Narthecium ossifragum* on mire ecosystems. *J. Ecol.* 60: 793–98.

Suneson, C. A. and G. A. Wiebe. 1942. Survival of barley and wheat varieties in mixtures. *J. Amer. Soc. Agron.* 34: 1052–56.

Sussman, A. S. and H. O. Halvorsen. 1966. *Spores: Their Dormancy and Germination.* New York: Harper and Row.

Suthers, R. A. 1970. Vision, olfaction, taste. In W. A. Wimsett (ed.), *Biology of Bats,* vol. 2. New York: Academic Press.

Swain, T. 1976. Nature and properties of flavonoids. In T. W. Goodwin (ed.), *Chemistry and Biochemistry of Plant Pigments,* vol. 1, 2d ed. New York: Academic Press.

Swihart, S. L. 1970. The neural basis of colour vision in the butterfly, *Papilio troilus. J. Insect Physiol.* 16: 1623–36.

Sybenga, J. 1975. *Meiotic Configurations.* Berlin, Heidelberg, and New York: Springer-Verlag.

Syvertsen, J. P., G. L. Nickell, R. W. Spellenberg, and G. L. Cunningham. 1976. Carbon reduction pathways and standing crop in three Chihuahuan Desert plant communities. *Southwest. Nat.* 21: 311–20.

Tahvanainin, J. O. and R. B. Root. 1972. The influence of vegetational diversity on the population ecology of a specialized herbivore, *Phyllotreta* (Coleoptera: Chrysomelidae). *Oecologia* 10: 321–46.

Tamm, C. O. 1972. Survival and flowering of perennial herbs. III. The behavior of *Primula veris* on permanent plots. *Oikos* 23: 159–66.

Taylor, S. E. 1975. Optimal leaf form. In D. M. Gates and R. B. Schmerl (eds.), *Perspectives of Biophysical Ecology,* New York: Springer-Verlag.

Taylor, S. E. and D. M. Gates. 1970. Some field methods for obtaining meaningful leaf diffusion resistances and transpiration rates. *Oecologia* 5: 105–13.

Taylor, S. E. and O. J. Sexton. 1972. Some implications for leaf tearing in Musaceae. *Ecology* 53: 143–49.

Taylorson, R. B. and H. A. Borthwick. 1969. Light filtration by foliar canopies: Significance for light-controlled weed seed germination. *Weed Sci.* 17: 48–51.

Teeri, J. A. and L. G. Stowe. 1976. Climatic patterns and the distribution of C_4 grasses in North America. *Oecologia* 23: 1–12.

Templeton, A. R. and E. D. Rothman. 1976. *Evolution and Fine-*

grained *Environmental Runs*. Technical Report no. 65, Univ. of Michigan, Dept. of Statistics.

Terborgh, J. 1973. On the notion of favorableness in plant ecology. *Amer. Nat.* 107: 481–501.

Tevis, L. 1958a. Germination and growth of ephemerals induced by sprinkling a sandy desert. *Ecology* 39: 681–88.

Tevis, L. 1958b. A population of desert ephemerals germinated by less than one inch of rain. *Ecology* 39: 688–95.

Thien, L. B., W. H. Heimermann, and R. T. Holman. 1975. Floral odors and quantitative taxonomy of *Magnolia* and *Liriodendron*. *Taxon* 24: 557–68.

Thoday, D. 1931. The significance of reduction in the size of leaves. *J. Ecol.* 19: 297–303.

Thomas, A. G. 1972. Autecological studies on *Hieracium* in Wellington County, Ontario. Ph.D. dissertation, University of Guelph, Ontario.

Thomas, A. G. and H. M. Dale. 1975. The role of seed reproduction in the dynamics of established populations of *Hieracium floribundum* and a comparison with that of vegetative reproduction. *Canad. J. Bot.* 53: 3022–31.

Thomas, A. G. and H. M. Dale. 1976. Cohabitation of three *Hieracium* species in relation to the spatial heterogeneity in an old pasture. *Canad. J. Bot.* 54: 2517–29.

Thompson, P. A. 1973. Seed germination in relation to ecological and geographical distribution. In V. H. Heywood (ed.), *Taxonomy and Ecology*. London and New York: Academic Press.

Thompson, P. A. 1975. Characterization of the germination responses of *Silene dioica* (L.) Clairv. populations from Europe. *Ann. Bot.* 39: 1–19.

Thrower, N. J. W. and D. Bradbury, eds. 1977. *Atlas of the Mediterranean Scrub Project (AES)*. Stroudsburg, Pa.: Dowden, Hutchinson, and Ross.

Tieszen, L. L. 1973. Photosynthesis and respiration in Arctic tundra grasses: Field light intensities and temperature responses. *Arct. Alp. Res.* 5: 239–52.

Tieszen, L. L. 1979. Photosynthesis. In J. Brown, F. Bunnell, S. MacLean, P. C. Miller, and L. L. Tieszen (eds.), *The Structure and Function of the Tundra Ecosystem; vol. 1: A U.S. Tundra Biome Synthesis*. Stroudsburg, Pa.: Dowden, Hutchinson, and Ross (in press).

564   LITERATURE CITED

Tieszen, L. L., P. C. Miller, M. C. Lewis, J. C. Mayo, F. S. Chapin, III, and W. C. Oechel. 1976. Processes of primary production in the tundra. In J. J. Moore (ed.), *The Ecology of Tundra and Related Habitats*. Cambridge: Cambridge University Press.

Tigerstedt, P. M. A. 1973. Studies on isozyme variation in marginal and central populations of *Picea abies*. *Hereditas* 75: 47–60.

Tinbergen, L. 1960. The natural control of insects in pine-woods. I. Factors influencing the intensity of predation by songbirds. *Arch. Neerl. Zool.* 13: 265–343.

Ting, I. P., H. B. Johnson, and S. Szarek. 1972. Net CO_2 fixation in crassulacean acid metabolism plants. In C. C. Black (ed.), *Net Carbon Dioxide Assimilation in Higher Plants*. Raleigh, N.C.: Cotton, Inc.

Tischler, G. 1935. Die Bedeutung der Polyploidie für die Verbreitung der Angiospermen, erlautert an den Arten Schleswig Holsteins, mit Ausblicken auf andere Florengebiete. *Bot. Jahrb.* 67: 1–36.

Titman, D. 1976. Ecological competition between algae: Experimental confirmation of resource-based competition theory. *Science* 192: 463–65.

Tobgy, H. A. 1943. A cytological study of *Crepis fuliginosa, C. neglecta* and their F_1 hybrid, and its bearing on the phylogenetic reduction in chromosome numbers. *J. Genet.* 45: 67–111.

Toole, E. H. and E. Brown. 1946. Final results of the Duvel buried seed experiment. *J. Agr. Res.* 72: 201–10.

Torres, A. M., U. Diedenhofen, and I. M. Johnstone. 1977. The early allele of alcohol dehydrogenase in sunflower populations. *J. Hered.* 68: 11–16.

Tramer, E. J. 1975. The regulation of plant species diversity on an early successional old-field. *Ecology* 56: 905–14.

Tregunna, E. B., B. N. Smith, J. A. Berry, and W. J. S. Downton. 1970. Some methods for studying the photosynthetic taxonomy of the angiosperms. *Canad. J. Bot.* 48: 1209–14.

Treharne, D. J. 1972. Biochemical limitations in photosynthetic rates. In A. Rees, K. Cockshull, and D. Hurd (eds.), *Crop Processes in Controlled Environment*. New York: Academic Press.

Trenbath, B. R. and J. L. Harper. 1973. Neighbour effects in the genus *Avena*. I. Comparison of crop species. *J. Appl. Ecol.* 10: 379–400.

Trimble, G. R., Jr. 1975. Summaries of some silvical characteristics of several Appalachian hardwood trees. USDA Forest Service General Technical Report NE-16.

Tripathi, R. S. and J. L. Harper. 1973. The comparative biology of *Agropyron repens* (L.) Beauv. and *A. caninum* (L.) Beauv. I. The growth of mixed populations established from tillers and from seeds. *J. Ecol.* 61: 353–68.

Troughton, A. 1968. Influence of genotype and mineral nutrition on the distribution of growth within plants of *Lolium perenne* L. grown in soil. *Ann. Bot.* 32: 411–23.

Troughton, A. and W. J. Whittington. 1969. The significance of genetic variation in root systems. In W. J. Whittington (ed.), *Root Growth*. London: Butterworths.

Tsunoda, S. 1972. Photosynthetic efficiency in rice and wheat. In *Rice Breeding*. Los Baños, Philippines: Rich Research Institute.

Tullock, G. 1971. The coal tit as a careful shopper. *Amer. Nat.* 105: 77–80.

Turesson, G. 1922a. The species and the variety as ecological units. *Hereditas* 3: 100–113.

Turesson, G. 1922b. The genotypical response of the plant species to the habitat. *Hereditas* 3: 211–350.

Turesson, G. 1925. The plant species in relation to habitat and climate. Contributions to the knowledge of genecological units. *Hereditas* 6: 147–236.

Turesson, G. 1931. The selective effect of the climate upon the plant species. *Hereditas* 14: 99–152.

Turkington, R. A. 1975. Relationships between neighbours among species of permanent grassland (especially *Trifolium repens* L.). Ph.D. dissertation, University College of North Wales, Bangor.

Turner, R. M., S. M. Alcorn, and G. Olin. 1969. Mortality of transplanted saguaro seedlings. *Ecology* 50: 835–44.

Turner, R. M., S. M. Alcorn, G. Olin, and J. A. Booth. 1966. The influence of shade, soil and water on saguaro seedling establishment. *Bot. Gaz.* 127: 95–102.

Uchichima, Z. 1962. Studies on the microclimate within the plant communities. I. On the turbulent transfer coefficient within plant layers. *J. Agr. Meteorol.* (Tokyo) 18: 1–9.

USDA. 1948. *The Yearbook of Agriculture*. U.S. Department of Agriculture, Washington, D.C.

Utech, F. H. and S. Kawano. 1975. Spectral polymorphism in angiosperm flowers determined by differential ultraviolet reflectance. *Bot. Mag. Tokyo* 88: 9–30.

van der Pijl, L. 1969a. *Principles of Dispersal in Higher Plants*. Berlin: Springer-Verlag.

van der Pijl, L. 1969b. Evolutionary action of tropical animals on the reproduction of plants. *Biol. J. Linn. Soc. (London)* 1: 85–92.

van der Pijl, L., and C. H. Dodson. 1966. *Orchid Flowers, Their Evolution and Pollination*. Coral Gables, Fla.: University of Miami Press.

Van Riper, W. 1960. Does a hummingbird find nectar through its sense of smell? *Sci. Amer.* 202: 157.

Vareschi, V. 1970. *Flora de los Páramos de Venezuela*. Merida: Universidad de los Andes.

Vasek, F. C. and J. Harding. 1976. Outcrossing in natural populations. V. Analysis of outcrossing, inbreeding, and selection in *Clarkia exilis* and *C. tembloriensis*. *Evolution* 30: 403–11.

Vieira da Silva, J. 1976. Water stress, ultrastructure and enzymatic activity. In O. L. Lange , L. Kappen, and E. D. Schulze (eds.), *Water and Plant Life*. Berlin: Springer-Verlag.

Vigue, C. and F. Johnson. 1973. Isozyme variability in *Drosophila*. VI. Frequency-property: Environment relationships of allelic alcohol dehydrogenases in *D. melanogaster*. *Biochem. Genet.* 9: 213.

Vogel, S. 1966. Parfums ammelnde Bienen als Bestauber von Orchidaceen und *Gloxinia*. *Osterr. Bot. Z.* 113: 302–61.

Volkens, G. 1887. Die Flora der aegyptisch-arabischen Wüste. *Grundlage anat. physiol. Forsch.* Berlin.

Waddington, C. H. 1957. *The Strategy of the Genes*. London: G. Allen.

Waddington, C. H. 1960. Evolutionary Adaptation. In S. Tax (ed.), *Evolution after Darwin; vol. 2: The Evolution of Life*. Chicago: University of Chicago Press.

Waggoner, P. E., and W. E. Reifsnyder. 1968. Simulation of temperature, humidity, and evaporation profiles in a leaf canopy. *J. Appl. Meteorol.* 7: 400–409.

Waggoner, P. E., G. M. Furnival, and W. E. Reifsnyder. 1969. Simulation of microclimate in a forest canopy. *For. Sci.* 15: 37–45.

Waldron, G. E., R. J. Hilton, and J. D. Ambrose. 1976. Evidence of microevolution in an escaped pear population. *Canad. J. Bot.* 54: 2857–67.

Wallace, D. H., J. L. Ozbun, and H. M. Munger. 1972. Physiological genetics of crop yield. *Adv. Agron.* 24: 97–146.

Wallace, J. W. and R. L. Mansell, eds. 1976. *Biochemical Interaction between Plants and Insects*. New York and London: Plenum Press.

Walter, H. 1973. Vegetation of the Earth. New York: Springer-Verlag.

Walter, H. and K. Kreeb. 1970. Die Hydratation und Hydratur des Protoplasmas der Pflanze und ihre ökophysiologische Bedentung. Weimar: Springer.

Ware, S. A. and E. Quarterman. 1969. Seed germination in cedar glade Talinum. Ecology 50: 137–40.

Wareing, P. F. 1966. Ecological aspects of seed dormancy and germination. In J. G. Hawkes (ed.), Reproductive Biology and Taxonomy of Vascular Plants. London: Pergamon.

Warembourg, F. P. and E. A. Paul. 1977. Seasonal transfers of assimilated ^{14}C in grassland: Plant production and turnover, soil and plant respiration. Soil Biol. Biochem. 9: 295–301.

Warren Wilson, J. 1966. Effect of temperature on net assimilation rate. Ann. Bot. 30: 753–61.

Watson, D. J. 1947. Comparative physiological studies on the growth of field crops. I. Variation in net assimilation rate and leaf area between species and varieties and within and between years. Ann. Bot. n.s. 11: 41–76.

Watt, A. S. 1919. On the causes of failure of natural regeneration in British oakwoods. J. Ecol. 7: 173–203.

Watt, A. S. 1923. On the ecology of British beechwoods, with special reference to their regeneration. J. Ecol. 11: 1–48.

Watt, A. S. 1955. Bracken versus heather: study in plant sociology. J. Ecol. 43: 490–506.

Watt, A. S. 1974. Senescence and rejuvenation in ungrazed chalk grassland (Grassland B) in Breckland: The significance of litter and of moles. J. Appl. Ecol. 11: 1157–71.

Webb, D. P. 1976. Root growth in Acer saccharum marsh seedlings: Effects of light intensity and photoperiod on root elongation rates. Bot. Gaz. 137: 211–17.

Webb, L. J. 1968. Environmental relationships of the structural types of Australian rain forest vegetation. Ecology 49: 296–311.

Weller, S. G. and R. Ornduff. 1977. Cryptic self-incompatibility in Amsinckia grandiflora. Evolution 31: 47–51.

Went, F. W. 1948. Ecology of desert plants. I. Observations on germination in the Joshua Tree National Monument, California. Ecol. 29: 242–53.

Went, F. W. 1949. Ecology of desert plants. II. The effect of rain and temperature on germination and growth. Ecology 30: 1–13.

Went, F. W. and N. Stark. 1968. Mycorrhiza. BioScience 18: 1035–39.

Went, F. W. and M. Westergaard. 1949. Ecology of desert plants. III. Development of plants in Death Valley National Monument, California. Ecology 30: 26–38.

Went, F. W., G. Juhren, and M. C. Juhren. 1952. Fire and biotic factors affecting germination. Ecology 33: 351.

Werner, E. E. 1977. Species packing and niche complementarity in three sunfishes. Amer. Nat. 111: 553–78.

Werner, E. E. and D. J. Hall. 1976. Niche shifts in sunfishes: Experimental evidence and significance. Science 191: 404–6.

Werner, P. A. 1975a. Predictions of fate from rosette size in teasel (Dipsacus fullonum L.). Oecologia 20: 197–201.

Werner, P. A. 1975b. A seed trap for determining patterns of seed deposition in terrestrial plants. Canad. J. Bot. 53: 810–13.

Werner, P. A. 1975c. The effects of plant litter on germination in teasel (Dipsacus sylvestris). Amer. Midl. Nat. 94: 470–76.

Werner, P. A. 1976. Ecology of plant populations in successional environments. Syst. Bot. 1: 246–68.

Werner, P. A. 1977. Colonization success of a "biennial" plant species: Experimental field studies in species colonization and replacement. Ecology 58: 840–49.

Werner, P. A. 1978. On the determination of age in Liatris aspera using cross-sections of corms. Amer. Nat. 112: 1113–20.

Werner, P. A. and H. Caswell. 1977. Population growth rates and age vs. stage-distribution models for teasel (Dipsacus sylvestris Huds.). Ecology 58: 1103–11.

Werner, P. A. and W. J. Platt. 1976. Ecological relationships of co-occurring goldenrods (Solidago: Compositae). Amer. Nat. 110: 959–71.

Werner, P. A. and R. Rioux. 1977. The biology of Canadian weeds. 24. Agropyron repens (L.) Beauv. Canad. J. Plant Sci. 57: 905–19.

Wesson, G. and P. F. Waring. 1969. The induction of light sensitivity in weed seeds by burial. J. Exp. Bot. 20: 414–25.

White, J., and J. L. Harper. 1970. Correlated changes in plant size and number in plant populations. J. Ecol. 58: 467–85.

Whitmore, T. C. 1975. Tropical Rain Forests of the Far East. Oxford: Clarendon Press.

Whittaker, R. H. 1969. Evolution of diversity in plant communities. Brookhaven Symp. Biol. 22: 178–96.

Whittaker, R. H. 1972. Evolution and measurement of species diversity. *Taxon* 21: 213–51.

Whittaker, R. H. and S. A. Levin, eds. 1976. *Niche: Theory and Application.* Stroudsburg, Pa.: Dowden, Hutchinson, and Ross.

Whittaker, R. H. and W. A. Niering. 1964. Vegetation of the Santa Catalina Mountains, Arizona. I. Ecological classification and distribution of species. *J. Ariz. Acad. Sci.* 3: 9–34.

Wiebe, H. H. 1978. The significance of the plant vacuole. *BioScience* 28: 327–31.

Wiens, J. A. 1976. Population responses to patchy environments. *Ann. Rev. Ecol. Syst.* 7: 81–120.

Wilbur, H. M. 1976. Life history evolution of seven milkweeds of the genus *Asclepias. J. Ecol.* 64: 223–40.

Wilbur, H. M. 1977. Propagule size, number, and dispersion pattern in *Ambystoma* and *Asclepias. Amer. Nat.* 111: 43–68.

Wilbur, H. M., D. W. Tinkle, and J. P. Collins. 1974. Environmental certainty, trophic level, and resource availability in life history evolution. *Amer. Nat.* 108: 805–17.

Willemsen, R. W. 1975a. Effect of stratification temperature and germination temperature on germination and the induction of secondary dormancy in common ragweed weeds. *Amer. J. Bot.* 62: 1–5.

Willemsen, R. W. 1975b. Dormancy and germination of common ragweed seeds in the field. *Amer. J. Bot.* 62: 639–44.

Williams, E. D. 1973. Seed germination of *Agrostis gigantea* Roth. *Weed Res.* 13: 310–24.

Williams, G. C. 1975. *Sex and Evolution.* Princeton, N.J.: Princeton University Press.

Williams, G. C. and J. B. Mitton. 1973. Why reproduce sexually? *J. Theor. Biol.* 39: 545–54.

Williams, J. 1974. Root density and water potential gradients near the plant root. *J. Exp. Bot.* 25: 669–74.

Williams, J. 1976. Dependence of root water potential on root radius and density. *J. Exp. Bot.* 27: 121–24.

Williams, N. H. and E. H. Dodson. 1972. Selective attraction of male euglossine bees to orchid floral fragrances and its importance in long distance pollen flow. *Evolution* 26: 84–95.

Williams, W. T., G. N. Lance, L. J. Webb, and M. B. Dale. 1969. Studies in the numerical analysis of complex rain-forest communities. III. Analysis of successional data. *J. Ecol.* 57: 515–35.

Willson, M. F. and B. J. Rathcke. 1974. Adaptive design of the floral display in *Asclepias syriaca* L. *Amer. Nat.* 92: 47–57.

Wilson, E. O. 1965. The challenge from related species. In H. G. Baker and G. L. Stebbins (eds.), *The Genetics of Colonizing Species*. New York and London: Academic Press.

Wilson, E. O. 1970. Chemical communication within animal species. In E. Sondheimer and J. B. Simeone (eds.), *Chemical Ecology*. New York: Academic Press.

Wilson, R. E. and E. L. Rice. 1968. Allelopathy as expressed by *Helianthus annuus* and its role in old field succession. *Bull. Torrey Bot. Club* 95: 432–48.

Winge, O. 1917. The chromosomes: Their numbers and general importance. *C. R. Trav. Lab. Carlsberg* 13: 131–275.

Winkler, H. 1916. Über die experimentelle Erzeugung von Pflanzen mit abweichenden Chromosomenzahlen. *Zeitschr. Bot.* 8: 471–531.

Winter, K., J. H. Troughton, and K. A. Card. 1976. ^{13}C values of grass species collected in the northern Sahara Desert. *Oecologia* 25: 115–23.

Wolf, L. L. 1969. Female territoriality in a tropical hummingbird. *Auk* 85: 490–504.

Wolf, L. L., F. R. Hainsworth, and F. B. Gill. 1975. Foraging efficiencies and time budgets in nectar-feeding birds. *Ecology* 56: 117–28.

Wolf, L. L., F. G. Stiles, and F. R. Hainsworth. 1976. Ecological organizations of a tropical highland hummingbird community. *J. Anim. Ecol.* 45: 349–79.

Wood, C. 1955. Evidence for the hybrid origin of *Drosera anglica*. *Rhodora* 57: 105–30.

Wood, O. M. 1938. Seedling reproduction of oak in southern New Jersey. *Ecology* 19: 276–93.

Woodell, S. R. J., H. A. Mooney, and A. J. Hill. 1969. The behaviour of *Larrea divaricata* (creosote bush) in response to rainfall in California. *J. Ecol.* 57: 37–44.

Woodwell, G. M., R. H. Whittaker, and R. A. Houghton. 1975. Nutrient concentrations in plants in the Brookhaven oak–pine forest. *Ecology* 56: 318–32.

Wooten, J. W. 1970. Experimental investigations of the *Sagittaria graminea* complex: Transplant studies and genecology. *J. Ecol.* 58: 233–41.

Wright, J. W. 1976. *Introduction to Forest Genetics.* New York: Academic Press.

Wright, S. 1931. Evolution in Mendelian populations. *Genetics* 16: 97–159.

Wright, S. 1932. The roles of mutation, inbreeding, crossbreeding and selection in evolution. *Proc. VI Int. Congr. Genet.* 1: 356–366.

Wright, S. 1938. Size of population and breeding structure in relation to evolution. *Science* 87: 430–31.

Wright, S. 1940. Breeding structure of populations in relation to speciation. *Amer. Nat.* 74: 232–48.

Wright, S. 1943. Isolation by distance. *Genetics* 28: 114–38.

Wright, S. 1946. Isolation by distance under diverse systems of matings. *Genetics* 31: 39–59.

Wright, S. 1948. On the roles of directed and random changes in gene frequency in the genetics of populations. *Evolution* 2: 279–95.

Wright, S. 1951. The genetical structure of populations. *Ann. Eugen.* 15: 323–54.

Wright, S. 1969. *Evolution and the Genetics of Populations,* vol. 2. Chicago: University of Chicago Press.

Wu, K. K. 1974. Ecogenetic studies on populations structure of *Bromus rubens* L. and *B. mollis* L. Ph.D. dissertation, University of California, Davis.

Wu, L. and J. Antonovics. 1976. Experimental genetics in *Plantago.* II. Lead tolerance in *Plantago lanceolata* and *Cynodon dactylon* from a roadside. *Ecology* 57: 205–8.

Wu, L., A. D. Bradshaw, and D. A. Thurman. 1975. The potential for evolution of heavy metal tolerance in plants. III. The rapid evolution of copper tolerance in *Agrostis stolonifera.* *Heredity* 34: 165–87.

Wuenscher, J. E. and T. T. Kozlowski. 1971. The response of transpiration resistance to leaf temperature as a desiccation resistance mechanism in tree seedlings. *Physiol. Plant.* 24: 254–59.

Wykes, G. R. 1952. The preferences of honeybees for solutions of various sugars which occur in nectar. *J. Exp. Biol.* 29: 511–18.

Wylie, R. B. 1952. The role of the bundle sheath extension in leaves of dicotylendons. *Amer. J. Bot.* 39: 645–51.

Wylie, R. B. 1954. Leaf organization of some woody dicotylendons from New Zealand. *Amer. J. Bot.* 41: 186–91.

Yarrington, G. A. and R. G. Morrison. 1974. Spatial dynamics of a primary succession: Nucleation. *J. Ecol.* 62: 417–28.

Yarrington, M. and G. A. Yarrington. 1975. Demography of a jack pine stand. *Canad. J. Bot.* 53: 310–14.

Yeaton, R. I. and M. L. Cody. 1974. Competitive release in island song sparrow populations. *Theor. Popul. Biol.* 5: 42–58.

Yeaton, R. I. and M. L. Cody. 1976. Competition and spacing in plant communities: The northern Mohave Desert. *J. Ecol.* 64: 689–96.

Yim, Y., H. Ogawa, and T. Kira. 1969. Light interception by stems in plant communities. *Jap. J. Ecol.* 19: 223–38.

Yocum, C. S. and P. W. Lommen. 1975. Mesophyll resistances. In D. M. Gates and R. B. Schmerl (eds.), *Perspectives of Biophysical Ecology*. New York: Springer-Verlag.

Yoda, K., T. Kira, H. Ogawa, and K. Hozumi. 1963. Self-thinning in overcrowded pure stands under cultivated and natural conditions. *J. Biol. Osaka City Univ.* 14: 107–29.

Yokoi, Y. 1976a. Growth and reproduction in higher plants. I. Theoretical analysis by mathematical models. *Bot. Mag. Tokyo* 89: 1–14.

Yokoi, Y. 1976b. Growth and reproduction in higher plants. II. Analytical study of growth and reproduction of *Erythronium japonicum*. *Bot. Mag. Tokyo* 89: 15–31.

Zabadal, T. J. 1974. Eco-physiological aspects of drought resistance. Ph.D. dissertation, Cornell University.

Zak, B. 1971. Characterization and classification of mycorrhizae of Douglas fir. II. *Pseudotsuga menziesii* + *Rhizopogon vinicolor*. *Canad. J. Bot.* 49: 1079–84.

Zelitch, I. 1971. *Photosynthesis, Photorespiration, and Plant Productivity*. New York: Academic Press.

Zimmerman, C. A. 1976. Growth characteristics of weediness in *Portulaca oleracea* L. *Ecology* 57: 964–74.

Zimmermann, M. H. 1974. Long distance transport. *Plant Physiol.* (Lancaster) 54: 472–79.

Zobel, R. W. 1975. The genetics of root development. In J. G. Torrey and D. T. Clarkson (eds.), *The Development and Function of Roots*. New York: Academic Press.

Zohary, M. 1962. *Plant Life of Palestine*. New York: Ronald Press.

Zohary, D. 1965. Colonizer species in the wheat group. In H. G.

Baker and G. L. Stebbins (ed.), *The Genetics of Colonizing Species.* New York and London: Academic Press.

Zouros, E., F. Golding, and T. MacKay. 1977. The effect of combining alleles into electrophoretic classes on detecting linkage disequilibrium. *Genetics* 85: 543–50.

PARTICIPANTS IN THE CONFERENCE OF PLANT POPULATION BIOLOGY HELD IN ITHACA, N.Y., JUNE 15–18, 1977

Mr. Mark Angevine, Section of Ecology & Systematics, Langmuir Labs, Cornell University, Ithaca, NY 14853

Dr. David Bates, Section of Ecology & Systematics, Bailey Hortorium, Cornell University, Ithaca, NY 14850

Dr. Brian Chabot, Section of Ecology & Systematics, Langmuir Labs, Cornell University, Ithaca, NY 14853

Dr. Robert Cook, Dept. of Biology, Harvard University, Cambridge, MA 02138

Mr. William Curtis, Gray Herbarium, Harvard University, Cambridge MA 02138

Mr. Thomas Ducker, Biological Laboratories, Harvard University, Cambridge, MA 02138

Ms. Mary Enama, Dept. of Biology, Washington University, St. Louis, MO 63130

Dr. Thomas Givnish, Dept. of Biology, Harvard University, Cambridge, MA 02138

Dr. Leslie Gottlieb, Dept. of Genetics, University of California, Davis, CA 95616

Dr. James Hamrick, Dept. of Botany, University of Kansas, Lawrence, Kansas 66045

Ms. Robin Hart, The Academy of Natural Sciences, Nineteenth and the Parkway, Philadelphia, PA 19103

Ms. Joan M. Herbers, Dept. of Biological Sciences, Northwestern University, Evanston, IL 60201

CONFERENCE PARTICIPANTS

Dr. James Hickman, Dept. of Biology, Swarthmore College, Swathmore, PA 19081

Dr. Henry Horn, Dept. of Biology, Princeton University, Princeton, NJ 08540

Dr. Subodh K. Jain, Dept. of Agronomy & Range Science, University of California, Davis, CA 95616

Dr. George Johnson, Dept. of Biology, Washington University, St. Louis, MO 63130

Mr. Bruce Karr, Dept. of Biology, Washington University, St. Louis, MO 63130

Dr. Dwight Kincaid, Gray Herbarium, Harvard University, Cambridge, MA 02138

Dr. Donald A. Levin, Dept. of Botany, The University of Texas at Austin, Austin, Texas 78712

Dr. Phillip Miller, Dept. of Biology, San Diego State University, San Diego, CA 92182

Dr. Harold A. Mooney, Dept. of Biological Sciences, Stanford University, Stanford, CA 94305

Dr. Harold Moore, Jr., Section of Ecology & Systematics, Bailey Hortorium, Cornell University, Ithaca, NY 14850

Dr. Sandra Newell, Gray Herbarium, Harvard University, Cambridge, MA 02138

Dr. Louis F. Pitelka, Dept. of Biology, Bates College, Lewiston, ME 04240

Dr. Peter H. Raven, Missouri Botanical Garden, 2315 Tower Grove Ave., St. Louis, MO 63110

Dr. Otto T. Solbrig, Gray Herbarium, Harvard University, Cambridge, MA 02138

Dr. G. Ledyard Stebbins, Dept. of Genetics, University of California, Davis CA 95616

Ms. Patricia Sullivan, Biological Laboratories, Harvard University, Cambridge, MA 02138

Dr. James Teeri, Dept. of Biology, University of Chicago, Chicago, IL 60637

Dr. John Tenhunen, Dept. of Botany, University of Michigan, Ann Arbor, MI 48109

Dr. Charles Uhl, Section of Ecology & Systematics, Cornell University, Ithaca, NY 14853

Dr. Natalie Uhl, Section of Ecology & Systematics, Bailey Hortorium, Cornell University, Ithaca, NY 14850

Dr. Patricia A. Werner, W. K. Kellogg Biol. Station, Michigan State University, Hickory Corners, MI 49060
Dr. Robert Whittaker, Section of Ecology & Systematics, Langmuir Labs, Cornell University, Ithaca, NY 14853

TAXON INDEX

Gossypium, 35
Gramineae, 356–65 passim, 372, 373
Gunnera, 386, 387
Gymnocladus dioica, 402
Gymnosperms, 94

Hakea, 391, 401
Helianthus, 183, 292; H. annuus, 112, 202
Heteromeles, 401, 402
Hieracium, 21, 292, 305; H. floribundum, 215
Hordeum jubatum, 112; H. vulgare, 112, 464
Horsetails, 389
Hymenopapus artemisiaefolius, 110; H. scabiosaeus, 110

Ilex, 399; I. opaca, 399
Imperata cylindrica, 369
Iris setosa var. interior, 23; I. versicolor, 23; I. virginica, 23

Jatropha deppeana, 349
Jepsonia, 120
Juglans, 402
Juniperus, 24; J. deppeana, 340, 348

Lantana camara, 369
Larix, 384; L. decidua, 108
Larrea divaricata, 197, 333, 340, 348, 349, 354
Leavenworthia, 123, 124, 172, 210; L. alabamica, 127; L. crassa, 127
Ledum procumbens, 434
Lepidodendron, 389
Liatris, 100, 102, 104, 147; L. cylindracea, 100, 109
Liliaceae, 358, 391
Limnanthes alba, 112, 177, 178; L. floccosa, 112, 177, 178
Linanthus parryae, 37, 38
Liriodendron, 398
Lithraea caustica, 452
Lolium multiflorum, 112
Lonicera, 307
Lotus corniculatus, 476

Lupinus nanus, 113, 476; L. subcarnosus, 113; L. succulentus, 113; L. texensis, 113
Lycopersicon cheesmanii, 109; L. chmielewskii, 109; L. esculentum, 109; L. esculentum var. cerasiformae, 109; L. parviflorum, 109
Lycopod, 389
Lythrum salicaria, 138; L. tribracteatum, 113

Macaranga, 404
Madiinae, 31, 32, 36
Malvaceae, 395
Marantaceae, 391
Medicago polymorpha, 173, 182; M. sativa L., 239
Melaleuca, 391
Mimulus guttatus, 101, 109
Minuartia, 214; M. uniflora, 258
Musaceae, 391
Musanga, 404
Myrtaceae, 386, 391

Neurospora, 118
Nicotiana, 29, 35, 37; N. digluta, 20
Nostoc, 386
Nuphar, 391
Nyctaginaceae, 358
Nymphaea, 391
Nymphaeaceae, 391
Nymphoides, 391

Ochroma, 404
Oenothera, 21, 22; O. argillicola, 110; O. biennis, 110; O. hookeri, 110; O. parviflora, 110; O. strigosa, 110
Oncopeltus, 472
Oxytropis, 24

Palicourea riparia, 204
Palms, 402
Papaver, 292
Paspalum distichum, 370
Paulownia, 404; P. tomentosa, 404
Phalaris arundinacea, 477
Phlox cuspidata, 113; P. drummondii, 113, 258, 470

Picea, 384; *P. abies*, 108; *P. engelman-
nii*, 108; *P. mariana*, 415
Pinus banksiana, 212; *P. longaeva*, 94,
95, 102, 103, 104, 108; *P. ponderosa*,
104, 105, 108, 340, 347, 348, 349; *P.
pungens*, 94, 108; *P. resinosa*, 84; *P.
rigida*, 351, 352, 353; *P. strobus*, 351,
352, 353; *P. sylvestris*, 108
Plantaginaceae, 391
Plantago, 214, 217, 292, 391; *P. major*,
217, 370; *P. media*, 217
Poa annua, 21, 101, 227, 370
Polygalaceae, 358
Polygonum, 236, 248–57 passim, 262,
292; *P. aniculare*, 370; *P. casca-
dense*, 249–57 passim; *P. kelloggii*,
238, 245, 250–57 passim
Polytrichum, 213
Populus, 26, 236, 464; *P. tremuloides*,
200
Portulacaceae, 358
Portulaca oleracea, 370
Potentilla, 21, 31; *P. glandulosa*, 29,
472
Primula kewensis, 20
Prodenia eridania, 330
Prosopis glandulosa, 321, 322
Proteaceae, 32, 391
Prunus ilicifolia, 401; *P. pennsylvanica*,
200
Pseudotsuga menziesii, 108, 425
Pteridium aquilinum, 476, 477

Quercus, 25; *Q. alba*, 386, 387, 462; *Q.
borealis*, 351, 352, 353; *Q. coccinea*,
387; *Q. dumosa*, 401; *Q. ilicifolia*,
386, 387; *Q. wislizenii*, 401

Ranunculaceae, 391
Ranunculus, 169, 214, 292, 391; *R.
alismaefolius*, 391; *R. lingua*, 391; *R.
repans*, 169, 215
Raphanobrassica, 20
Rhododendron, 68, 126
Rhus, 307, 402; *R. integrifolia*, 402; *R.
laurina*, 402; *R. ovata*, 201, 402, 441,
450, 452; *R. trilobata*, 402
Rodigia, 30

Rosaceae, 402
Rubus, 21; *R. hispida*, 303
Rumex, 292
Rutaceae, 402

Salix gilliesii, 450; *S. pulchra*, 434
Sambucus, 403
Sanchus oleraceus, 370
Sassafras, 307, 398
Scorzonera, 391
Scrophulariaceae, 358
Sedum, 214
Senecio, 227; *S. vulgaris*, 203
Setoria verticillata, 370
Sida aggregata, 395
Silene dioica, 192; *S. maritima*, 109
Simmondsia chinensis, 340, 348, 349
Solanum nigrum, 20, 370
Solidago, 159, 236, 288, 292, 293, 300,
301, 302, 304, 308; *S. caesia*, 304; *S.
canadensis*, 200, 293–304 passim; *S.
gigantea*, 304; *S. graminifolia*, 293–
99 passim, 304; *S. houghtonii*, 304;
S. missouriensis, 293–99 passim,
304; *S. nemorales*, 293–304 passim;
S. remota, 304; *S. riddellii*, 304; *S.
rigida*, 304; *S. rugosa*, 304; *S. sempervirens*, 304; *S. speciosa*, 293, 294,
296, 297, 299, 304; *S. uliginosa*, 304
Sorghum halepense, 369
Stellaria media, 370
Stephanomeria, 159, 470; *S. exigua*
ssp. *coronaria*, 113, 159, 266–86
passim; *S. malheurensis*, 266–86
passim

Taraxacum, 21, 464; *T. officinale*, 158,
292, 293, 370
Taxodium, 384
Thuja occidentalis, 464
Tilia, 200
Toxicodendron, 402
Tradescantia, 24
Tragopogon, 391; *T. dubius*, 110; *T.
mirus*, 110; *T. miscellus*, 110; *T. por-
rifolius*, 110; *T. pratensis*, 110
Trifolium, 292, 476; *T. hirtum*, 173; *T.
repens*, 292; *T. subterraneum*, 464

SUBJECT INDEX